Single Molecule Biology

Single Molecule Biology

Alex E. Knight
(National Physical Laboratory)

AMSTERDAM • BOSTON • HEIDELBERG • LONDON
NEW YORK • OXFORD • PARIS • SAN DIEGO • SAN FRANCISCO
SINGAPORE • SYDNEY • TOKYO
Academic Press is an imprint of Elsevier

ELSEVIER

Academic Press is an imprint of Elsevier
525 B Street, Suite 1900, San Diego, CA 92101-4495, USA
30 Corporate Drive, Suite 400, Burlington, MA 01803, USA
32, Jamestown Road, London NW1 7BY, UK
360 Park Avenue South, New York, NY 10010-1710, USA

Library of Congress Cataloging-in-Publication Data
A catalog record for this book is available from the Library of Congress

British Library Cataloguing in Publication Data
A catalogue record for this book is available from the British Library

ISBN: 978-0-12-374227-8

For information on all Academic Press publications
visit our website at www.elsevierdirect.com

Typeset by Charon Tec Ltd., A Macmillan Company
www.macmillansolutions.com

Printed and bound by CPI Group (UK) Ltd, Croydon, CR0 4YY

Transferred to Digital Print 2011

To Jacky and Lewis

Contents

Preface

This book is intended to act as an overview of the ways in which "single molecule" methods have contributed to our understanding of biological systems and processes. The chapters have been written specially for the book and are aimed at the level of a final year undergraduate or a first-year PhD student. The hope, therefore, is that the book should be accessible to readers from a wide variety of backgrounds, as I feel is essential for this field of research, which is intrinsically interdisciplinary. Some biological knowledge, however, will be a benefit. The book is by no means comprehensive – nor could it hope to be   but I hope that it will provide a primer, and a starting point for further exploration.

In the first chapter, I have striven to give some background to the reader new to the field. The subsequent chapters are all written by leaders in their fields, and each covers a biological system that has been illuminated by the single molecule approach. Finally, the Appendix is intended to provide a useful reference on abbreviations, symbols and units that are commonly encountered in the field; in particular, as a scientist working at the UK's national measurement institute, I wanted to include some notes on the SI and its use in biology.

Alex Knight
National Physical Laboratory, Teddington
August 2008

Acknowledgments

Thanks are due to many people for helping me to put this book together. First of all, at the National Physical Laboratory, I must thank Marc Bailey for support and encouragement and Anna Hills for reviewing drafts. My work on this book has been supported by NPL's Strategic research programme, and also by the National Measurement System of the Department for Innovation, Universities and Skills (DIUS).

Elsewhere, thanks are also due to Edward Bittar for suggesting the book in the first place; to Justin Molloy for his encouragement, and for providing such a striking cover image; and to the team at Academic Press/Elsevier including Luna Han, Gayle Luque and April Graham.

List of Contributors

Colin Echeverría Aitken   Biophysics Program, Stanford University School of Medicine, Stanford, CA, USA

Richard M. Berry   Clarendon Laboratory, Department of Physics, University of Oxford, Oxford, UK

Laurence R. Brewer   Department of Chemical Engineering and Bioengineering, Center for Reproductive Biology, Washington State University, Pullman, WA, USA

David Colquhoun   Department of Pharmacology, University College London, London, UK

Magdalena Dorywalska   Department of Structural Biology, Stanford University School of Medicine, Stanford, CA, USA

Rachel E. Farrow   Division of Physical Biochemistry, MRC National Institute for Medical Research, The Ridgeway, Mill Hill, London, UK

Jeremy C. Fielden   Division of Physical Biochemistry, MRC National Institute for Medical Research, The Ridgeway, Mill Hill, London, UK

Samir M. Hamdan   Department of Biological Chemistry and Molecular Pharmacology, Harvard Medical School, Boston, MA, USA

Lydia M. Harriss   Chemistry Research Laboratory, University of Oxford, Oxford, UK

Thomas Haselgrübler   Biophysics Institute, Johannes Kepler University Linz, Linz, Austria

Jan Hesse   Center for Biomedical Nanotechnology, Upper Austrian Research GmbH, Linz, Austria

Lukas C. Kapitein   Department of Physics and Astronomy, VU University Amsterdam, Amsterdam, The Netherlands; Department of Neuroscience, Erasmus Medical Center, Rotterdam, The Netherlands

Alex E. Knight   Biotechnology Group, National Physical Laboratory, Teddington, Middlesex, UK

Hiroaki Kojima   Kobe Advanced ICT Research Center, National Institute of Information and Communications Technology, 588-2 Iwaoka, Nishi-ku, Kobe, Japan

Remigijus Lape   Department of Pharmacology, University College London, London, UK

Sanford H. Leuba Department of Cell Biology and Physiology, University of Pittsburgh School of Medicine, Petersen Institute of NanoScience and Engineering, Department of Bioengineering, Swanson School of Engineering, 2.26g Hillman Cancer Center, University of Pittsburgh Cancer Institute, Pittsburgh, PA, USA

R. Andrew Marshall Department of Chemistry, Stanford University School of Medicine, Stanford, CA, USA

Justin E. Molloy Division of Physical Biochemistry, MRC National Institute for Medical Research, The Ridgeway, Mill Hill, London, UK

Kazuhiro Oiwa Kobe Advanced ICT Research Center, National Institute of Information and Communications Technology, 588-2 Iwaoka, Nishi-ku, Kobe, Japan; Graduate School of Life Science, University of Hyogo, Harima Science Park City, Hyogo, Japan

Erwin J.G. Peterman Department of Physics and Astronomy, VU University Amsterdam, Amsterdam, The Netherlands

Joseph D. Puglisi Department of Structural Biology, Stanford University School of Medicine, Stanford, CA, USA; Stanford Magnetic Resonance Laboratory, Stanford University School of Medicine, Stanford, CA, USA

Gerhard J. Schütz Biophysics Institute, Johannes Kepler University Linz, Linz, Austria

Lucia Sivilotti Department of Pharmacology, University College London, London, UK

Yoshiyuki Sowa Clarendon Laboratory, Department of Physics, University of Oxford, Oxford, UK

Antoine M. van Oijen Department of Biological Chemistry and Molecular Pharmacology, Harvard Medical School, Boston, MA, USA

Mark I. Wallace Chemistry Research Laboratory, University of Oxford, Oxford, UK

Christian Wechselberger Center for Biomedical Nanotechnology, Upper Austrian Research GmbH, Linz, Austria

Introduction: The "Single Molecule" Paradigm

Alex E. Knight

Biotechnology Group, National Physical Laboratory, Hampton Road, Teddington, Middlesex TW11 0LW, UK

Summary

A new experimental paradigm, based on the detection of individual molecules, has been making great strides in the dissection of biomolecular function in vitro in the past two decades. A technological convergence – of improved detectors, probes, microfluidics and other tools – is leading both to an explosion of this area of research and its development into a tool for investigating processes in living cells.

Key Word

single molecule detection

The "Single Molecule" Paradigm

Imagine a busy motorway, packed with all kinds of vehicles. Now imagine that you are trying to describe the traffic on that motorway (see Figure I.1). You could try to summarize it by a single number; the average speed of the traffic would be a good example. This gives a good indication as to whether the traffic is flowing or obeying the speed limit, but it does not tell you much more. Sports cars may be tearing along in the outside lane, more cautious drivers cruising in the center lane, while trucks rumble along in the slow lane. Indeed, some vehicles may be pulled over on the hard shoulder. What's more, vehicles will occasionally change lanes, slow down, or accelerate. We don't get a full picture of this diversity from a single number, but this is the kind of measurement of molecular properties, quantities, or behavior that we usually make in the life sciences.

For example, if we measure the properties of a molecule by a spectroscopic technique, such as fluorescence spectroscopy, we are likely to be measuring the average characteristics of

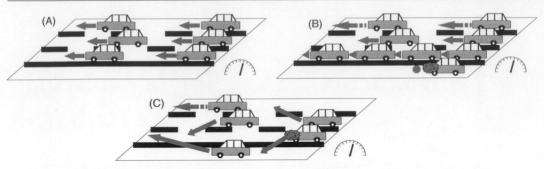

Figure I.1: Single molecule detection can unravel differences between molecules in a population. This figure illustrates the "motorway" analogy of single molecule experiments used in the text. When a single number is used to describe the properties of a population of molecules – represented by cars on a motorway – it gives no information about how the properties vary within that population. For example, if we know only the average speed of the cars on the motorway, we do not know if all the cars are moving at the same speed (A) or whether their speeds differ (B). This is known as *static heterogeneity*. Furthermore, it may be that the cars are changing speed – or stopping and starting – and again this is not apparent from the average speed (C). This is known as *dynamic heterogeneity*

a very large ensemble of molecules. If our cuvette holds 3 ml, and our sample is of a protein at 1 mg/ml, then for a typical protein of a molecular weight 50 000 Da we have 60 nmol of protein in the cuvette. This may sound like a relatively small amount, but it corresponds to 36 000 000 000 000 000 individual molecules. This huge number arises because Avogadro's constant (N_A), the number of entities in a mole, is such a huge number – approximately 6×10^{23}. Viewed from this perspective, we are looking at a very large sample indeed!

So in most techniques, even if the quantities are, in molar terms, tiny, any measurement we make is an average across many millions or billions of molecules. The usual approach is to assume that all the molecules are the same. But this is often not the case, particularly for the complex molecules that are found in biology; the molecules may have different properties (sports cars, trucks – or breakdowns) and indeed, these properties can change over time (switching lanes) – and moreover in a random (or stochastic) fashion. Sometimes the ensemble, "averaged" measurement is good enough. But at other times, we need much more understanding of the molecules – in fact, we need a whole new approach.

This new approach is one that has been developing steadily over the past two decades, and now appears to be undergoing something of an explosion. This is the "single molecule" approach. A rather inelegant name, perhaps, but this describes a philosophy where molecules are thought of – and measured – as individual entities.[1] It is important not to get too fixated on the word "single" – even if you measure a single molecule, it does not tell you much; after all, how can you be sure that it is representative? So even single molecule experiments may characterize hundreds or thousands of molecules, for as in other fields of biology, good statistics are vital. Indeed, often what we are interested in is the shape of the *distribution* of our property of interest.

This is the key point, then: not that we analyze a sample at the absolute limit of detection (although we do), but that we treat all the molecules in that sample as individual entities.

When we consider such tiny samples, the conventional units of quantity become somewhat ungainly. A single molecule is approximately 1.66 yoctomoles[2]; a zeptomole corresponds to approximately 600 molecules. Therefore in this type of work, experimenters tend to report on numbers of molecules rather than molar quantities[3] – see Figure I.2.

Why Single Molecules?

So what are the advantages of observing or measuring single molecules? The reaction of many, on hearing about single molecule detection, is to assume that the benefit is in the ability to detect and even quantitate very small amounts of material. While this is true up to a point, it misses the main advantages of the single molecule approach, as will be shown later. Another common (but somewhat more acute) reaction is that one cannot infer much from looking at a single molecule: how does one know this molecule is typical? This is an excellent point, but in reality, "single molecule" experiments are never really done with *single* molecules. In fact, it is the name that is misleading – really we are interested in performing *discrete* molecule experiments, that is, experiments where we observe a group of molecules as a population of discrete individuals rather than as an undifferentiated *ensemble*. This implicitly requires that we can, in some sense, detect a single molecule but this alone would never make sense as an experimental design.

The continuous improvements in analytical science have pushed detection limits to extraordinarily low levels – picomoles or femtomoles, for example – so it is natural that single molecule detection techniques, where we are reaching the ultimate detection

---

[1] Bustamante has suggested the term *in singulo* to denote "single molecule" experiments, contrasted with *in multiplo* to denote "bulk" or "ensemble" measurements (Bustamante, 2008).

[2] The less familiar SI prefix yocto- indicates a factor of 10^{-24}, whereas zepto- indicates 10^{-21}. See appendix.

[3] Moerner (1996) has wittily suggested the adoption of a new unit, the guacamole, corresponding to a single molecule, where the prefix guaca- corresponds to 1/Avocado's number.

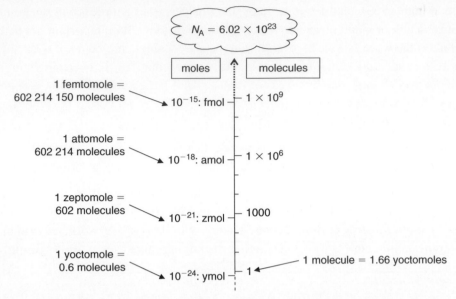

Figure I.2: The scale of single molecule detection. Historically, we tend to quantify molecules in terms of the mole, where a mole contains N_A molecules (where $N_A = 6.02 \times 10^{23}$) – a very large number (see appendix). For large biological molecules, we tend to deal with much smaller quantities – submultiples of the mole – and as measurement techniques increase in sensitivity we are dealing in smaller and smaller quantities. Once we get into the subattomole range, it becomes more convenient to think in terms of numbers of molecules. This logarithmic scale provides a quick comparison between moles and molecules. The less familiar SI prefixes of zepto- and yocto- are brought into play to express these tiny quantities; a zeptomole is approximately 600 molecules, whereas a yoctomole is less than 1 molecule. Since molecules are discrete entities, at this scale they are best quantified by a counting approach and expressed as numbers of molecules

limit of yoctomole sensitivity (Figure I.2), should be seen in this light. However, as the sample sizes become smaller, the number of molecules likewise becomes smaller, until the random statistical variations in the numbers of molecules counted – known as "shot" or "Poisson" noise – become a significant factor. A more significant problem with the real-world application of these techniques simply for detection and quantitation is the sampling issues. The detection volumes for most single molecule techniques are typically very small; how can we be sure that our detection volume accurately reflects the concentration of the molecule in the larger sample? Also, as with many microscale techniques, there are questions about purification and handling of the sample and losses due to the molecule of interest being retained in matrices or on surfaces. This is not to say that quantitative results on numbers of molecules cannot be obtained, within stated uncertainty limits, but rather that this is not the true strength of this approach.

Let us now try to summarize the reasons why single molecule approaches are useful:

1. Static heterogeneity: By identifying *subpopulations* of molecules within a sample, we may be able to understand more about the characteristics of the molecule and its mechanism. For example, different molecules may experience different local microenvironments and thus exhibit different activities; or there may be a variety of different conformational states. Obtaining detailed statistics is a benefit of the single molecule approach.

2. Dynamic heterogeneity: Often the behavior of interest concerns the *transitions* of the molecule between different states; for example, where an enzyme is binding to a substrate molecule, we may wish to know the rate of release of the product. These transitions are often lost in ensemble measurements because of the intrinsic averaging, or special tricks have to be used to "synchronize" the molecules. We may also be interested in rare or transient states of the molecule, which are obscured in bulk measurements.

3. Microscopic properties: The molecules may have properties which are key to their function, but which can only be measured at a microscopic, single molecule scale. For example, the activity of myosin motor proteins in muscles can be measured on a larger scale, such as a whole-muscle fiber or a myofibril, because they are organized into arrays that integrate their forces and displacements. In contrast, many other sorts of myosins, such as myosin V or myosin I (see Chapter 1, this volume), operate as individuals and their activity can thus only be measured at the single molecule level. Another example is when we are looking for a change in the *orientation* of molecules (Figure I.3). Where the molecules are randomly orientated, changes in the orientation of individual molecules make no difference to the orientation of the population. This is important in the study of rotary motors such as the F_1 ATPase, where direct single molecule observations proved the hypothesis that these were rotary motors, and have since enabled a detailed dissection of the mechanism (Noji et al., 1997).

4. Trace detection: Notwithstanding the comments above, an advantage of the single molecule approach is that very small amounts of material are typically needed. This is obviously a benefit where samples are difficult to obtain (e.g., a low-abundance protein) but could also be a benefit where large numbers of experiments are required,

Figure I.3: Single molecule detection can reveal characteristics obscured in the bulk. This figure illustrates how some properties can only be meaningfully measured at the single molecule level. At top left, a population of molecules (circles) is randomly oriented (arrows). By single molecule approaches, the orientation of each molecule may be determined. At right, a bulk method will only conclude that there is no net orientation in the sample. In the lower panels, the sample is remeasured at a different point in time. The single molecule approach detects changes in the orientations of the individual molecules, but the bulk approach reports no change

as in some high-throughput screening methods, for example, in genomics, drug discovery, or systems biology. For example, microarray or microfluidic devices can be used to screen large numbers of samples (see, e.g., Chapter 10, this volume).

5. Spatial information: In many (but not all) approaches, images of molecules or precise localizations or distances are obtained. This spatial information can be of great

benefit. For example, single molecule FRET can be used to observe conformational changes in molecules or complexes; single molecules can be localized in cells or the colocalization of molecules can be observed.

6. "Digital" detection: The discrete (or "digital") nature of single molecule detection is very different from the conventional "analogue" approach. It can lead to more accurate measurements because "noise" or "background" signals are more readily distinguished from the molecules of interest or their behaviors (depending on the type of experiment). For example, molecules can be counted or steplike "digital" state changes can be observed.

7. Direct approach: Bulk methods often require that the behavior of molecules is *inferred* rather than measured directly. Often this inference relies upon a model or assumptions about the system. In contrast, single molecule approaches are more direct and typically less model dependent. However, single molecule approaches may also introduce artifacts (e.g., due to the introduction of labels), which must be carefully allowed for in the experimental design.

In summary, the single molecule approach has many advantages. We should remember, though, that most problems in biology are solved through the application of a variety of complementary tools, and indeed this is demonstrated in the subsequent chapters.

Life as a Molecule

To perform and interpret single molecule experiments, we must have some understanding of the microscopic world in which the molecules exist. In some ways it is surprisingly similar to our familiar macroscopic world and in others, startlingly different.

Let us start out by thinking about the *scale* of the phenomena we are measuring. Biological molecules are the natural world's equivalent of nanotechnology (and usually far superior in their capabilities); their characteristic dimensions are typically of the order of nanometers.[4] Similarly, the forces that molecules can exert are of the order of

---

[4] Some biological molecules can, of course, be much larger in one dimension. The obvious example is DNA, with a diameter of 2 nm and a variable, but often very great, length. For example, the genome of λ phage, often used in laboratory experiments, is approximately 16 μm long: an aspect ratio of approximately 8000. Many DNA molecules are far, far longer than this; for example, human chromosome 1 if fully extended would have a contour length of approximately 84 mm. Some proteins can also reach dimensions of the order of micrometers, such as titin, an important component of muscle. Still greater lengths can be achieved by filaments assembled from many smaller protein subunits, such as the cytoskeletal filaments actin and tubulin.

piconewtons (Howard, 2001). On this molecular scale, the environment that molecules experience differs from our more familiar macroscopic world in two critical respects: *Brownian motion* and *Reynolds number*.

Brownian, or thermal, motion is the random motion of microscopic objects due to thermal energy (as a consequence of their constant bombardment by water molecules). The mean thermal energy of an object is $k_B T/2$ per degree of freedom, where k_B is Boltzmann's constant $(1.4 \times 10^{-23}\,\mathrm{J\,K^{-1}})$ and T is the absolute temperature. At room temperature, $k_B T \approx 4\,\mathrm{pN\,nm}$ (1 pN nm = 1 zeptojoule or 10^{-21} J). (We can see that pN nm is a convenient unit for expressing energy at the molecular scale.) Although this expression gives the mean value of the thermal energy, in fact this varies randomly, with an exponential distribution. The actual motion of an object depends not just on the thermal forces exerted on the object, but also on any constraints on its motion. For example, there may be viscous drag from the medium or from a membrane it is embedded in; or there may be a restraining stiffness from a tether to a surface, or perhaps (in an experiment) from an optical trap or an atomic force microscope (AFM) tip. The same physical constraints will apply to any force exerted by the molecule under study or by the experimenter.

The other important aspect in which the microscopic environment differs is expressed by the *Reynolds number*, R_e, which expresses the ratio between inertial and viscous forces. For macroscopic objects, $R_e \gg 1$, inertial forces are dominant. For example, a rowing boat continues to move through the water for some distance after you stop rowing. At the microscopic scale, $R_e \ll 1$, viscous drag, and not inertia, is most important. This means that objects stop moving very rapidly when a force is no longer applied. This can have some counterintuitive consequences; for example, large beads (several micrometers in diameter) are sometimes used as tags or handles for motor proteins. It might be thought that the motors would be incapable of generating sufficient force to move such objects, many times larger than themselves; however, the absence of significant inertial forces means that the motors are quite capable of moving them.

These features of the microscopic world are most important when we are performing mechanical measurements – of force, stiffness position, or velocity – but diffusion rates are, of course, significant in many types of experiment.

One might expect at the microscopic scale that the behavior we observe might be different in other ways from our everyday experience. Perhaps surprisingly, the behavior of many of the systems can in fact be modeled in quite familiar ways – for example, the mechanics encountered in most experiments can be well described by sets of springs and dampers (or "dashpots"). For a deeper insight into the topics covered in this section, see the excellent book by Howard (2001).

Single Molecule Techniques

The focus of this book is very much on the *results* of single molecule research and the insights that have been gained from it, rather than the techniques that are used. However, it is impossible to describe the experiments without touching upon the techniques, as most of the chapters of this book demonstrate. As each technique has many variations and can be applied in a variety of ways, only a brief introduction to the methods will be given here. For a recent review see Walter et al. (2008). Figure I.4 illustrates a simple scheme by which we might classify single molecule methods.

Mechanical Techniques

There are a number of techniques by which the mechanical properties or behavior of a molecule can be observed (Greenleaf et al., 2007; Neuman and Nagy, 2008; Walter et al., 2008). Measurables might include the step size of a motor, the force it generates, or the stiffness or length of a molecule. One of the most popular techniques is the *optical tweezers* or *optical trap* (Knight et al., 2005). Here, microscopic particles may be manipulated with a focused laser beam, typically within an optical microscope. These particles are often "handles" that are attached to a molecule of interest, such as a cytoskeletal filament, a molecular motor, or a nucleic acid. A position sensor may be added to the system, which permits measurements of force and displacement to be made. By the addition of a feedback loop, between the sensor and the trap, measurements may be made of position at a constant force, or force at a fixed position. Many elaborations are possible, for example, with multiple traps or through combinations with other methods. For examples of the application of this technique, see Chapters 1–3, 5–7, this volume. Other ways of manipulating microscopic objects include magnetic tweezers and hydrodynamic flow (Chapter 6, this volume, Fig. 1) and electrorotation (Chapter 4, this volume, Fig. 3). The scanning probe microscopies, particularly AFM, have also been widely used to study protein folding/unfolding and intramolecular interactions (Greenleaf et al., 2007; Neuman and Nagy, 2008).

Electrical Techniques

Patch clamping was the first single molecule technique to be widely applied (see Chapter 8, this volume) and is still an important technique today. Typically what is measured is

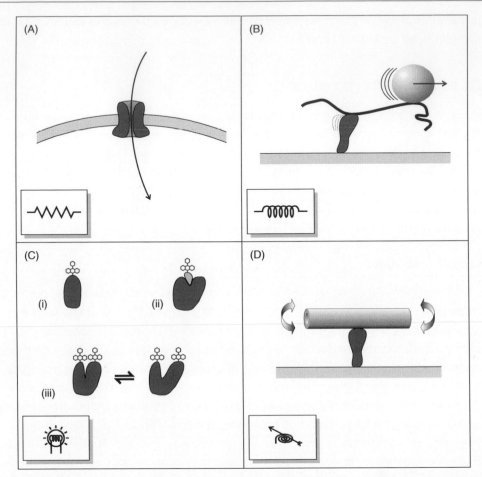

Figure I.4: Types of single molecule measurement. These cartoons illustrate the main types of single molecule measurement. In (A), electrical measurements of channels or pores are shown. Here, a *current* of ions through an individual channel is measured. The inset indicates that the molecule being measured is analogous to a resistor in a conventional electrical circuit. Typically, observations are of the channel current and of the opening and closing (gating) of the channel. In (B), single molecule mechanical measurements are illustrated. Typically, a relatively large object, such as an AFM tip or a microbead manipulated by an optical or magnetic tweezers, is connected to the molecule of interest. Often the measurement may be of the *stiffness* of the molecule (hence the inset showing a spring) or a more comprehensive analysis of the molecule's force-extension curve. Alternatively, forces exerted by the molecule or changes in position may be measured, particularly for molecular motors. In (C), some example of single molecule fluorescence measurement are shown.

the ion current through the pore, which corresponds to an electrical conductance. Often the parameters of interest are the distributions of "open" and "closed" lifetimes, which can yield detailed mechanistic information. More recently, the measurement of current through nanopores has been applied in a variety of sensing applications (Bayley and Cremer, 2001).

Optical Techniques

There are many optical techniques in use for single molecule measurements today (Walter et al., 2008), and examples are discussed in most of the chapters of this book. Many of these techniques are based on variants of fluorescence detection, as this is one of the most sensitive detection techniques available, because a single fluorescent molecule can release many thousands of photons. Conventional epifluorescence microscopy uses a mercury lamp and filters to image samples. While it is possible to image single molecules using this approach, it is very challenging to reduce the levels of background fluorescence and stray excitation light sufficiently. Most single molecule fluorescence methodologies rely on lasers to excite fluorescence because of their typically greater stability, single wavelength, and collimated (parallel) beams. In some approaches, such as TIRF (Axelrod et al., 1999; Knight et al., 2005), the laser beam illuminates the whole field of view and a sensitive camera is used to record an image. In other approaches, the laser beam is focused to a tight spot and detection is through a confocal pinhole with a point detector (such as a photomultiplier or an avalanche photodiode). The spot may be scanned to

In (i), a fluorescent group is attached to the molecule of interest. The fluorescence signal can then be monitored using a camera or other detector, as an indicator of the molecule's position, orientation, or conformation, for example. In (ii), the fluorescent group is attached to a ligand (e.g., an enzyme substrate) so that binding and dissociation of the ligand may be monitored. In (iii), FRET (see appendix) may be used to monitor the distance between two points in the molecule. This can be used to monitor conformational changes, for example. In (D), a general approach of using a larger object (which can be resolved by light microscopy) as a marker for the behavior of a molecule is illustrated. The macroscopic object may be a bead or other particle, or a cytoskeletal polymer such as an actin filament or a microtubule (illustrated). These can be used to monitor position or orientation of molecules. Often thermal or Brownian motion is used to "probe" the system. In this example, the thermal rotation of a cytoskeletal filament is used to measure the torsional stiffness of a molecule

produce an image (scanning confocal microscopy); or the behavior of a molecule within the spot can be tracked over time. Alternatively, the fluctuations in signal caused by the diffusion of molecules through the beam may be monitored (as in FCS).

In many of these techniques, the central challenge is to reduce the volume in which fluorescence is excited. In TIRF, this is achieved by exciting fluorescence in a very thin layer near a surface; in confocal methods, the laser is focused down to a diffraction-limited volume. The excited volume may be further reduced by using multiphoton excitation. Another approach used in near-field scanning microscopy, for example, is to excite fluorescence only in the vicinity of a scanned probe tip.

Much of the research described in this book relates to single molecule measurements in vitro, but single molecule fluorescence detection is also possible in cells (Mashanov et al., 2006) where it can be used to study intramolecular interactions, localization, and transport of molecules, among many other phenomena.

Various different aspects of the fluorescence from single molecules may be monitored to obtain different sorts of information. This may include collection of spectra, measurement of fluorescence lifetime, or measurement of polarization, for example. Variations also arise from the probes used and the way in which they are attached to the molecules of interest; for example, FRET can be used to detect interactions or to measure distance at the nanometer scale (Roy et al., 2008).

An exciting recent development in microscopy has been the development of so-called "super-resolution" methods, which allow the conventional resolution limits of microscopy to be exceeded. In some cases, these are single molecule methods. (For more details see Chapter 9, this volume, Table 9.1; and Walter et al., 2008.)

"Handles" and Passive Observation

Information about the behavior of a molecule of interest can sometimes be gleaned by attaching a large "handle" as a mechanical probe of its properties. By watching the behavior of the visible "handle," we can infer the properties or behavior of the system of interest. Sometimes we can simply let thermal motion probe the system for us. For example, "tethered particle motion" has been used to monitor the length of DNA molecules during transcription (Finzi and Gelles, 1995; Schafer et al., 1991; Tolic-Norrelykke et al., 2004; Yin et al., 1994). Here, the thermal motion of a bead is constrained by a DNA tether to a surface; the range of motion indicates the length of the tether. The handle can also be a cytoskeletal filament. For example, Hunt and Howard (1993) used a microtubule's thermal oscillations to measure the torsional stiffness of a kinesin molecule; Nishizaka et al. (2000)

used a similar technique to measure the torsional stiffness of myosins. Perhaps most strikingly, an actin filament was used to demonstrate the rotational motion of the F_1 ATPase (Noji et al., 1997). The use of thermal motion is also a component of some other techniques; for example, changes in the amplitude of thermal motion of an actin filament can be used to detect interactions with myosin (Molloy et al., 1995; Chapter 1, this volume).

Overview of Single Molecule Biology

The chapters of this book present the dizzying diversity of biological systems that have been investigated by single molecule techniques. In most cases, the application of these techniques has resulted in profound insights into the function or mechanism of these systems.

Probably the first single molecule technique to come into widespread use was patch clamping (see Chapter 8, this volume) – first introduced in the 1970s – which is used to study ion channels in membranes. The impact of this particular single molecule method is beyond dispute – it has revolutionized many aspects of neurobiology, and is an important tool in the development of new drugs, as many of them target membrane channels. Recently there has been much interest in the applications of membrane pores in a variety of sensing applications, even for single molecule DNA sequencing (Bayley, 2006; Bayley and Cremer, 2001; Sabanayagam et al., 2005; Soni and Meller, 2007). Other membrane proteins and lipids have been tracked by single molecule fluorescence approaches (see Chapter 9, this volume). This has provided fundamental insights into the structures of biological membranes, and the interactions and mechanisms of membrane proteins such as receptors and other signaling molecules.

One of the most characteristic properties of living organisms is movement – of organisms, cells, and subcellular components. In many cases, such movements are driven by biological motors and these have been one of the most fruitful fields of study by the single-molecule tool-kit. The most familiar motors are the myosins, which generate force in skeletal, cardiac, and smooth muscle – and also within cells – by interacting with actin filament tracks. The progress that has been made in elucidating the mechanism and functions of these proteins is reviewed in Chapter 1, this volume. The myosins are *linear motors*, that is, they operate on straight "tracks" that are composed of polymerized cytoskeletal proteins – in the case of myosin, the tracks are actin filaments. Many of the pioneering applications of single molecule techniques were in the field of myosin research.

The other class of cytoskeletal tracks that motors run on are the microtubules. There are two types of microtubule-based motors: the kinesins and the dyneins. Chapter 2, this volume, introduces the kinesin family of proteins and their diverse functions in intracellular transport and the organization of the cytoskeleton. A variety of single molecule methods, including optical tweezers and single molecule fluorescence imaging, have illuminated various aspects of how these motors function, such as how big a step they take along the microtubule, the fact that one ATP molecule is consumed per step, and the way in which the two motors in a dimeric kinesin are coordinated with each other. They have also revealed the existence of surprising new modes of motility in some of the kinesins, including "reverse" motility and lattice diffusion. Finally, light has been cast on the mechanisms by which these motors are regulated and bind to their "cargo."

The dyneins were the first class of microtubule-based motors to be discovered, but their investigation has progressed more slowly than our knowledge of the kinesins because of their greater complexity (Chapter 3, this volume). Dyneins drive the motility of cilia and flagella on eukaryotic cells and are also found in intracellular forms that are important for the organization and function of the mitotic spindle and the position of some organelles within the cell. Optical tweezers and fluorescence measurements have determined the step size and forces exerted by single dynein molecules, and models for the mechanism of motility have been developed. Like muscle myosins, flagellar dyneins operate in large, coordinated arrays to fulfil their function, and one of the most interesting questions is how this is achieved; the combination of single molecule and single-flagellum measurements is beginning to address this conundrum.

There is another type of "track" within cells upon which linear motors operate; these tracks are nucleic acids such as DNA and RNA. Many of the enzymes and enzyme complexes that interact with nucleic acids can also be regarded as molecular motors. Indeed, many classes of nucleic acid enzymes have now been studied by single molecule approaches, for example, DNA and RNA polymerases, exonucleases, helicases, and topoisomerases; these are all discussed in Chapter 6, this volume. That chapter also discuss single molecule measurements of the nucleic acids themselves, which turn out to have some intriguing properties. As mentioned above, the DNA molecules encountered in cells are many orders of magnitude longer than the cell diameter, which raises the fascinating question of how they are packaged up into such a small volume; this is discussed in Chapter 5, this volume. The ribosome translocates along a messenger RNA as it synthesizes a polypeptide; so in a sense the ribosome is also a motor. The function of the ribosome has also begun to be dissected in exquisite detail by single molecule techniques, as is set out in Chapter 7, this volume.

Although not covered in this book, recent work has exposed the stochastic (random) nature of gene expression in single cells. Several recent papers describe the monitoring of individual mRNA molecules in living cells (Fusco et al., 2003; Golding and Cox, 2004; Shav-Tal et al., 2004). Two of them (Fusco et al., 2003; Shav-Tal et al., 2004) monitor the movement of complexes of mRNA and proteins (mRNPs) in mammalian cells and conclude that it is mostly diffusive in nature. Golding and Cox (2004) similarly followed the behavior of mRNA in *Escherichia coli* cells. They found that most of the mRNA molecules they observed remained tethered in one location, consistent with transcripts remaining tethered to the DNA (as one might expect in a prokaryote). They also measured the number of transcripts per cell (under repressed conditions) and found a Poisson distribution where most cells had no transcripts and a few had one or more.

Two papers from the laboratory of Xie at Harvard describe the detection of individual protein molecules as they are expressed (Cai et al., 2006; Yu et al., 2006). Interestingly, these papers describe radically different approaches. In the paper by Cai et al. (2006), an "indirect" approach is used to follow the number of enzyme molecules expressed in a cell. The fluorescent products of the enzyme accumulate in a microfluidic chamber, and the rate of increase is determined by differentiating the fluorescence signal. Steps in the rate of synthesis of the fluorescent compound correspond to the expression of individual enzyme molecules. The authors found that enzyme molecules were expressed in "bursts." They interpreted each burst as corresponding to a transcription event and showed that the number of enzymes produced per burst followed an exponential distribution, which was consistent with a competition between mRNA degradation and translation.

The second paper (Yu et al., 2006) describes a "direct" approach to detecting protein expression. Here, the reporter protein molecules – natively fluorescent proteins – are detected directly by fluorescence microscopy. To prevent their images being "smeared out" by diffusion, they are expressed as a fusion with a membrane-anchoring protein. This approach permits a more detailed analysis of the mechanics of gene expression and the following of protein expression through cell lineages.

Many of the molecules that have been studied by single molecule approaches are enzymes. Indeed, enzymes as a broad class of molecules have also been the subject of intensive single molecule research (English et al., 2006; Xie, 2001; Xie and Lu, 1999). These studies have revealed many intriguing features, including both the dynamic and static types of heterogeneity discussed above. They have also revealed that enzyme molecules apparently retain a "memory" of their recent past.

Many of the molecules discussed above are also linear motors. However, there are at least two *rotary* motors that are of central importance in biology and have received fundamental insights from single molecule (or at least single motor) investigations.

The bacterial flagellum (a completely different structure to the eukaryotic flagellum) is driven by such a rotary motor, which in this case is driven by an ion gradient across the bacterial cell membrane. Single molecule techniques have now begun to reveal the details of this extremely large and complex motor (see Chapter 4, this volume). Perhaps one of the most fascinating rotary motors which has been revealed by single molecule methods is the F_1-F_o ATPase. This enzyme is (under normal physiological conditions) responsible for using the energy of the proton gradient across the inner mitochondrial membrane to convert ADP into ATP (Kinosita et al., 2004; Noji and Yoshida, 2001). In fact, it is a complex of two rotary motors: the F_o component is driven by the ion gradient (rather like a turbine), and the rotation is transferred by a rotating shaft to the F_1 portion, which catalyzes the synthesis of ATP (this portion can also be run "backwards," consuming ATP and driving rotation) (Adachi et al., 2007; Noji et al., 1997).

Finally, it is interesting to see that single molecule methods are beginning to be developed for applications outside the research lab. These typically involve the manipulation and detection of trace quantities of biological molecules. One approach is the use of microfluidics, as is seen, for example, in Chapter 5, this volume. A popular method for the detection and quantitation of biomolecules in complex mixtures is the microarray. In Chapter 10, this volume, it is shown how this approach can be pushed to the ultimate limit through the use of single molecule detection. Detection of specific mRNAs is demonstrated with a detection limit of 112 molecules and almost five orders of magnitude dynamic range. Also discussed are applications to measuring DNA methylation and DNA fragment sizes, DNA mapping, and sequencing.

Conclusions

The pace of development of single molecule experiments continues to accelerate, in terms of both the physical techniques used and the systems they are applied to. This book aims to provide a snapshot or cross section of the current state of play in the field. It can never be comprehensive, and rapid progress will continue; but I hope that the reader will be left with a strong impression of the profound changes that single molecule measurements are making to our understanding of biological systems.

Acknowledgments

The writing of this chapter was supported by NPL's Strategic research program, and by the National Measurement System of the Department for Innovation, Universities and Skills (DIUS).

References

Adachi, K., Oiwa, K., Nishizaka, T., Furuike, S., Noji, H., Itoh, H., Yoshida, M., and Kinosita, K. (2007). Coupling of rotation and catalysis in F-1-ATPase revealed by single molecule imaging and manipulation. *Cell* 130, 309–321.

Axelrod, D., Sund, S. E., Johns, L. M., and Holz, R. W. (1999). Total internal reflection fluorescence microscopy: applications to cell biology. *J Gen Physiol* 114, 15.

Bayley, H. (2006). Sequencing single molecules of DNA. *Curr Opin Chem Biol* 10, 628–637.

Bayley, H. and Cremer, P. S. (2001). Stochastic sensors inspired by biology. *Nature* 413, 226–230.

Bustamante, C. (2008). In singulo biochemistry: when less is more. *Annu Rev Biochem* 77, 45–50.

Cai, L., Friedman, N., and Xie, X. S. (2006). Stochastic protein expression in individual cells at the single molecule level. *Nature* 440, 358–362.

English, B. P., Min, W., van Oijen, A. M., Lee, K. T., Luo, G., Sun, H., Cherayil, B. J., Kou, S. C., and Xie, X. S. (2006). Ever-fluctuating single enzyme molecules: Michaelis–Menten equation revisited. *Nat Chem Biol* 2, 87.

Finzi, L. and Gelles, J. (1995). Measurement of lactose repressor-mediated loop formation and breakdown in single DNA molecules. *Science* 267, 378–380.

Fusco, D., Accornero, N., Lavoie, B., Shenoy, S. M., Blanchard, J. M., Singer, R. H., and Bertrand, E. (2003). Single mRNA molecules demonstrate probabilistic movement in living mammalian cells. *Curr Biol* 13, 161–167.

Golding, I. and Cox, E. C. (2004). RNA dynamics in live *Escherichia coli* cells. *Proc Natl Acad Sci U S A* 101, 11310–11315.

Greenleaf, W. J., Woodside, M. T., and Block, S. M. (2007). High-resolution, single molecule measurements of biomolecular motion. *Annu Rev Biophys Biomol Struct* 36, 171–190.

Howard, J. (2001). *Mechanics of Motor Proteins and the Cytoskeleton*. Sinauer Associates, Sunderland, MA.

Hunt, A. J. and Howard, J. (1993). Kinesin swivels to permit microtubule movement in any direction. *Proc Natl Acad Sci U S A* 90, 11653–11657.

Kinosita, K., Adachi, K., and Itoh, H. (2004). Rotation of F-1-ATPase: how an ATP-driven molecular machine may work. *Annu Rev Biophys Biomol Struct* 33, 245–268.

Knight, A., Mashanov, G., and Molloy, J. (2005). Single molecule measurements and biological motors. *Eur Biophys J* 35, 89.

Mashanov, G. I., Nenasheva, T. A., Peckham, M., and Molloy, J. E. (2006). Cell biochemistry studied by single molecule imaging. *Biochem Soc Trans* 34, 983–988.

Moerner, W. E. (1996). High-resolution optical spectroscopy of single molecules in solids. *Acc Chem Res* 29, 563–571.

Molloy, J. E., Burns, J. E., Kendrick-Jones, J., Tregear, R. T., and White, D. C. S. (1995). Movement and force produced by a single myosin head. *Nature* 378, 209–212.

Neuman, K. C. and Nagy, A. (2008). Single molecule force spectroscopy: optical tweezers, magnetic tweezers and atomic force microscopy. *Nat Methods* 5, 491–505.

Nishizaka, T., Seo, R., Tadakuma, H., Kinosita, K., and Ishiwata, S. (2000). Characterization of single actomyosin rigor bonds: load dependence of lifetime and mechanical properties. *Biophys J* 79, 962–974.

Noji, H., Yasuda, R., Yoshida, M., and Kinosita, K. (1997). Direct observation of the rotation of F-1-ATPase. *Nature* 386, 299–302.

Noji, H. and Yoshida, M. (2001). The rotary machine in the cell, ATP synthase. *J Biol Chem* 276, 1665–1668.

Roy, R., Hohng, S., and Ha, T. (2008). A practical guide to single molecule FRET. *Nat Methods* 5, 507–516.

Sabanayagam, C. R., Eid, J. S., and Meller, A. (2005). Long time scale blinking kinetics of cyanine fluorophores conjugated to DNA and its effect on Forster resonance energy transfer. *J Chem Phys* 123, 224708.

Schafer, D. A., Gelles, J., Sheetz, M. P., and Landick, R. (1991). Transcription by single molecules of RNA polymerase observed by light microscopy. *Nature* 352, 444–448.

Shav-Tal, Y., Darzacq, X., Shenoy, S. M., Fusco, D., Janicki, S. M., Spector, D. L., and Singer, R. H. (2004). Dynamics of single mRNPs in nuclei of living cells. *Science* 304, 1797–1800.

Soni, G. V. and Meller, A. (2007). Progress toward ultrafast DNA sequencing using solid-state nanopores. *Clin Chem* 53, 1996–2001.

Tolic-Norrelykke, S. F., Engh, A. M., Landick, R., and Gelles, J. (2004). Diversity in the rates of transcript elongation by single RNA polymerase molecules. *J Biol Chem* 279, 3292–3299.

Walter, N. G., Huang, C. Y., Manzo, A. J., and Sobhy, M. A. (2008). Do-it-yourself guide: how to use the modern single molecule toolkit. *Nat Methods* 5, 475–489.

Xie, X. S. (2001). Single molecule approach to enzymology. *Single Mol.* 2, 229–236.

Xie, X. S. and Lu, H. P. (1999). Single molecule enzymology. *J Biol Chem* 274, 15967–15970.

Yin, H., Landick, R., and Gelles, J. (1994). Tethered particle motion method for studying transcript elongation by a single RNA polymerase molecule. *Biophys J* 67, 2468–2478.

Yu, J., Xiao, J., Ren, X. J., Lao, K. Q., and Xie, X. S. (2006). Probing gene expression in live cells, one protein molecule at a time. *Science* 311, 1600–1603.

CHAPTER 1

Single Molecule Studies of Myosins

Rachel E. Farrow
Division of Physical Biochemistry, MRC National Institute for Medical Research,
The Ridgeway, Mill Hill, London NW7 1AA, UK

Jeremy C. Fielden
Division of Physical Biochemistry, MRC National Institute for Medical Research,
The Ridgeway, Mill Hill, London NW7 1AA, UK

Alex E. Knight
Biotechnology Group, National Physical Laboratory, Hampton Road, Teddington,
Middlesex TW11 0LW, UK

Justin E. Molloy
Division of Physical Biochemistry, MRC National Institute for Medical Research,
The Ridgeway, Mill Hill, London NW7 1AA, UK

Summary

Myosins are motor proteins that are present in virtually all eukaryotic cells; indeed the recent explosion in genomic information shows that there are at least 24 different families of myosin in higher organisms. The first myosin to be isolated, now called myosin II, is the protein in muscles that, together with actin, brings about contraction. It is a molecular machine that converts the chemical free energy of adenosine triphosphate (ATP) to mechanical force and movement. Studies, by both "traditional" and "single molecule" approaches, have been focused on understanding the mechanism of this energy conversion and the diversity of functions of myosins in cells.

Keywords

myosin; motor protein; actin; motility assay; optical tweezers

Introduction

Myosins have been categorized into a large number of different families based on their sequence similarities, although the exact number of families is somewhat controversial, ranging from a relatively modest 24 (Foth et al., 2006) to 35 (Odronitz and Kollmar, 2007) or even 37 (Richards and Cavalier-Smith, 2005). Conventionally, the classes are numbered with roman numerals in their approximate order of discovery. Humans have 39 myosin genes, 9 of which encode myosins responsible for muscle contraction and the rest responsible for a wide variety of other cell motilities. In recent years, much interest has turned to these newly discovered, "nonmuscle" myosins. All myosins share the same basic "body plan" and have three main functional regions:

- A force-producing *motor* region that (a 90 kDa prolate spheroid of 7 nm × 5 nm) contains the active site responsible for adenosine triphosphate (ATP) hydrolysis and a binding site for the cytoskeletal (actin) filament.

- A *regulatory* region that binds up to six "light chains" (low-molecular-weight proteins of the calmodulin family). This region can work as a "lever arm" to amplify (and sometimes reverse) the action of the motor. The motor and regulatory regions together form the so-called *myosin head*.

- Finally, there is a highly divergent *tail* region that enables different myosins to bind different "cargoes" and therefore to perform a wide variety of cellular functions.

Atomic structures of several myosins have been determined, and we know that the functional regions listed above are themselves composed of several structural domains.

Filamentous actin (F-actin) forms the polymer track that myosin moves along. It consists of molecules of globular actin (a 42 kDa oblate spheroid of 2.5 nm × 5 nm) arranged as a two-start, helical chain with a repeat distance of around 72 nm (or 14 monomers) and a pseudo-repeat of about 36 nm (Schroder et al., 1993). Each actin monomer contacts four neighboring molecules and the same face of each monomer projects outward from the filament backbone. Every monomer has the same axial rise of 2.75 nm, but each is rotated azimuthally by ~166° for each successive monomer in the chain, giving rise to a so-called 13/6 helix, which repeats every 14 monomers. Because the monomers pack together with their outer faces all pointing in the same direction, the filament is polar with a so-called "plus end" (which polymerizes rapidly) and a "minus end" (slow growing). The myosin motor region binds in a stereospecific manner to actin and therefore takes up a defined orientation with respect to the filament axis, and this sets the geometry of the system (Figure 1.1) (Houdusse et al., 2000). The system of interacting actin and

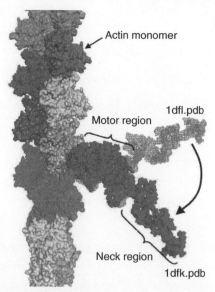

Figure 1.1: The structure of F-actin (PDB entry 1 mvw) and a bound myosin II head (1 dfk and 1 dfl superimposed) have been aligned using 1 mvw as a template structure. Note the helical arrangement of actin monomers and the "lever" action of myosin demonstrated by the two crystal structures of myosin: with nucleotide analogues bound (1 dfl), and in the apo- or rigor state (1 dfk) (Houdusse et al., 2000). The barbed or plus end of actin is at the bottom of the figure. The swing of the regulatory domain through an angle of about 60° would generate a movement of about 5–10 nm as the myosin molecule moves through its catalytic cycle

myosin molecules is referred to as *actomyosin*. Nearly all myosins move toward the plus end of actin but there is at least one exception (myosin VI), which has a small additional structure within its motor region that causes it to move backwards.

The myosin head is the functional core of the molecule and it is highly conserved across the myosin classes, suggesting that all myosins work by a similar molecular mechanism. The current view is that myosin binds to F-actin with the products of ATP hydrolysis (ADP and phosphate, P_i) bound to its catalytic site. Then, as the products are sequentially released (P_i first, then ADP), myosin changes conformation to produce a translational movement or "working stroke." After the products have been released, the catalytic site is vacant and myosin forms a tightly bound "rigor" complex with actin. Binding of a new ATP molecule to myosin causes the rigor complex to dissociate and subsequent ATP hydrolysis resets the original myosin conformation so that the cycle can be repeated. This catalytic cycle is shown schematically in Figure 1.2.

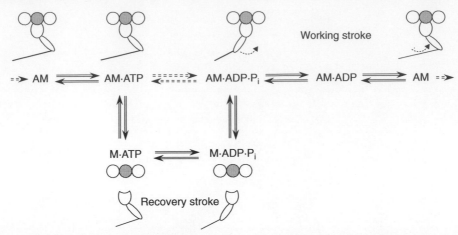

**Figure 1.2: The actomyosin ATPase cycle. Actin and myosin form a tightly bound "AM,"
or rigor state, in the absence of ATP. When ATP binds, the actomyosin complex quickly
dissociates and ATP is hydrolyzed to ADP and P_i. The products remain trapped in the
catalytic site until myosin rebinds actin. When bound to actin, the products are rapidly
released and myosin undergoes a conformational change that produces force and movement.
The cycle is driven by the free energy liberated by ATP hydrolysis (10^{-21}J or 100 pN nm), and
about half of this free energy is converted to external mechanical work**

In muscle, there is a near-crystalline arrangement of myosin-containing thick filaments and
actin-containing thin filaments within the sarcomeres of each muscle cell (or fiber). This
highly ordered structure enables individual molecules to work together as a team and generate
huge forces and rapid velocities of shortening. Much of what we know about the mechanism
by which myosin produces force and movement stems from early structural, mechanical, and
biochemical studies performed using single muscle fibers from frogs and purified proteins
from rabbits (Huxley, 1969; Huxley and Simmons, 1971; Lymn and Taylor, 1971).

The development of in vitro motility assays (Kron and Spudich, 1986; Yanagida et al.,
1984) allowed myosin-driven movement of individual actin filaments to be observed
using fluorescence light microscopy. More recently, biophysicists have devised methods to
reduce the number of interacting components so that the underlying molecular motions are
revealed. The ultimate experiment has been to perform mechanical studies on individual
actomyosin interactions as a single molecule of ATP is broken down (Finer et al., 1994;
Molloy et al., 1995). Results from single molecule experiments have provided great insights
into the mechanism of force generation by actomyosin and allowed us to make mechanical
measurements on nonmuscle myosins that are available in only minute quantities within

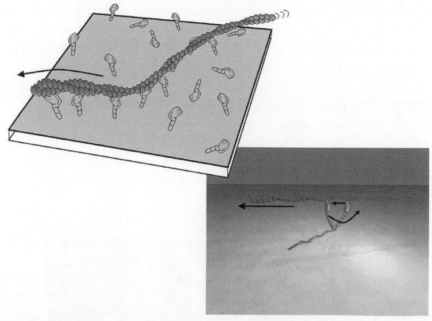

Figure 1.3: The actomyosin in vitro motility assay. A microscope coverslip is precoated with a hydrophobic material (such as a silane reagent or nitrocellulose solution); then a dilute myosin-containing solution is added and myosin adheres to the surface. Rhodamine-phalloidin-labeled actin filaments are then added in a buffered solution containing Mg·ATP. Actin filament sliding can be observed using fluorescence microscopy and video images are stored on computer to allow determination of sliding velocity. Some myosin types (such as myosin V, shown in the lower panel) can move along actin in a processive manner. These myosins can work individually to drive actin filaments over the surface. The activity of single processive myosin motors can therefore be viewed by this simple fluorescence microscopy–based assay

the cell and that often perform their physiological function as individual molecules (rather than the large molecular ensembles involved in muscle contraction). The single molecule techniques originally devised to study actomyosin from muscle have opened up an entire field of science that allows biophysicists to make major contributions to our understanding of cell motility and cell biology.

Motility Assays

In vitro motility assays (Scholey, 1993) enable actomyosin motility to be reconstituted from purified component molecules (see above and Figure 1.3). Rhodamine-phalloidin-labeled actin filaments can be visualized by fluorescence microscopy as they move on a microscope coverslip surface that has been coated with myosin or one of its proteolytic subfragments (Margossian and Lowey, 1982). In the absence of Mg·ATP, the filaments bind tightly to the surface, but when ATP is added, the filaments start sliding over the surface. A sensitive video camera is used to record the filaments (which appear as bright, red-colored threads) as they snake their way across the surface. Images of individual filaments that appear on sequential video frames are captured by a computer frame grabber and can then be tracked using image analysis software (Figure 1.4). The speed and direction of filament sliding can then be determined. Under typical conditions, rabbit skeletal myosin moves actin at a velocity of around 5–9 μm s^{-1}, which is similar to the maximum shortening velocity of the sarcomeres within intact muscle.

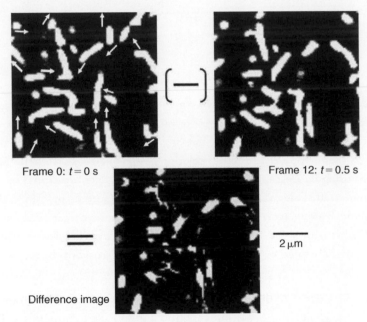

Frame 0: $t = 0$ s   Frame 12: $t = 0.5$ s

2 μm

Difference image

Figure 1.4: Actin filament sliding velocity can be quantified by recording sequential video images acquired using a camera coupled to the fluorescence microscope. Images are stored on computer and then individual filaments are tracked by image analysis software. The upper two images show video frames 0.5 s apart. The lower panel is the difference image (i.e., one image subtracted from the other). This reveals how the filaments have moved between successive frames. Tracking many hundred of filaments automatically requires sophisticated software, and in the early experiments, this was a significant computational challenge. It is a trivial task for a modern computer

Because in vitro assays reduce the system to its simplest functional components, much can be learned because of the ease with which the reaction conditions can be manipulated. These conditions include the chemical composition of the bathing solution, temperature, type of myosin or actin being used, and the mode of surface attachment of the proteins. In the context of this chapter, in vitro motility assays not only paved the way to single-molecule actomyosin studies but, along the way, gave rise to several important mechanistic findings. We list a few of these below:

1. The minimal requirements for motility are purified actin, the myosin head, and a buffered Mg·ATP solution (Toyoshima et al., 1987).

2. The actin filament polarity determines the direction of movement. Furthermore, myosins heads can swivel (presumably at the neck–tail junction) to realign themselves with actin (Sellers and Kachar, 1990; Toyoshima et al., 1989).

3. The maximum velocity of actin filament sliding is determined by the type of myosin (Sellers and Kachar, 1990).

4. The velocity of actin sliding varies with the length of the myosin regulatory domain, reinforcing the idea that this region acts as a "lever arm" to amplify the conformational change in the motor domain (Anson et al., 1996).

5. Actin rotates slowly about its long axis as it is moved by myosin (Nishizaka et al., 1993).

6. Some myosins interact intermittently with actin (such as muscle myosin II) whereas others are processive (such as myosin V), and a single molecule is sufficient to produce actin sliding over long distances (Howard, 1997; Mehta et al., 1999).

7. Some myosins (e.g., myosin VI) move actin in the "reverse" direction and are known as "minus-end-directed" myosin motors (Wells et al., 1999).

8. By genetically reengineering the myosin neck region so that it points back across the myosin head, the directionality of a plus-end-directed myosin can be reversed (Tsiavaliaris et al., 2004).

Why Work with Single Molecules?

It is likely that the molecular mechanism deduced from modeling the biochemical and mechanical behavior of an ensemble of molecules is correct (Huxley and Simmons, 1971; Lymn and Taylor, 1971). But, there are important subtleties that can only be resolved by directly studying the turnover of individual molecules. Furthermore, the mechanical properties of the recently discovered cellular (or "unconventional") myosins can *only* be studied using single molecule methods. To summarize:

- Single molecule experiments can give unequivocal information about how myosins work and can provide new insights into their mechanism.

- Sequential steps that make up the biochemical pathway can be observed directly, so that the chemical trajectory of an individual myosin can be followed in space and time.

- There is no need to synchronize a population of myosins in order to study their biochemical kinetics.

- Single molecule data sets can be treated in a wide variety of ways – for example, one can specifically look for heterogeneity in behavior.

- Molecules that perform their physiological function as individual entities – such as myosin V – often need to be studied as single molecules.

Molecular Mechanics

At the scale of single molecules, "life" is dominated by thermal energy. At absolute zero, molecular vibrations cease and chemical reactions stop. But, at room temperature, thermal energy (k_BT; Boltzmann's constant $\times$ absolute temperature $= 4\,\mathrm{pN\,nm}$) does sufficient mechanical work to bend and distort molecules to allow chemical reactions to proceed [see Eq. (1.2)]. During the myosin ATPase cycle, shape changes occurring at the catalytic site propagate across the molecule affecting, at one end, its affinity for actin and, at the other, the disposition of the regulatory region (lever arm). If prevented from moving, the molecule becomes distorted and force is generated. By storing mechanical work as elastic strain energy, the actomyosin complex captures the sudden changes (nanosecond timescale) in chemical potential associated with steps in the biochemical cycle so that external work can be performed on a slower timescale (milliseconds to seconds) – this can be either as a muscle fiber shortens or as a vesicle is being transported through the cytoplasm. If one were able to observe a single myosin motor at work, one would expect it to produce sudden (jerky) movements as it tugs on the actin filament.

Evolution of Single Molecule Actomyosin Experiments

Part of the "central dogma" of the actomyosin force-producing cycle (described above) is that each myosin molecule works independently and generates force as a stochastic, square-wave pulse that arises each time it binds, pulls, and then releases actin. The amplitude of force fluctuations produced by a small ensemble of myosins should depend upon the square root of the number of molecules, whereas the total force that they produce should be linear with the number (known as Rice's law). This points to a form of statistical analysis that could be applied to force signals produced by small numbers of myosin and which might give insights into the underlying molecular mechanism. In fact, this idea of "noise analysis" originates from electrical studies of the acetylcholine receptor at the nerve synapse (Katz and Miledi, 1970). To recapitulate this type of study with actomyosin, one might think of trying to measure the force fluctuations produced by a muscle fiber due to individual myosins binding and pulling. However, because a single muscle fiber contains about 10^{12} molecules, the expected fluctuations in force would be just one millionth of the total force. So, early efforts focused on using single myofibrils (containing about 10^7 myosin heads) in the hope of measuring force fluctuations of about one-thousandth the total force signal. Unfortunately, these heroic early attempts basically failed to detect mechanical fluctuations from individual myosin heads (or "cross-bridges" as they are known in the muscle field) (Borejdo and Morales, 1977; Iwazumi, 1987). In the early 1990s, a new method was developed in which individual actin filaments were held using a glass microneedle that was positioned above a microscope coverslip that had been coated with myosin molecules (Ishijima et al., 1991). When the actin filament landed on the surface, it was pulled by a small number (less than 50) of myosins and the microneedle was deflected. The workers found that when loaded in this way it fluctuated in position in a manner that depended upon the square root of the total force signal being recorded. This is a characteristic signature for mechanical noise being produced by individual myosin molecules. These early experiments heralded the start of single molecule studies of actomyosin.

The critical issue in making mechanical recordings from an individual muscle myosin (so-called muscle myosin II) is that the amount of energy change involved in each interaction is pitifully small: ATP hydrolysis produces a maximum of 100 pN nm or 10^{-21} J of energy, and of this, only about 50% is converted to mechanical work. This is just 10-fold greater than thermal energy (4 pN nm) and the expected forces and movements will be in the piconewton and nanometer range; that is, of the same order of magnitude as thermal fluctuations. Furthermore, because muscle myosin II interacts with actin in an intermittent fashion, one needs to hold both actin filament and myosin molecule in place during the recording period, otherwise they will simply diffuse away from each other. This presents a significant technical challenge in terms of holding a relatively flexible, filamentous protein (actin) and a single myosin at fixed positions with nanometer stability. To achieve this, a completely new type of apparatus was required to make the measurements. In the early 1990s, several laboratories exploited a new method, which had been developed around that time, by Ashkin called *optical tweezers* (Ashkin, 1970; Ashkin et al., 1986) which will be discussed in detail below.

Technologies

To measure the force and movement produced by a single myosin, a mechanical *transducer* (a device that produces an electrical signal proportional to movement or load) of the correct stiffness and sensitivity is required. Three approaches have been used for actomyosin studies: glass microneedles (Figure 1.5A), atomic force microscopy (AFM, Figure 1.5B), and optical tweezers (Figure 1.6).

Microneedles

Ishijima et al. (1991) constructed an apparatus that used an ultrathin glass microneedle that was attached to the end of a single actin filament. An image of the microneedle was cast onto a split photodetector that enabled mechanical recordings to be made from a small ensemble of myosin molecules interacting with actin (Figure 1.5A). Low microneedle stiffness ($\sim 1\,\mathrm{pN\,nm^{-1}}$ deflection at the tip) and exceptional sensitivity of the split-photocell detection system allowed forces produced by very small numbers of myosin (~ 10) to be measured. They found that the force signal exhibited large amplitude, stochastic fluctuations with a noise spectrum that "rolled-off" at around $5\,\mathrm{Hz}$. The amplitude of the force fluctuations was proportional to the square root of the mean force. Both observations are consistent with actomyosin cycling, since the ATP turnover, and hence mean cycle time, is around $5\,\mathrm{s^{-1}}$. Detailed analysis of the noise spectrum showed that the average force produced by a single myosin head was about $1\,\mathrm{pN}$.

AFM

Kitamura et al. (1999) developed an instrument based on the principle of the AFM whereby a single myosin molecule was attached to a very fine probe (made of a zinc oxide spicule) and this was then brought into contact with an actin filament bundle and displacements and forces determined from movements of the probe tip (Figure 1.5B). They found that a muscle myosin II head produces a series of $5\,\mathrm{nm}$ steps (resembling a staircase) from a single ATP molecule.

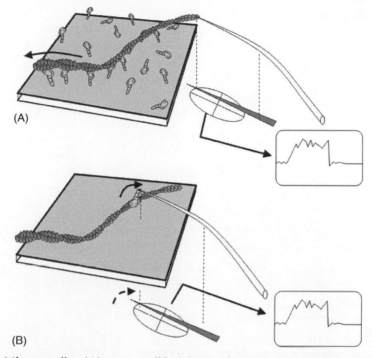

Figure 1.5: Microneedles (A) or a modified form of atomic force microscopy (AFM) (B) has been used to measure the force produced by small numbers of myosin molecules and individual molecules, respectively. In (A), a single actin filament is attached specifically to a glass microneedle (made using a micropipette puller device). When the actin filament touches a microscope coverslip coated with myosin molecules, it is pulled and causes the needle to deflect. Deflections are recorded using a sensitive split photodiode detector system with a sensitivity <1 nm. In (B), a single myosin molecule is captured on the tip of a zinc oxide whisker coupled to a microneedle; when brought close to a bundle of actin filaments fixed to the microscope coverslip, the myosin binds and moves, causing the needle to deflect so that individual forces can be measured (as in (A))

The "upside-down" geometry adopted by this group (compared to the in vitro motility assay arrangement) meant that the actin filament (or a bundle of actin filaments) is fixed and myosin moves along the top of it. This geometry has been adopted more recently for studies of processive myosins such as myosin V and myosin VI (see section "Studies of the Processive Motor, Myosin V").

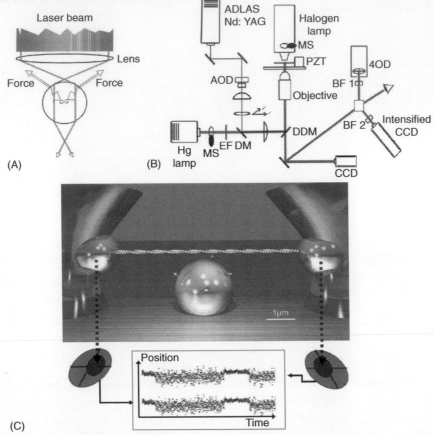

Figure 1.6: Microscopic particles (on the order of the wavelength of light ~1 μm diameter) can be captured and manipulated using a tightly focused beam of laser light by an effect called optical tweezers. An oil immersion microscope objective lens (NA >1.2, magnification >40×) and a Gaussian single mode laser (>5 mW power) are required. Panel B shows how various optical components are arranged around a light microscope so that optical tweezers can be produced by a near-infrared laser (Nd:YAG). The beam path is steered using acousto-optic deflectors (AODs) and combined with green light using a dual dichroic mirror (DDM), so that rhodamine-labeled actin filaments can be visualized using a CCD camera and manipulated by attaching it at either end to an optically trapped bead. Panel C shows the experimental arrangement adopted to make mechanical recordings from single myosin molecules that are adhered to a glass bead attached to the microscope coverslip surface. Bead deflections are monitored using four-quadrant photodiode detectors (4QD) and signal recorded by computer. In most experimental setups, the entire experiment is computer controlled. Closed-loop feedback can be applied to servo the bead positions by sending suitable control signals from the computer to the piezoelectric stage (PZT) or AOD devices (Knight et al., 2005)

Optical Tweezers

Optical tweezers (otherwise known as optical traps) harness the photon pressure produced by an intense beam of laser light to hold and manipulate micrometer-sized particles. Each photon of light carries energy $h\nu$ and momentum $h\nu/c$, so if absorbed by an object the momentum transferred from a light beam of power P gives a reaction force F on the object, given by[1]:

$$F = \frac{nP}{c} \tag{1.1}$$

If light is reflected by an object then the momentum transferred is double that of Eq. (1.1), and if it is refracted or diffracted to take a new path at an angle θ to the incident beam then the resulting force will be equal to $F \cos(\theta)$ (Ashkin, 1970). Optical tweezers work by arranging the direction and intensity of the incoming beams of light to be such that the object is held fixed in three dimensions. One might imagine that this would require a very complicated optical arrangement. However Ashkin discovered that all that is required is a high numerical aperture lens and an input beam of Gaussian intensity profile (Ashkin et al., 1986). A microscope objective lens and a laser pointer of a few milliwatts power output is all that is required to capture and manipulate micrometer-sized glass or plastic beads suspended in water (Figure 1.6). Optical tweezers have been adapted to measure the minute forces and movements produced by individual myosin molecules by monitoring the position of the trapped particle with nanometer precision. Most optical tweezers–based devices are constructed around a light microscope combined with a number of external optical components, required to align and direct an infrared (Nd:YAG, $TEM_{0,0}$) laser beam toward the microscope objective lens. By splitting the input laser beam into two paths or by rapidly chopping the laser beam between two sets of x,y coordinates, two diffraction-limited spots of light can be created and moved independently within the $x-y$ object plane of the microscope (Figure 1.6B). Optical tweezers produce a restoring force to hold the micrometer-sized plastic microsphere at the focus. This restoring force is linearly related to displacement (i.e., it is Hookean) over a distance of about ± 250 nm. The bead position can therefore be used to measure

---

[1] In these equations, h is Planck's constant and ν is the frequency; c is the speed of light; n is the ratio of the refractive indices of the particle and the medium.

both force and movement. In most apparatus, the bead position is measured using a four-quadrant photodiode position sensor, and either the scattered laser light emanating from the trapping laser or a bright field image of the bead is imaged onto the sensor. A calibrated, electrical signal is captured using analogue-to-digital converters and saved by a computer. If scattered light from the optical trap is used, the signal is proportional to displacement from the trap center (i.e., force), whereas if the bead is imaged directly onto the sensor, using bright- or dark-field microscopy the signal measures the absolute position of the bead. In both cases, the sensor determines the centroid of the bead position with a resolution of better than 0.5 nm. The most common geometrical arrangement used to make actomyosin mechanical measurements is the so-called three-bead arrangement (Figure 1.6C) (Finer et al., 1994). In the experiment, an actin filament is tensioned between two beads, held in independent optical tweezers, and this bead–actin–bead assembly is positioned close to a third bead that is glued to the microscope coverslip. The third bead is sparsely coated with myosin molecules, so that when a single myosin head binds to actin and pulls, the beads are displaced by a small distance (a few nanometers). The motions produced by repeated actomyosin interactions can then be recorded using the position sensor (Figure 1.7). This arrangement is ideal for studying myosins that interact in an intermittent fashion with actin because time series data show repeated, stochastic, binding interactions between an individual myosin and actin. Binding events are interspersed with periods during which the actin is free. Because of the low optical tweezers stiffness used in these experiments, thermal motion of the actin filament carries it back and forth past the myosin, so there is positional noise that needs to be factored out from the measurements. This complication to the measurement led to an early controversy in estimating the size of the working stroke from such records, which is discussed in the following section.

The Myosin Working Stroke

To determine the displacement or "working stroke" produced by a single actomyosin interaction, the stiffness of the apparatus must be much *less than* that of the actomyosin complex (so that myosin may undergo its full working stroke unhindered). To measure displacement, most workers use optical tweezers with a stiffness (κ) of ≈ 0.05 pN nm^{-1}. At such low stiffness, the transducer (the bead held in the optical tweezers) necessarily exhibits large amounts of Brownian motion. The mean squared amplitude, $\langle x \rangle^2$, can be readily calculated from the principle of equipartition of energy:

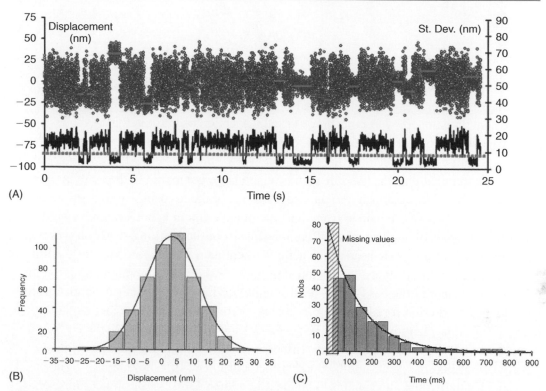

Figure 1.7: Single molecule mechanical interactions between rabbit skeletal muscle myosin II (specifically, the two-headed proteolytic subfragment HMM (Margossian and Lowey, 1982)) and rabbit F-actin measured using optical tweezers. The upper trace shows the position of one of the optically trapped beads measured along the axis of the actin filament (defined as the x-axis); it is noisy because background thermal vibration causes about 50 nm peak-to-peak motion of the bead within the optical trap. The lower trace is a plot of the local variance of the positional data. When myosin binds actin system, stiffness increases and the amplitude of thermal vibration is reduced (variance falls). Individual events can be detected by thresholding the variance data (dotted line) so that individual events can then be identified in the original position data (solid lines). Panel B shows the distribution of binding event amplitudes measured from the mean resting position of the bead, scored for several hundred events. The fit is a Gaussian, with a mean of 4.5 nm and standard deviation of 13 nm. Panel C shows the distribution of event lifetimes, with a fit to a single exponential with a rate constant of $7\,\mathrm{s}^{-1}$). The results are from different experiments (performed at 23°C, $1\,\mu$M Mg·ATP (panel A and B) and $2\,\mu$M Mg·ATP (panel C))

$$\frac{1}{2}\kappa\langle x\rangle^2 = \frac{1}{2}k_\mathrm{b}T \tag{1.2}$$

Because the optically trapped beads are held by a springlike stiffness within aqueous solution, variations in position should follow a Gaussian distribution, and the motion will be heavily damped by viscous drag. Visual inspection of unfiltered data shows peak-to-peak motions of about 60 nm (4 pN nm/0.05 pN nm^{-1} = 80 nm^2 = 9 nm standard deviation) with a noise spectrum that extends to a few hundred hertz. Thus, movements of the actin filament past the fixed myosin head are large compared to the actin monomer spacing and very large compared to the expected size of the myosin working stroke. So, interpretation of records obtained from this type of experiment is not straightforward, and different types of analysis give different estimates of the working stroke. It was noted that attachment of a myosin head to the actin filament increases the stiffness of the link between the actin filament and "mechanical ground" with a corresponding reduction in thermal movement of the bead–actin–bead assembly. Changes in variance can therefore be used as the criterion for identifying individual myosin attachments to actin (Molloy et al., 1995); this can most simply be achieved by plotting the "running variance" of the data (proportional to the instantaneous stiffness of the system) alongside the original position data (Figure 1.7A) – although other methods exist. Thresholding the variance data enables individual attachments to be identified, their level (or amplitude) measured, and the results plotted as a histogram. The spread of the histogram of myosin-bound levels closely matches the random positions explored by actin due to thermal motion, but its mean position is shifted in one or other direction by the size of the myosin working stroke (e.g., +5 nm). If the actin filament polarity is reversed, then the mean position of the histogram is also reversed (e.g., changing to −5 nm).

This observation implies that, when one averages results from several experiments, myosin can bind at any position along the actin filament. However, if data are recorded from just a single actin filament, then a different pattern emerges. One finds "hot spots" for myosin binding on a minor repeat distance of 5 nm (the actin monomer periodicity) and a major repeat of 36 nm (the repeat of the actin pseudo-helix). Experiments made using an optical tweezers system that had extremely good mechanical stability (<2 nm drift over several minutes) enabled the actin filament structure (the geometrical origin of the actin-binding sites) to be "mapped" (Steffen et al., 2001). Steffen et al. found that myosin-binding interactions marked the position of individual actin-binding sites along the filament. Furthermore, because the optical tweezers are elliptical in the z-axis, the trapped microbeads (although nominally of spherical and isotropic form) took up a defined orientation, so that actin was unable to rotate about its long axis, and only monomers with the correct azimuth could bind myosin. The fact that myosin binds preferentially to actin monomers of the correct azimuthal orientation has interesting implications for how myosin might work within the muscle sarcomere, and also how a two-headed, processive myosin might "walk" along the filament.

Since the individual power stroke made by myosin can be measured, one might then ask whether myosin moves actin instantaneously as soon as it binds, or whether it first binds and then rocks forward, in order to produce its working stroke. In other words, does the working stroke occur in a single phase, or is it composed of several, discrete motions associated with sequential conformational rearrangements of the actomyosin complex (corresponding to the transitions in Figure 1.2)?

To answer these more detailed questions, we need to consider the expected timescale of the events we wish to measure and the time resolution of our experimental measurements. From biochemical studies, we know that ADP and phosphate leave the myosin active site within a few milliseconds after actin binding, whereas the time resolution of the optical tweezers–based transducers, limited by viscous drag on the beads, is around 10 ms. However, time resolution can be improved by applying a small-amplitude, high-frequency oscillation to the transducer (e.g., a 1 kHz oscillation to the optical tweezer), so that by monitoring pick up of this oscillation, changes in stiffness can be measured within 1 ms. Using this approach, it was found that some myosins (myosin I and smooth muscle myosin II) produce their working stroke in two discrete phases (Molloy and Veigel, 2003; Schmitz et al., 2002; Veigel et al., 1999). These mechanical results are consistent with cryo-electron microscopy studies made using the same myosins (Jontes et al., 1995) which showed a smaller lever-arm swing or rotation of 32° than had been previously observed (see Figure 1.1). More recent studies, using fast skeletal muscle myosin II, also indicate that the working stroke occurs in two phases (Capitanio et al., 2006). So, it appears that myosin binds to actin, generates about 80% of its working stroke within the time resolution of the measurement, and completes the final 20% of the movement following a finite time delay.

The working stroke size depends upon the type of myosin and the length of the regulatory domain. For instance, a single head of skeletal muscle myosin II (which has two light chains) produces a working stroke, d, of about 5 nm, but this can be increased or decreased by lengthening or shortening the regulatory region respectively (Ruff et al., 2001). Similarly, the working stroke for myosin V (which has six calmodulin light chains) is around 25 nm (Veigel et al., 2002), but this can be shortened to 5 nm by removing light chains (Sakamoto et al., 2003).

Force and Stiffness

Accurate determination of the force developed by myosin under isometric conditions (where the molecule is prevented from moving) requires that the stiffness of the apparatus (*transducer*) be much *greater than* that of the actomyosin complex. Under such conditions, myosin is prevented from moving, so it then generates its maximum possible force. To measure force requires a transducer with a stiffness of $\sim 10\,\mathrm{pN\,nm^{-1}}$ and the ability to measure its movement with subnanometer precision. To achieve this, negative feedback can be applied to *servo* the optical tweezer and compensate for bead movements as measured by the position detector. When myosin binds and pulls on the actin filament, the force is immediately counterbalanced by moving the optical tweezer in the opposite direction. Knowing the stiffness of the optical trap and the distance it has been moved, the opposing force applied can then be calculated. However, the situation becomes complicated because of compliance[2] at the bead-to-actin connection (Dupuis et al., 1997). This compliant linkage is extended (or "pulled out") as myosin exerts force. Because the extension of this linkage is unknown, the maximum force that can be generated by myosin tends to be underestimated. For this reason, published values cover a wide range from around 1 to 5 pN. More recent work, using a closed-loop feedback system that servos the position of one bead by moving the other bead (Takagi et al., 2006), obviates the series compliance problem and gave an estimate for the maximum force, F, produced by a single myosin of about 10 pN.

Correction for series compliance can also be made by applying a large-amplitude, low-frequency oscillation to one of the optical tweezers so that myosin is pushed and pulled several times while it is attached to actin. The resulting force-extension diagrams enable the stiffness of different components to be extracted. Various groups have estimated the myosin stiffness, κ_m, in this way to be around 1–$2\,\mathrm{pN\,nm^{-1}}$ (Veigel et al., 1998). So if we combine this finding with the estimate of the working stroke (5 nm for myosin II), the expected maximum force would be about 10 pN (similar to that of Takagi et al.). We can now estimate the total work done to be either $1/2\kappa_m d^2 = 25\,\mathrm{pN\,nm}$, or, if under constant load, $Fd = 50\,\mathrm{pN\,nm}$.

Kinetics

In order to measure the kinetics of the actomyosin interaction, one needs to measure the lifetimes of the bound events and then plot the distribution of lifetimes as a histogram. Single molecule experiments are often performed at limiting concentrations of ATP so that

---

[2] Compliance is simply the inverse (or reciprocal) of stiffness.

the cycle is artificially slowed down and the bound lifetimes of the actomyosin complex depend upon the rate of ATP binding. For most myosins this is around $10^6 \, M^{-1} \, s^{-1}$, so at $1 \, \mu M$ ATP, the rate of binding will be $1 \, s^{-1}$ and the average bound lifetimes will be around $1 \, s$. Notice, however, that the individual lifetimes will be stochastic in nature, since each binding event is terminated by the probabilistic binding of ATP to the rigor complex. To get a feeling for what the distribution of lifetimes will look like, let us suppose that the chance of an ATP molecule binding to an actomyosin rigor complex during a $1 \, s$ interval (our chosen bin size) is 50%. We can see that if we observed a large number of events, after $1 \, s$, half of all myosins will have bound ATP and detached from actin (i.e., 50% of the events terminate). Importantly, the remaining molecules, which are still bound to actin, have no knowledge of time. So, over the next $1 \, s$ time period, half of them will bind ATP and detach. Over each subsequent $1 \, s$ time interval, half the remainder will detach. Thus, we expect an exponential function with a characteristic "half-time," τ, of $1 \, s$ (the average time taken for 50% of the molecules to bind ATP). So, if the event lifetime is determined by a single probabilistic process, then the lifetime distribution will fit to a single exponential decay function, $A_t = A_0 e^{-t/\tau}$. Here A_0 is the amplitude at time 0 and A_t the amplitude at time t, where the amplitude corresponds to the number of molecules remaining attached to actin. Note that for a single exponential function, the average of the bound lifetimes, τ_{av}, equals τ. However, if the kinetic pathway is more complex and consists of several sequential biochemical steps (as in Figure 1.2), then we might expect more complicated lifetime distributions. By varying the concentration of different ligands (e.g., ATP, ADP, and P_i) and then curve fitting the lifetime distributions, it is possible to deduce the lifetime of different biochemical steps (Rief et al., 2000; Veigel et al., 1999).

The lifetime distributions that we create by pooling data from many hundreds of molecules, or from many hundreds of events recorded from an individual molecule, give rise to ensemble behavior. So we can see that by drawing histograms of data, we can move from single molecule behavior back to the more familiar ensemble behavior (rate constants, amplitudes, etc.). The great advantage of having the raw, single molecule, data sets is that we can extract far more information – we can ask whether the lifetime distributions correlate with the amount of force exerted on the myosin while it is bound to actin; if the average position of the myosin is different at the start and end of an event; or whether the size of the working stroke depends upon the average lifetime of the event (i.e., do short-lived events produce more or less movement?).

We should note at this point that there is a technical complication to the histogram method, as we need to account for missing events at short times because there is increasing difficulty in detecting short-lived binding interactions. One can either model the detection probability and correct for that in our observations, or define a minimum threshold value, called t_{min}, and then curve fit only for values greater than t_{min}, or use the formula $\tau = (t_{avmin} - t_{min})$, where t_{avmin} is the average of all values greater than t_{min} (Colquhoun and Sigworth, 1995) (see Figure 1.7C and the discussion of this problem in Chapter 8, this volume).

Studies of the Processive Motor, Myosin V

Myosin V is a two-headed, processive motor (Mehta et al., 1999; Rief et al., 2000; Walker et al., 2000) involved in vesicle and organelle trafficking (Cheney et al., 1993; Reck-Peterson et al., 2000). It produces successive 36 nm steps as it steps along the actin filament pseudo-repeat. This means that myosin V walks approximately parallel to the axis of the actin filament (Ali et al., 2002). Kinetic studies show that ADP release from the actomyosin V complex is the rate-limiting step in the ATPase cycle so it spends >70% of its ATPase cycle bound tightly to actin (De La Cruz et al., 1999). This is in contrast to muscle myosin II, which is an intermittent motor that spends just 5% of its time bound to actin. The two heads of myosin V can therefore move processively along actin, provided there is some level of cooperativity or coordination between its two heads, to ensure that at least one head always remains attached to actin (Veigel et al., 2002).

Myosin VI is also involved in intracellular transport (Buss et al., 2001) and is the only myosin shown consistently to move toward the minus end of actin (Wells et al., 1999). Reverse directionality of this motor is thought to arise because of a unique 53-amino-acid insert within the "converter" domain that links the motor domain to the neck. Kinetic studies of myosin VI show that ADP release is rate limiting (De La Cruz et al., 1999) and, like myosin V, it appears to be a processive motor (Nishikawa et al., 2002; Rock et al., 2001). However, it seems that this myosin can exist either as a nonprocessive, single-headed species in the cell (Lister et al., 2004) or in its dimeric form (Park et al., 2006).

Myosin V is now the best-studied myosin at the single molecule level. The fact that it "walks" processively along a single actin filament means that a wide variety of exciting new techniques have been developed in order to study its mode of action. The very first studies involved optical trapping using the methods developed for myosin II (the "three-bead" geometry described above and see Figure 1.8) (Mehta et al., 1999; Moore et al., 2001; Veigel et al., 2002). This work showed that a single (dimeric) myosin V molecule could bind actin and then move along it in several (up to about 10) 36 nm steps before detaching. This finding, combined with negatively stained electron micrograph observations (Walker et al., 2000), indicated that the two heads of myosin V could span the actin helical repeat distance (36 nm) and walk in a hand-over-hand fashion.

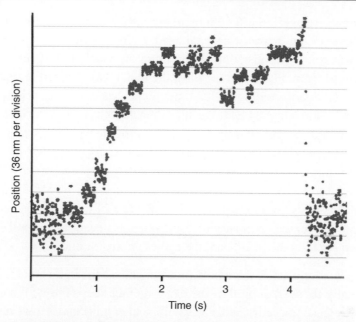

Figure 1.8: An optical tweezer's mechanical record showing the processive movement exhibited by full-length (dimeric) myosin V. Because this, two-headed, molecule can "walk" along actin, the beads held in the optical tweezer (see Figure 1.6) move in a "staircase" fashion as myosin pulls the attached actin forward in a series of steps. Each step is caused by the rearward head moving forward to become the new leading head. The molecule pauses after each step until ADP is released from the rear head so that ATP can bind and the next step can commence. This experiment was performed at 100 μM ATP, 23°C (Figure reproduced from Veigel et al., 2002)

Because of its processive nature, a simpler, "upside-down" trapping geometry was developed to study the movement of myosin V along actin. A single actin filament was fixed to a microscope coverslip and then a single myosin, bound to an optically trapped bead, was observed as it walked along, pulling the bead forward in a stepwise manner (Rief et al., 2000). This single-bead optical trapping geometry allows feedback control in which the bead displacement from the trap center position is kept constant (and therefore the force is also constant) so that walking kinetics can be examined under a range of different loads. The motivation behind this general type of experiment is to try and understand the detailed process by which the myosin heads move in a coordinated fashion, one after the other. Another approach has been to use a recombinant, monomeric form of the molecule (truncated after the regulatory domain so it cannot dimerize) in order to study the working stroke of a single head. Using the three-bead geometry, the working stroke was found to consist of a rapid initial movement of 20 nm followed by a further movement of 5 nm. Furthermore, when the intact, two-headed molecule walked along actin, the stiffness of attachment fell dramatically immediately prior to each step (Veigel et al., 2002). This implies that one head detaches completely from actin and then swings forward toward its new binding site, but because the working stroke is smaller than the step size, the new lead head must diffuse forward by about 10 nm in order to bind onto its target actin monomer. This process strains the molecule internally, causing the rear head to detach more rapidly and making the lead head stay bound for longer. This means that their kinetic cycles are no longer independent of one another. Internal load has a synchronizing effect on the chemistry of the two heads because as it pulls forward on the rear head, and back on the lead head, the cycles are maintained 180° out of phase.

As myosin V walks along actin, the whole molecule advances in discrete 36 nm steps, pausing periodically for ATP to bind (if the experimenter keeps the ATP concentration low) or for ADP release to occur (under physiological conditions). However, if it works by placing one head in front of the other (like the movement of our feet as we walk), then the individual heads should move by 72 nm for each 36 nm step taken (Figure 1.8). Since these distances are large, they should be measurable using optically based single molecule localization methods (Gelles et al., 1988; Schmidt et al., 1996), without the need for external mechanical intervention or use of macroscopic probes (such as plastic microspheres). In fact, the stepping motion of myosin V has been studied using a wide variety of so-called "super-resolution" optical techniques, many of which have been devised and pioneered for these specific studies. It is likely that their future impact and broad application across biology will be significant (see, e.g., Chapter 9, this volume, in particular Table 9.1).

Single Myosin Optical Studies

To study individual actomyosin interactions using imaging methods, one can either use *microscopic* probes (e.g., fluorophores, quantum dots, or nanogold) or *macroscopic* probes (e.g., latex beads, actin filaments, or microtubules) to label the molecules. The optical tweezers experiments discussed earlier relied upon imaging the centroid of a latex bead with high spatial precision, using a four-quadrant photodetector. The same approach can be used to track microscopic or macroscopic probes so long as they remain in the focal plane. Myosin V walking along a surface-attached actin filament is an ideal specimen. Movement can be tracked over long distances using camera-based detection systems. Often, rather exotic cameras are chosen by the experimenter based on sensitivity, quantum efficiency, noise level, or speed. In the experiments, an image of the probe is cast over several pixels of the camera and then a sequence of video images are saved to computer so that the centroid positions can be calculated either by fitting the pattern to a functional form (e.g., 2-D Gaussian) or by a center-of-mass calculation to give the position, intensity, or angle of the probe.

The assumption is that the probe accurately reports the position or angle of the molecule so that structural information can be directly obtained. In other experiments, the probe might be attached to a ligand (e.g., ATP) so that the timing of the chemical reaction can be followed in a temporal and spatial manner. One critical issue, which we need to consider when making high time resolution spatial measurements, is the amplitude and bandwidth of the independent thermal motions of the probe. If the probe is coupled tightly to the protein either by short chemical linkers or using bifunctional attachment chemistry, then independent movement of the probe should not be a problem. Also, if a microscopic probe is used, its independent thermal motion should be fast compared to the observation time window, and so movements will average to zero. However, if a macroscopic probe is used (such as a 1 μm diameter plastic microsphere), then we need to carefully consider both the amplitude $\langle x^2 \rangle$ and bandwidth, f_c, of its thermal motion, as this may interfere with our measurement just as we found for the optical trapping studies earlier. Thermal motion depends on the stiffness of attachment, κ, and viscous drag on the particle, which is given by $\beta = \pi \eta a$, where η is the viscosity of the medium and a is the radius of the bead. The amplitude is given by Eq. (1.2), and the bandwidth is shown in Figure 1.9 and Eq. (1.3).

$$f_c = \frac{\kappa}{2\pi\beta} \tag{1.3}$$

Single fluorophores and quantum dots might at first seem to be ideal because they are physically very small and unlikely to interfere with the mechanical behavior of the myosin. However, because they do not emit much light and their image will be shot noise (photon count) limited, we need to consider how accurately we can measure their position. We know that photons emitted by a single fluorophore will be imaged as a diffraction-limited spot with an approximately Gaussian intensity profile, having a full width at half maximum of 350 nm. So, our ability to determine the centroid (or mean position) of such an image is given by photon counting statistics; the standard error of the mean is $\sigma/\sqrt{n}$, where n is the number of photons collected and σ is the standard deviation of the distribution (175 nm) (see Figure 1.11). This means we require around 10 000 photons to detect the

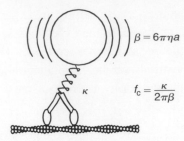

$\beta = 6\pi\eta a$

κ

$f_c = \dfrac{\kappa}{2\pi\beta}$

Figure 1.9: When a macroscopic probe is tracked using high-resolution video-tracking-based methods, one needs to consider both the amplitude (Eq. (1.2)) and the bandwidth of the independent probe motion. The bandwidth depends upon the viscous drag, and hence the size of the probe. As a rule of thumb, for very small probes such as single fluorophores, any motion will be over a wide bandwidth and so will average to zero, but for large probes (on the micrometer length scale), motion might occur over a period of several seconds

position with nanometer precision. But there is a confounding problem, in that single fluorophores typically photobleach after emitting about 10^7 photons. This means that excitation intensity must be carefully adjusted to optimize the photon emission rate. If the excitation power is too high, then the fluorophore will photobleach rapidly (<1 s); if it is too low, the fluorophore emission rate will be insufficient to measure above background. The distribution of fluorophore longevities will be exponentially distributed, so some fluorophores will last much longer than others. For some studies, one can simply retain the long-lived events and ignore the short-lived events, while for others, data sets will have to be corrected for photobleaching. Basically, the optimization process depends upon the type of measurement being made and the sort of information that one is interested in collecting.

The first convincing report of single fluorophore imaging in aqueous solution was made using myosin II from rabbit muscle (Funatsu et al., 1995). Individual myosin molecules that had been covalently tagged with a fluorescent dye called Cy5 were observed bound to a microscope coverslip surface, and individual ATP turnover events were followed as Cy3-labeled ATP bound and released from the myosin catalytic sites. In these early studies, the myosin was immobilized to the microscope coverslip so that no motility was observed. However, in later work, the same group combined this approach with the optical tweezers technique so that movement and ATP activity could be followed simultaneously (Ishijima et al., 1998). These two reports popularized use of a technique called total internal reflection fluorescence microscopy (TIRFM) that had been pioneered originally by Axelrod (1981). The principle of TIRFM is to direct a beam of light toward a microscope slide or coverslip such that its incident angle exceeds the critical angle

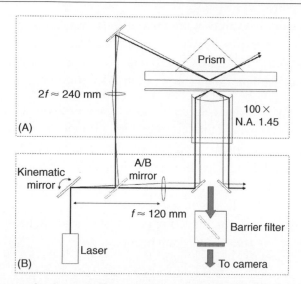

Figure 1.10: Two methods to realize total internal reflection fluorescence microscopy (TIRFM) within a research-grade inverted microscope. Panel A shows the prism-coupled method, whereby the laser beam is focused via a prism at the glass–water interface on the lower side of the microscope slide. Panel B shows an alternative arrangement, whereby laser light is coupled via a high numerical aperture, oil immersion objective lens (where N.A. >1.45). In both cases, the incident laser beam angle must exceed the critical angle for total internal reflection (~63°) adjusted using the kinematic mirror mount. Fluorescence emission is collected via the normal imaging light path using a suitable barrier filter to remove stray laser light. Light paths are selected by moving "mirror A/B" in or out of the light path

required for total internal reflection at the glass–water interface (~63°). The incident beam is then totally internally reflected but an evanescent field permeates the aqueous solution (to ~100 nm depth) and this will excite only the fluorophores in solution that come close to, or bind at, the surface. The signal-to-background noise ratio is sufficiently high that at nanomolar concentrations of fluorophore, individual molecules become visible as individual spots of light. In free solution, the fluorophores diffuse rapidly and are barely detectable, but when bound to the surface, they persist as bright and immobile spots of light until they eventually photobleach and disappear in a single rapid event.

TIRFM can be realized by introducing a laser beam into in a fluorescence light microscope either by a prism-coupling or by objective lens–coupling methods (Mashanov et al., 2003) (Figure 1.10). Once aligned, the objective lens–coupled method is much the easiest and most reliable method to use, so most groups have adopted this technique for visualizing myosin molecules at work in vitro or within living cells (Mashanov et al., 2004, 2006). To discover how myosin V moves its two heads as it walks along actin, a number of experimenters have attached various probes to one or both heads and then followed the motion of the probes using light microscopy (often TIRF illumination is used to do this).

1. By substituting a Cy3-labeled calmodulin light chain onto one of the myosin heads, Yildiz et al. (2003) discovered that myosin V moves in a "hand-over-hand" fashion. They showed that each myosin V head moves 72 nm each time the molecule advances by 36 nm (see Figure 1.11 and Molloy and Veigel, 2003). Nanometer tracking of individual fluorophores enabled the detailed motions of one myosin head to be resolved, and a simple kinetic argument showed that two ATPs were required for each 72 nm step; so we can infer that the other (unlabeled) myosin head must detach and then reattach before the next visible step occurs.

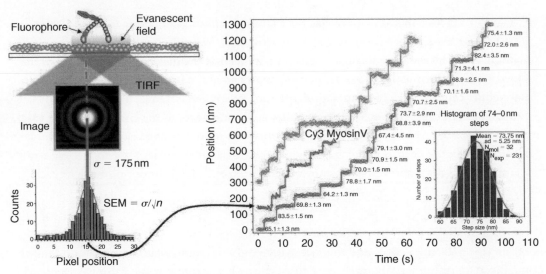

Figure 1.11: (Left panel) A single fluorophore bound, for instance, to one of the calmodulin light chains of myosin V (top) can be imaged onto a camera by TIRFM. The image appears as a diffraction-limited spot of half width 175 nm (center). The ability to detect the centroid of such an image depends upon the photon counting statistics shown in the histogram (bottom). (Right panel) As the myosin V walks along actin, the fluorophore moves in a stepwise fashion. Each step is separated by a distance of 73 nm, and at low ATP concentration, the dwell times are limited by the rate of ATP binding to the rigor head, which is usually the rear head, so that the myosin motor (and associated fluorophore) steps unidirectionally along actin (Yildiz et al., 2003)

2. By attaching two, differently colored quantum dots to either myosin head (using an N-terminally biotinylated recombinant myosin V molecule), Warshaw et al. (2005) were able to confirm that the two heads do indeed alternate in position between leading and trailing positions (like human walking) and so must move in a hand-over-hand fashion.

3. By labelling one of the myosin V heads with a fluorescent tag, Forkey et al. (2003) found that the angle of the angle of the light chain binding region first points forwards then backwards as the actin-attached myosin head changed from leading to trailing positions. This indicates that the step is driven by a structural rearrangement of the regulatory domain, such that it points first back and then forward along the axis of the actin filament.

4. To complement and extend Forkey's work, Shiroguchi and Kinosita (2007) investigated the angle changes of both the free (detached) and actin-bound (attached) myosin V heads during the stepping process. To do this, they fixed an elongated probe to one of the myosin regulatory domains, using a chimeric form of calmodulin that included a mutant kinesin head that stays tightly bound to a microtubule. By substituting chimeric calmodulin light chains onto the regulatory domain of myosin V, they were able to attach a microtubule that aligned itself along the length of the regulatory domain (like strapping a very long beanpole to someone's leg). They found that the myosin regulatory domain performs a rapid swing (power stroke) while the head is bound to actin, and makes random angular fluctuations when free, consistent with the free head making a diffusive search for the next actin-binding site.

5. Dunn and Spudich (2007) used a 40 nm gold particle attached to one of the myosin heads and viewed its position using a high-speed camera (>3000 frames per second). They found that while the myosin head is bound to actin it takes up a defined position, but after it detaches and moves toward its next binding site, it fluctuates in position, making a diffusional search for its next binding site along actin.

6. Ali and others attached an asymmetric particle (a doublet of 1 μm diameter latex beads) to the myosin V molecule and found that it walks around the actin filament in a left-handed spiral path as it advances (Ali et al., 2002; Yildiz et al., 2003). The motor made one full revolution for every 2 μm advance along actin, indicating that the average step size, 34.8 nm, is slightly less than the true pseudo-repeat distance, 36 nm, so that the leading head often "falls short" of the on-axis, 13th, monomer along actin, presumably binding to the 11th monomer instead.

We can see that single molecule and single-particle imaging methods have given immense insights into the detailed molecular mechanism of myosin V. The scientific literature in this area is rapidly increasing, especially as many of the approaches are now being applied to other members of the myosin family. Although it is likely that common mechanistic themes will emerge from these studies, we also expect many surprises, especially as the scope, time resolution, and precision of the measurements increases.

The critical technical innovations presented in this section have been to use optical signals to

- localize an object or individual fluorophore with subnanometer precision;

- measure its angular disposition, either from polarization or directly from the orientation of macroscopic "tags"; and

- visualize individual fluorescently tagged ligand molecules (such as Cy3 ATP "fuel").

Conclusions

The aim of this chapter has been to introduce the reader to some of the background to single molecule studies of actomyosin and specifically to illustrate this large body of work by some specific examples including work with myosin II and myosin V. Single molecule studies of actomyosin have been at the forefront in terms of technology development and applying new and emerging techniques to address highly specific questions of how this particular molecular motor works. Hopefully, we have been able to give sufficient background to the actomyosin system that the general reader understands the motivation for using and developing new single molecule techniques, and also how the specific findings have contributed to our basic knowledge and understanding of intermittent and processive myosin types. We are obtaining an increasingly clear picture of how myosin works at the level of molecular rearrangements during the ATPase cycle but there is still much to learn. Single molecule approaches have made a major impact in this particular area of research and we believe that in the future many of the methods developed will be broadly applicable to other fields.

Acknowledgments

The authors' work is supported by the Medical Research Council, UK (REF, JCF, and JEM), and the Chemistry and Biology Knowledge Base program of the UK National Measurement System, funded by the Department for Innovation, Universities and Skills (AEK).

References

Ali, M. Y., Uemura, S., Adachi, K., Itoh, H., Kinosita, K., and Ishiwata, S. (2002). Myosin V is a left-handed spiral motor on the right-handed actin helix. *Nat Struct Biol* 9, 464–467.

Anson, M., Geeves, M. A., Kurzawa, S. E., and Manstein, D. J. (1996). Myosin motors with artificial lever arms. *EMBO J* 15, 6069–6074.

Ashkin, A. (1970). Acceleration and trapping of particles by radiation pressure. *Phys Rev Lett* 24, 156–159.

Ashkin, A., Dziedzic, J. M., Bjorkholm, J. E., and Chu, S. (1986). Observation of a single-beam gradient force optical trap for dielectric particles. *Opt Lett* 11, 288–290.

Axelrod, D. (1981). Cell substrate contacts illuminated by total internal reflection fluorescence. *J Cell Biol* 89, 141–145.

Borejdo, J. and Morales, M. F. (1977). Fluctuations in tension during contraction of single muscle fibres. *Biophys J* 20, 315–334.

Buss, F., Arden, S. D., Lindsay, M., Luzio, J. P., and Kendrick-Jones, J. (2001). Myosin VI isoform localized to clathrin-coated vesicles with a role in clathrin-mediated endocytosis. *EMBO J* 20, 3676–3684.

Capitanio, M., Canepari, M., Cacciafesta, P., Lombardi, V., Cicchi, R., Maffei, M., Pavone, F. S., and Bottinelli, R. (2006). Two independent mechanical events in the interaction cycle of skeletal muscle myosin with actin. *Proc Natl Acad Sci U S A* 103, 87–92.

Cheney, R. E., O'shea, M. K., Heuser, J. E., Coelho, M. V., Wolenski, J. S., Espreafico, E. M., Forscher, P., Larson, R. E., and Mooseker, M. S. (1993). Brain myosin-V is a 2-headed unconventional myosin with motor-activity. *Cell* 75, 13–23.

Colquhoun, D. and Sigworth, F. J. (1995). *Single-Channel Recording*. Plenum, New York.

De La Cruz, E. M., Wells, A. L., Rosenfeld, S. S., Ostap, E. M., and Sweeney, H. L. (1999). The kinetic mechanism of myosin V. *Proc Natl Acad Sci U S A* 96, 13726–13731.

Dunn, A. R. and Spudich, J. A. (2007). Dynamics of the unbound head during myosin V processive translocation. *Nat Struct Mol Biol* 14, 246–248.

Dupuis, D. E., Guilford, W. H., Wu, J., and Warshaw, D. M. (1997). Actin filament mechanics in the laser trap. *J Muscle Res Cell Motil* 18, 17–30.

Finer, J. T., Simmons, R. M., and Spudich, J. A. (1994). Single myosin molecule mechanics: piconewton forces and nanometre steps. *Nature* 368, 113–118.

Forkey, J. N., Quinlan, M. E., Shaw, M. A., Corrie, J. E. T., and Goldman, Y. E. (2003). Three-dimensional structural dynamics of myosin V by single molecule fluorescence polarization. *Nature* 422, 399–404.

Foth, B. J., Goedecke, M. C., and Soldati, D. (2006). New insights into myosin evolution and classification. *Proc Natl Acad Sci U S A* 103, 3681–3686.

Funatsu, T., Harada, Y., Tokunaga, M., Saito, K., and Yanagida, T. (1995). Imaging of single fluorescent molecules and individual ATP turnovers by single myosin molecules in aqueous solution. *Nature* 374, 555–559.

Gelles, J., Schnapp, B. J., and Sheetz, M. P. (1988). Tracking kinesin-driven movements with nanometre-scale precision. *Nature* 331, 450–453.

Houdusse, A., Szent-Gyorgyi, A. G., and Cohen, C. (2000). Three conformational states of scallop myosin S1. *Proc Natl Acad Sci U S A* 97, 11238–11243.

Howard, J. (1997). Molecular motors: structural adaptations to cellular functions. *Nature* 389, 561–567.

Huxley, A. F. and Simmons, R. M. (1971). Proposed mechanism of force generation in muscle. *Nature* 233, 533–538.

Huxley, H. E. (1969). The mechanism of muscular contraction. *Science* 164, 1356–1366.

Ishijima, A., Doi, T., Sakurada, K., and Yanagida, T. (1991). Sub-piconewton force fluctuations of actomyosin in vitro. *Nature* 352, 301–306.

Ishijima, A., Kojima, H., Funatsu, T., Tokunaga, M., Higuchi, H., Tanaka, H., and Yanagida, T. (1998). Simultaneous observation of individual ATPase and mechanical events by a single myosin molecule during interaction with actin. *Cell* 92, 161–171.

Iwazumi, T. (1987). High-speed ultrasensitive instrumentation for myofibril mechanics measurements. *Am J Physiol* 252, C253–C262.

Jontes, J. D., Wilsonkubalek, E. M., and Milligan, R. A. (1995). A 32-degrees tail swing in brush-border myosin-I on ADP release. *Nature* 378, 751–753.

Katz, B. and Miledi, R. (1970). Membrane noise produced by Acetylcholine. *Nature* 226, 962–963.

Kitamura, K., Tokunaga, M., Iwane, A. H., and Yanagida, T. (1999). A single myosin head moves along an actin filament with regular steps of 5.3 nanometres. *Nature* 397, 129–134.

Knight, A., Mashanov, G., and Molloy, J. (2005). Single molecule measurements and biological motors. *Eur Biophys J* 35, 89.

Kron, S. J. and Spudich, J. A. (1986). Fluorescent actin filaments move on myosin fixed to a glass surface. *Proc Natl Acad Sci U S A* 83, 6272–6276.

Lister, I., Schmitz, S., Walker, M., Trinick, J., Buss, F., Veigel, C., and Kendrick-Jones, J. (2004). A monomeric myosin VI with a large working stroke. *EMBO J* 23, 1729–1738.

Lymn, R. W. and Taylor, E. W. (1971). The mechanism of adenosine triphosphate hydrolysis by actomyosin. *Biochem* 10, 4617–4624.

Margossian, S. S. and Lowey, S. (1982). Preparation of myosin and its subfragments from rabbit skeletal muscle. *Methods Enzymol* 85, 55–71.

Mashanov, G. I., Nenasheva, T. A., Peckham, M., and Molloy, J. E. (2006). Cell biochemistry studied by single molecule imaging. *Biochem Soc Trans* 34, 983–988.

Mashanov, G. I., Tacon, D., Knight, A. E., Peckham, M., and Molloy, J. E. (2003). Visualizing single molecules inside living cells using total internal reflection fluorescence microscopy. *Methods* 29, 142–152.

Mashanov, G. I., Tacon, D., Peckham, M., and Molloy, J. E. (2004). The spatial and temporal dynamics of pleckstrin homology domain binding at the plasma membrane measured by imaging single molecules in live mouse myoblasts. *J Biol Chem* 279, 15274–15280.

Mehta, A. D., Rock, R. S., Rief, M., Spudich, J. A., Mooseker, M. S., and Cheney, R. E. (1999). Myosin-V is a processive actin-based motor. *Nature* 400, 590–593.

Molloy, J. E., Burns, J. E., Kendrick-Jones, J., Tregear, R. T., and White, D. C. S. (1995). Movement and force produced by a single myosin head. *Nature* 378, 209–212.

Molloy, J. E. and Veigel, C. (2003). Myosin motors walk the walk. *Science* 300, 2045–2046.

Moore, J. R., Krementsova, E. B., Trybus, K. M., and Warshaw, D. M. (2001). Myosin V exhibits a high duty cycle and large unitary displacement. *J Cell Biol* 155, 625–635.

Nishikawa, S., Homma, K., Komori, Y., Iwaki, M., Wazawa, T., Iwane, A. H., Saito, J., Ikebe, R., Katayama, E., Yanagida, T., and Ikebe, M. (2002). Class VI myosin moves processively along actin filaments backward with large steps. *Biochem Biophys Res Commun* 290, 311–317.

Nishizaka, T., Yagi, T., Tanaka, Y., and Ishiwata, S. (1993). Right-handed rotation of an actin filament in an in vitro motile system. *Nature* 362, 269–271.

Odronitz, F. and Kollmar, M. (2007). Drawing the tree of eukaryotic life based on the analysis of 2,269 manually annotated myosins from 328 species. *Genome Biol* 8, R196.

Park, H., Ramamurthy, B., Travaglia, M., Safer, D., Chen, L. Q., Franzini-Armstrong, C., Selvin, P. R., and Sweeney, H. L. (2006). Full-length myosin VI dimerizes and moves processively along actin filaments upon monomer clustering. *Mol Cell* 21, 331–336.

Reck-Peterson, S. L., Provance, D. W., Mooseker, M. S., and Mercer, J. A. (2000). Class V myosins. *Biochim Biophys Acta* 1496, 36–51.

Richards, T. A. and Cavalier-Smith, T. (2005). Myosin domain evolution and the primary divergence of eukaryotes. *Nature* 436, 1113–1118.

Rief, M., Rock, R. S., Mehta, A. D., Mooseker, M. S., Cheney, R. E., and Spudich, J. A. (2000). Myosin-V stepping kinetics: a molecular model for processivity. *Proc Natl Acad Sci U S A* 97, 9482–9486.

Rock, R. S., Rice, S. E., Wells, A. L., Purcell, T. J., Spudich, J. A., and Sweeney, H. L. (2001). Myosin VI is a processive motor with a large step size. *Proc Natl Acad Sci U S A* 98, 13655–13659.

Ruff, C., Furch, M., Brenner, B., Manstein, D. J., and Meyhofer, E. (2001). Single molecule tracking of myosins with genetically engineered amplifier domains. *Nat Struct Biol* 8, 226–229.

Sakamoto, T., Wang, F., Schmitz, S., Xu, Y. H., Xu, Q., Molloy, J. E., Veigel, C., and Sellers, J. R. (2003). Neck length and processivity of myosin V. *J Biol Chem* 278, 29201–29207.

Schmidt, T., Schutz, G. J., Baumgartner, W., Gruber, H. J., and Schindler, H. (1996). Imaging of single molecule diffusion. *Proc Natl Acad Sci U S A* 93, 2926–2929.

Schmitz, S., Wang, F., Sellers, J. R., Molloy, J., and Veigel, C. (2002). The gait of myosin V constructs. *Biophys J* 82, 1819.

Scholey, J. M. (1993). *Motility Assays for Motor Proteins*. Academic Press Inc., New York.

Schroder, R. R., Manstein, D. J., Jahn, W., Holden, H., Rayment, I., Holmes, K. C., and Spudich, J. A. (1993). 3-Dimensional atomic model of F-actin decorated with dictyostelium myosin-S1. *Nature* 364, 171–174.

Sellers, J. R. and Kachar, B. (1990). Polarity and velocity of sliding filaments: control of direction by actin and of speed by myosin. *Science* 249, 406–408.

Shiroguchi, K. and Kinosita, K. (2007). Myosin V walks by lever action and Brownian motion. *Science* 316, 1208–1212.

Steffen, W., Smith, D., Simmons, R., and Sleep, J. (2001). Mapping the actin filament with myosin. *Proc Natl Acad Sci U S A* 98, 14949–14954.

Takagi, Y., Homsher, E. E., Goldman, Y. E., and Shuman, H. (2006). Force generation in single conventional actomyosin complexes under high dynamic load. *Biophys J* 90, 1295–1307.

Toyoshima, Y. Y., Kron, S. J., McNally, E. M., Niebling, K. R., Toyoshima, C., and Spudich, J. A. (1987). Myosin subfragment-1 is sufficient to move actin filaments in vitro. *Nature* 328, 536–539.

Toyoshima, Y. Y., Toyoshima, C., and Spudich, J. A. (1989). Bi-directional movement of actin filaments along tracks of myosin heads. *Nature* 341, 154–156.

Tsiavaliaris, G., Fujita-Becker, S., and Manstein, D. J. (2004). Molecular engineering of a backwards-moving myosin motor. *Nature* 427, 558–561.

Veigel, C., Bartoo, M. L., White, D. C. S., Sparrow, J. C., and Molloy, J. E. (1998). The stiffness of rabbit skeletal actomyosin cross-bridges determined with an optical tweezers transducer. *Biophys J* 75, 1424–1438.

Veigel, C., Coluccio, L. M., Jontes, J. D., Sparrow, J. C., Milligan, R. A., and Molloy, J. E. (1999). The motor protein myosin-I produces its working stroke in two steps. *Nature* 398, 530–533.

Veigel, C., Wang, F., Bartoo, M. L., Sellers, J. R., and Molloy, J. E. (2002). The gated gait of the processive molecular motor, myosin V. *Nat Cell Biol* 4, 59–65.

Walker, M. L., Burgess, S. A., Sellers, J. R., Wang, F., Hammer, J. A., Trinick, J., and Knight, P. J. (2000). Two-headed binding of a processive myosin to F-actin. *Nature* 405, 804–807.

Warshaw, D. M., Kennedy, G. G., Work, S. S., Krementsova, E. B., Beck, S., and Trybus, K. M. (2005). Differential labeling of myosin V heads with quantum dots allows direct visualization of hand-over-hand processivity. *Biophys J* 88, L30–L32.

Wells, A. L., Lin, A. W., Chen, L. Q., Safer, D., Cain, S. M., Hasson, T., Carragher, B. I., Milligan, R. A., and Sweeney, H. L. (1999). Myosin VI is an actin-based motor that moves backwards. *Nature* 401, 505–508.

Yanagida, T., Nakase, M., Nishiyama, K., and Oosawa, F. (1984). Direct observations of motion of single F-actin filaments in the presence of myosin. *Nature* 307, 58–60.

Yildiz, A., Forkey, J. N., McKinney, S. A., Ha, T., Goldman, Y. E., and Selvin, P. R. (2003). Myosin V walks hand-over-hand: single fluorophore imaging with 1.5-nm localization. *Science* 300, 2061–2065.

CHAPTER 2

Single Molecule Experiments and the Kinesin Motor Protein Superfamily: Walking Hand in Hand

Lukas C. Kapitein

Department of Physics and Astronomy, VU University Amsterdam,
De Boelelaan 1081, 1081HV Amsterdam, The Netherlands
Department of Neuroscience, Erasmus Medical Center, Dr. Molewaterplein 50,
3015 GE Rotterdam, The Netherlands

Erwin J.G. Peterman

Department of Physics and Astronomy, VU University Amsterdam, De Boelelaan 1081,
1081HV Amsterdam, The Netherlands

Summary

Kinesins are ATP-dependent motor proteins that can generate force and displacement along microtubules. They play key roles in mitosis and in the trafficking of organelles and vesicles. Shortly after kinesin's discovery in 1985, techniques were developed to measure the properties of individual proteins. Over the years, these techniques have been widely applied to kinesin, and it is justified to say that kinesin research has gone hand in hand with the coming of age of single molecule methods. Here, we review the properties of kinesin, with special emphasis on how single molecule approaches, such as optical tweezers and fluorescence microscopy, have helped to reveal kinesin's motility mechanism. We focus not only on the mechanical properties of the prototypical kinesin, Kinesin-1, but also on those of other related kinesins. We furthermore address diffusion along the microtubule lattice, a property of many kinesins and the regulation of motor activity, an area of growing attention.

Key Words

kinesin; motor protein; microtubule; motility assay; optical tweezers; single molecule fluorescence spectroscopy

Introduction

After the discoveries of myosin and dynein, the kinesin family was the last family of cytoskeletal motors to be discovered. First, myosin was found to be the protein that drives muscle contraction by interacting with actin filaments (Szent-Györgyi, 1945). Later, dynein was found to be the protein responsible for the beating of sperm tails and cilia, by interacting with microtubules (Gibbons and Rowe, 1965). The discovery of these two motor proteins provided a molecular explanation for many large-scale motility phenomena in organisms. On the other hand, intracellular motility had remained largely elusive and it was unclear as to what extent active transport played a role inside the cell. For very large and extended cells like neurons, however, there were several indications for active transport and the giant axon of squid had emerged as a model system to explore transport phenomena.

The development of video-enhanced differential interference contrast (VE-DIC) microscopy (Allen et al., 1981; Inoue, 1981) opened up the fascinating world of intracellular motility for exploration. This new technique allowed visualization of objects substantially smaller than the wavelength of light inside living cells and revealed that, in the giant axon, a dazzling amount of vesicles moves back and forth like traffic on a rush-hour highway. In addition, VE-DIC made it possible to image individual microtubules in vitro, which was an essential prerequisite for the reconstitution of motility using purified components. In 1985, this resulted in the discovery of a novel force-generating protein involved in motility along microtubules, for which the name "kinesin" was coined (Vale et al., 1985a).

In the following years, many proteins related to kinesin have been discovered. The human genome codes for 41 kinesin-like proteins (Dagenbach and Endow, 2004). The sequences of all these proteins contain a homologous globular domain that can hydrolyze ATP and interact with microtubules. On the basis of phylogeny, the kinesin superfamily has been subdivided into 14 families. From in vivo and in vitro experiments, it was found that the members of the kinesin superfamily not only function in the transport of vesicles and other cargo (Kinesin-1, Kinesin-2, Kinesin-3), but also play crucial roles in organizing the microtubule cytoskeleton during cell division and in interphase. Kinesin family members can regulate the growth and shrinkage of microtubules (Kinesin-4, Kinesin-8,

Kinesin-14), generate polarity-specific microtubule arrays (Kinesin-5, Kinesin-6, Kinesin-13), and mediate interactions between chromosomes and microtubules (Kinesin-7).

Understanding the precise function of these different motors requires knowledge of how they act, and for Kinesin-1, experiments that allow for the observation and manipulation of individual motor proteins have played an important role in achieving this. For most other kinesins, such single molecule studies have just begun, profiting greatly from the tools of molecular biology that allow overexpression, modification, affinity purification, and specific labeling of truncated and full-length protein complexes. In this chapter, we will highlight key experiments that have helped in addressing the motility mechanism of Kinesin-1 and also focus on novel assays and findings for several other kinesins.

Overall Mechanical Parameters of Kinesins

Multimotor Surface Gliding and Bead Assays

To understand how kinesins convert the chemical energy obtained from the hydrolysis of ATP into the mechanical work of force generation and/or translocation, well-controlled in vitro motility assays are required. These assays permit precise observation of the motor's mechanical activity, while environmental factors such as temperature, buffer conditions, and ATP concentrations can be controlled. As already mentioned in the Introduction, the discovery and isolation of Kinesin-1 from axoplasm of squid giant axons depended strongly on these assays (Vale et al., 1985a). The axoplasm was fractionated in different ways, and fractions that showed motility in these assays were selected for further purification, until only the two subunits of Kinesin-1 (the heavy chain and the light chain) remained. Two distinct motility assays were used, surface gliding assays and bead assays (Howard, 2001; Vale et al., 1985c), both based on assays developed to monitor myosin motion along actin (Sheetz and Spudich, 1983).

In "surface gliding assays," motors are adsorbed onto a glass surface using specific or nonspecific interactions (Figure 2.1B). Upon addition of microtubules and ATP, the action of the motor proteins can be observed in the gliding of the microtubules over the surface, just as a crowd surfer at a rock concert is passed overhead by other spectators, from person to person, driven by the muscle power of the individual spectators. In the surface gliding assay, the microtubules can be observed using VE-DIC microscopy or, when fluorescently tagged tubulin is used, wide-field fluorescence microscopy. Microtubules marked for polarity (with axoneme fragments or differently labeled microtubules as seeds to nucleate plus-end elongation) can be used to determine the directionality of the motors.

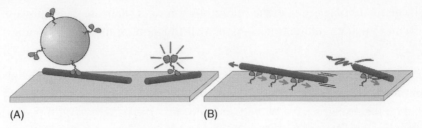

Figure 2.1: *In vitro* assays commonly used for kinesin. (A) Walking assays in which motors that are coated onto a bead or fluorescently labeled walk along surface-immobilized microtubules or axonemes. (B) Gliding assays in which motors that are adsorbed onto the surface drive the motility of microtubules landing from solution. When the surface concentration of motors is low enough, individual processive motors can drive motility, resulting in the pivoting of microtubules around a single anchor point

Another frequently applied assay is the "bead assay" (Figure 2.1A). In this assay, microtubules are stuck to the coverslip surface using antibodies, silanes with positively charged amino groups, or nonspecific interactions. A glass or polystyrene bead with a size of the order of a micrometer is coated with motors, using either specific or nonspecific interactions. When these beads are added to the sample in the presence of ATP, their binding to and their motion along microtubules can be observed using bright-field microscopy methods.

Using these two motility assays, it was demonstrated that Kinesin-1 is a motor that moves toward the microtubule plus end (Vale et al., 1985b). Furthermore, it was shown that its ATP-dependent velocity obeys Michaelis–Menten kinetics, with a maximal velocity of ~600 nm/s, at saturating ATP concentrations (Howard et al., 1989). It should be noted that these two motility assays are often used under such conditions that many motors are interacting with a single microtubule. As such, the activity of the ensemble of motors is measured, not that of a single motor.

Kinesin-1 is a Processive Motor

To study the mechanical properties of individual Kinesin-1 motor proteins, surface gliding assays were performed with a decreasing number of motors attached to the surface (Howard et al., 1989). Even at very low surface concentrations of kinesin, microtubules were still landing and subsequently moving at similar velocity. At these low kinesin concentrations, microtubules were typically attached to the surface at only one point, around which they swiveled, while moving for several micrometers (Figure 2.1B).

These observations indicate that single Kinesin-1 motors can move continuously for several micrometers along a microtubule. This property of a motor is called "processivity." Similar results were obtained using bead assays in which the number of motors absorbed onto the beads was titrated down until no more than half of the beads still bound to a microtubule (Block et al., 1990; Gelles et al., 1988). Under those conditions, only a single motor is expected to interact with a microtubule, yet continuous motion of the bead over a distance of 1.4 μm was observed (Block et al., 1990). In these experiments, the use of an optical trap to "park" the beads on a microtubule was important to accurately determine the mobile fraction of beads.

The motion of single kinesin motors along microtubules can be observed more directly by using "single molecule fluorescence microscopy" (Vale et al., 1996). The assay used in these experiments is similar to the bead assay, except that fluorescently tagged motors are used instead of motor-coated beads (Figure 2.1A). Kinesins can be made fluorescent either by fusion with an autofluorescent protein such as green fluorescent protein (GFP) or by chemically labeling the protein with a fluorescent dye. Chemical labeling has the advantage that smaller and more photostable dyes can be used (such as Cy3 or Alexa555) and targeted to specific amino acid residues. This requires, however, the tedious preparation of kinesin mutants with a single cysteine residue in a specific location, to which maleimide- or iodoacetamide-modified dyes can be attached with high specificity (Peterman et al., 2004; Pierce and Vale, 1998).

Using laser excitation and wide-field epi- or total-internal reflection illumination, the motion of individual fluorescent spots can be detected with a sensitive CCD camera. From these recordings, the velocity and run length of individual motors can be determined, although it should be noted that in these experiments the observed end of a run can be due to either detachment or photobleaching of the attached fluorophores.

In addition, the intensities of fluorescent spots can be calibrated to directly report on the number of fluorophores and, if the stoichiometry is known, on the number of proteins. Individual fluorophores tend to bleach in abrupt steps and the observation of, for example, two such steps is an indication that the fluorescent spot contained two fluorophores. In this way, it was shown that a dimeric construct of Kinesin-1 with two motor domains is required for processive motility (Vale et al., 1996).

Kinesin-1 Walks with 8 nm Steps and can Work against Loads of ~6 pN

"Optical tweezers" have been the method of choice to measure the forces of individual walking kinesin motors or apply controlled loads to them. Such experiments use the single-motor bead assay as described earlier, but leaving the laser trap on when the bead interacts with the microtubule. The motor will then pull the bead out of the center of the trap such that it experiences an increasing counteracting force (Figure 2.2A) (Svoboda et al., 1993). This force will pull the elastic connection between the bead and the motor taut, and will decrease the Brownian motion of the bead. Under these conditions, stepwise motion of the bead due to kinesin's stepping along the microtubule can be observed (Figure 2.2A). These steps of the bead reflect the center-of-mass motion of the kinesin. The step size of Kinesin-1 as measured in such assays with a calibrated optical trap is 8 nm, independent of the load applied and the ATP concentration (Svoboda et al., 1993). This 8 nm step size corresponds to the length of an α–β tubulin dimer, the building block of a microtubule (Amos and Klug, 1974), indicating that kinesin steps between consecutive tubulin dimers.

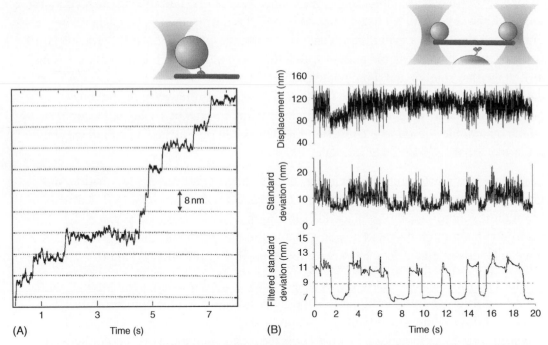

Figure 2.2: Kinesin-1's stepping and Kinesin-14's power strokes. (A) A single-bead assay as used in optical tweezers studies on processive motors, together with an example of a recording that clearly shows 8 nm steps. Adapted from Schnitzer and Block (1997) by permission from Macmillan Publishers Ltd: Nature, Copyright 1997. (B) A three-bead assay used to study nonprocessive motors with optical tweezers, together with an example recording. Binding events restrict the motion of the filament and can be determined by calculating the windowed standard deviation. Adapted from deCastro et al. (2000) by permission from Macmillan Publishers Ltd: Nature Cell Biology, Copright 2000

The step size of Kinesin-1 does not depend on the applied load, whereas the velocity of the motor decreases with load until the movement is stalled at a force of ~6 pN (Svoboda and Block, 1994). From the product of step size and maximum force, the maximum mechanical work per Kinesin-1 step can be calculated to be about 48 pN nm, corresponding to 12 kT, approximately 50% of the free energy associated with the hydrolysis of a single ATP in cellular conditions (~25 kT) (see Introduction, Chapter 1; Howard, 2001).

Kinesin-1 has been studied in great detail, much more so than the other members of the kinesin superfamily. Before addressing more advanced properties of the mechanism of Kinesin-1, we will first address other kinesin-like motor proteins whose motility has been studied with single molecule methods.

Other Processive Kinesins

"Nkin," a Kinesin-1 from the fungus *Neurospora crassa*, has been studied in detail using single molecule techniques. Unlike the motor domain, the sequence of this motor's neck is quite different from that of other well-characterized, but phylogenetically distant, animal Kinesin-1's. Like these motor proteins, Nkin is processive and makes 8 nm steps in the microtubule plus-end direction. Nkin is, however, more than twice as fast, with a velocity of ~2 µm/s (Lakämper et al., 2003). The molecular basis of this velocity difference remains unclear. Another kinesin, Kip3p, a fungal Kinesin-8, which plays a key role in regulating the length of cytoskeletal and mitotic spindle microtubules by promoting their depolymerization from the plus end, has recently been shown to be a slow (~50 nm/s) processive motor with an unusually long run length of more than 10 µm (Varga et al., 2006). It has been proposed that this high processivity helps Kinesin-8 to regulate the length of microtubules, since it would lead to an enrichment of depolymerizing motors on the plus ends of longer microtubules (Varga et al., 2006). A heterodimeric Kinesin-2, KIF3A/B, has also been shown to be processive in surface gliding assays, moving microtubules with a speed of 184 nm/s (Zhang and Hancock, 2004). Dimeric constructs of the Kinesin-3 motor Unc104 (from *Caenorhabditis elegans*) have been shown to be highly processive, with a run length of several micrometers and a velocity of ~2 µm/s (Tomishige et al., 2002). However, Kinesin-3 isolated from the tissue is mostly in a monomeric form. We will address later in this chapter the distinctive motility of these monomers and the suggested role of cargo-induced dimerization in the regulation of the activity of this motor. The tetrameric, bipolar Kinesin-5 Eg5 (from vertebrates) is a processive motor (Kwok et al., 2006; Valentine et al., 2006) that can cross-link and slide microtubules apart by moving on both microtubules it cross-links (Kapitein et al., 2005). Furthermore, the motility of this motor has unique characteristics that will be discussed below.

ncd is a Minus-End-Directed, Nonprocessive Kinesin-14

Kinesin-14 motors are unique in the kinesin superfamily because their motor domain is on the C-terminus of their polypeptide chains. Like many kinesins, Kinesin-14s are dimeric, but unlike other kinesins, their dimerizing coiled coil is not attached to the C-terminus of the catalytic motor domain but to the N-terminus, leading to a uniquely different connection of the two motor domains, potentially leading to different mechanical properties of these motors. Indeed, these motors are the only kinesins known to drive motility toward the microtubule's minus end, as observed in surface gliding assays using polarity-marked microtubules and ncd, a Kinesin-14 from *Drosophila melanogaster* (McDonald et al., 1990; Walker et al., 1990). In single molecule fluorescence motility assays, no processive motility was observed for GFP-tagged Ncd constructs, under conditions when Kinesin-1 showed reliable processive motion (Case et al., 1997). These results suggested that Ncd is not processive and walks in steps along a microtubule but behaves like myosin II and binds only for a short period, makes a single power stroke, and then releases from the track (Howard, 2001). More direct confirmation of ncd being nonprocessive came from optical trapping experiments (deCastro et al., 2000). The single-bead trapping assay as depicted in Figure 2.2A is not well suited to study nonprocessive motors. The short, isolated nonprocessive interactions between the motor and the track can be more efficiently and reliably observed using the so-called dumbbell or three-bead assay (Figure 2.2B), which was pioneered in studies of the mechanism of myosin II (Finer et al., 1994, Chapter 1; Molloy et al., 1995, Chapter 1). Using this assay for ncd, data as displayed in Figure 2.2B were observed: noisy displacement time traces, reflecting the Brownian motion of the optically trapped microtubule, which contain periods in which the motion of the filament is restricted due to attachment of the motor. The binding events can be detected by calculating the standard deviation of the data within a short, moving window. The duration of the binding events is ATP-concentration dependent, indicating that ATP binding takes place when a motor domain is bound to the microtubule. Individual binding events can be aligned and ensemble averaged, yielding traces with much higher signal-to-noise ratio. These traces show that ncd binds to a microtubule, most probably in the ADP-bound state, followed by ADP release, without significant conformational change. After an ATP-concentration-dependent delay, ATP binds and ncd moves the microtubule ~9 nm followed by release, indicating that ncd works with a power stroke model, swinging its "lever arm" just like the nonprocessive, muscle myosin (Howard, 2001). In later surface gliding assays and electron microscopy of ncd constructs with varying "lever arm" length, velocities and displacements of the tip of the "lever arm" were measured to be proportional to its length (Endres et al., 2006), confirming that ncd's motility can be described with a power stroke model.

It is unclear whether any kinesins from other Kinesin families are nonprocessive. For some plus-end directed motors, for example, NcKin3 (a dimeric, fungal Kinesin-3) (Adio et al., 2006a) and Xklp1 (a Kinesin-4 from *Xenopus laevis*) (Bringmann et al., 2004), gliding motility assays and kinetic arguments suggest nonprocessivity. But to be conclusive, single molecule studies are required.

Advanced Mechanochemistry of Kinesin-1

Kinesin-1 Makes One Step per ATP Hydrolyzed

In the previous section, we have discussed that Kinesin-1 makes 8 nm steps during which it can produce mechanical work corresponding to about half the free energy obtained from the hydrolysis of a single ATP molecule. On the basis of this insight, it was hypothesized that Kinesin-1 might hydrolyze one ATP per 8 nm step, and that stepping and hydrolysis are strongly coupled. Convincing evidence for this idea came from optical trapping experiments at low ATP concentrations, when ATP binding is the single rate-limiting step (Schnitzer and Block, 1997). Under these conditions, displacement time traces were recorded for single Kinesin-1s. Since a single motor makes steps separated by randomly distributed dwell times, the motion is not smooth and the motor's displacement exhibits a time-dependent variance. The variance measured at limiting ATP concentrations revealed that the dwell times between individual steps are due to a single rate-limiting transition, which is, under these conditions, the binding of a single ATP. Several other transitions become rate limiting when the same measurements are performed at higher ATP concentrations. This last observation does not fully exclude the possibility that the number of ATP hydrolyzed per step per Kinesin-1 is 1 at limiting ATP concentrations, but higher at higher ATP concentrations. This possibility was excluded by comparing the ATP-concentration-dependent velocity in single-motor bead motility assays with ATPase rate determinations (Hua et al., 1997). In a later study, both these assays were even performed on exactly the same sample: single Kinesin-1s attached to beads (Coy et al., 1999). These experiments showed that, at all ATP concentrations, dimeric Kinesin-1 hydrolyzes one ATP molecule per step.

Kinesin-1 Walks Hand over Hand

Two mechanisms have been postulated to explain how two motor domains can drive Kinesin-1's processive motion. In the "hand-over-hand" model, the two motor domains alternate in their role of leading and trailing heads, and both heads step, in turn, 16 nm, resulting in a center-of-mass motion of 8 nm (Figure 2.3). In the other model, the

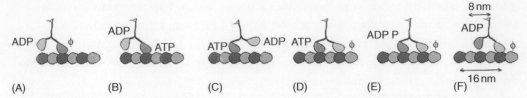

Figure 2.3: Model for processive Kinesin-1 stepping. Binding of one of kinesin's two motor domains to a microtubule induces the release of ADP from the bound motor domain (A). The subsequent binding of ATP to this bound motor domain (B) drives the formation of a conformation in which the unbound motor domain is positioned in front of the attached head (C), primed to bind to the microtubule. Binding of the unbound motor domain induces the release of ADP from this motor domain (D) and ATP hydrolysis in the rear motor domain (E), followed by phosphate release and detachment of this rear motor domain (F). Model based on Cross (2004) and Guydosh and Block (2006)

"inchworm" model, only one of the motor domains is catalytically active and leading all the time, making 8 nm steps, while the other motor domain follows, without using ATP. A number of single molecule experiments addressing the exact choreography of Kinesin-1's two motor domains have shown that Kinesin-1 walks hand over hand. Optical tweezers recordings from a heterodimeric Kinesin-1 construct with one motor domain slowed down by a mutation in the ATP-binding pocket convincingly showed alternating short and long dwell times between 8 nm steps (Kaseda et al., 2003). This result indicates that both motor domains hydrolyze ATP and step one after the other, a slow step due to the mutated motor domain followed by a fast step due to the wild-type motor domain. Similar experiments on certain homodimeric constructs showed that some of the motors "limp," exhibiting subsequent fast and slow steps (Asbury et al., 2003). Careful analysis of the dwell times led to the same conclusion: the two motor domains step in succession, although the cause of this asymmetric stepping behavior of these homodimeric constructs has remained unclear.

In another approach, a dimeric Kinesin-1 construct was fluorescently labeled in the motor domain (Yildiz et al., 2004). Using wide-field single molecule fluorescence imaging, the location of the fluorophore can be determined with an accuracy of better than a few nanometers, by fitting the Gaussian point-spread function to the image. For high accuracy, this approach requires a relatively large amount of detected photons and consequently long acquisition times (330 ms), demanding slowed-down motion of the motors at limiting ATP concentrations, when ATP binding is the single rate-limiting process. In this way, the stepping of individual motor domains was observed with an average step size of 16 nm, as expected in the hand-over-hand model. Furthermore, the distribution of the dwell times between these 16 nm steps could not be described by a single exponential distribution, but by a convolution of two identical exponential processes. Again, this is in full agreement with the hand-over-hand model since it requires the sequential hydrolysis of two ATP molecules (one by each motor domain) before the same motor domain can step again.

A Conformational Change of the Neck Linker Drives Kinesin's Motility

In Kinesin-1, the "neck linker," a chain of 15 amino acid residues connecting the motor domains with the coiled-coil neck, has been identified as a key mechanical element driving motility. Using a combination of electron microscopy, electron paramagnetic resonance, and (bulk) Förster resonance energy transfer (FRET) measurements on monomeric constructs, it was observed that during kinesin's catalytic cycle this neck linker switches between two states: in the ATP and ADP-Pi bound states it is "docked" on the motor domain, while in the ADP-bound and no-nucleotide states it is unzipped and flexible (Rice et al., 1999), in agreement with X-ray structural models of Kinesin-1 (Vale and Milligan, 2000) and other kinesins (Kikkawa et al., 2001). In dimeric constructs, similar conformational changes of the neck linker were observed using electron microscopy (Skiniotis et al., 2003).

To confirm that the neck linkers of dimeric Kinesin-1 switch conformations during processive motion, single molecule fluorescence assays were performed. First, it was shown that locking the neck linker in place by introduction of disulfide links between engineered cysteines prevented processive motion, indicating that flexibility of this structural element is a prerequisite for Kinesin-1's motility (Tomishige and Vale, 2000). Processive motion of these "chained" motors could be switched on again by reduction of the disulfide bonds. Later, it was shown using single molecule fluorescence polarization measurements on walking Kinesin-1 that the neck linker does indeed undergo the predicted conformational reorientation (Asenjo et al., 2006). In these measurements, a bifunctional fluorescent probe was attached to two engineered cysteines in the neck linker such that its orientation is fixed with respect to the structural element it was attached to. Conformational switching of the neck linker was later confirmed in single molecule experiments that dynamically measured the distance between the neck linker and a specific location on the motor domain in dimeric Kinesin-1 (Tomishige et al., 2006). In this study, FRET was measured between probes attached to engineered cysteines in the motor domain and the neck linker in one of the polypeptide chains of a heterodimeric Kinesin-1 construct.

Models for Kinesin-1 Motility Based on Single Molecule Experiments

On the basis of single molecule and other experiments, a consistent picture is emerging of the molecular basis of Kinesin-1's processive motility (Figure 2.3) (Asbury, 2005; Cross, 2004; Schnitzer et al., 2000). Key ingredients are as follows:

1. The affinity of the motor domains for microtubules is coupled to their nucleotide states such that the motor domains switch between strongly (ATP, ADP-Pi, and no nucleotide) and weakly (ADP) microtubule-bound states.

2. The neck linkers switch between docked (ATP and ADP-Pi) and undocked, disordered states (ADP and no nucleotide) (Asenjo et al., 2006; Rice et al., 1999; Tomishige and Vale, 2000; Tomishige et al., 2006).

3. The two motor domains step in a hand-over-hand fashion, with each motor domain making, in turn, 16 nm steps (Asbury et al., 2003; Kaseda et al., 2003; Yildiz et al., 2004).

4. The motor domains keep each other out of phase such that when one is weakly bound, the other is strongly bound.

These ingredients lead to models like the one depicted in Figure 2.3. The fourth element implies that the heads "sense" the state of the other head, such that they decrease the probability of both being in a weakly microtubule-bound state at the same time, which would lead to the end of a run. How could this sensing mechanism work? The most obvious mechanism would be mechanical strain between the motor domains reducing or increasing the rate of specific reaction steps. An attractive mechanism could be that the ATP affinity of the leading motor domain is decreased when the trailing motor domain is bound (increased strain, Figure 2.3D and E) compared to when it is not bound (less strain, Figure 2.3A and F). Evidence for such a mechanism comes from optical trapping measurement using a trace of a nonhydrolyzable ATP analogue (Guydosh and Block, 2006). The analogue was shown to induce long pauses in the runs, with the analogue bound to the trailing motor domain. Stepping could only resume after a single backward step, indicating that the nucleotide affinity is lower for the leading motor domain than that for the trailing one.

Lattice Diffusion as an Additional Motility Mode

In the previous section, we have addressed the motility properties of several kinesin motor proteins and shown how single molecule experiments have contributed to distinguishing between the possible motility schemes. In this section, we will give examples of new, unanticipated motility modes revealed by single molecule experiments.

Biased Diffusion along Microtubules by KIF1A

As discussed earlier, Kinesin-1 needs two motor domains for processive motility. Kinesin-1 maintains attachment during stepping because the heads keep each other out of phase in such a way that while one is in a weakly microtubule-bound nucleotide state, the other is strongly bound. Therefore, the observation that certain monomeric constructs of the Kinesin-3 motor KIF1A appeared to "pile up" at the microtubule plus end came as quite a surprise (Okada and Hirokawa, 1999). Closer inspection with single molecule fluorescence microscopy revealed that individual monomeric motors did indeed move directionally along the microtubule lattices, although in a far more irregular fashion than Kinesin-1 (Figure 2.4A) (Okada and Hirokawa, 1999).

The mean squared displacement, as obtained from the KIF1A motility time traces, could not be described by only a quadratic term and an additional linear term was required, indicating that the motility of KIF1A can be understood as a biased random walk (Okada and Hirokawa, 1999). An object that moves with a constant speed in one direction will have a mean squared displacement that is proportional to the square of the time interval. In an unbiased one-dimensional random walk, a stepping object can step backwards and forward with equal probability, resulting in a stochastic trajectory with, on average, no specific direction. The mean squared displacement of these trajectories is expected to increase linearly with time, i.e., when many objects are released from one point at the same time, they will spread out in both directions from the origin and the width of the distribution of the objects (proportional to the square root of the variance) will increase with the square root of time. In the case of a biased random walk, the objects can still step backwards and forward, but with probabilities that are not equal. The mean displacement, as obtained by averaging over many time traces, will increase linearly with time. The mean squared displacement will increase with time by a combination of a quadratic (reflecting the directionality) and a linear term (reflecting the diffusiveness).

Further experiments have shown that the KIF1A motor domains cycle through a weakly bound state and a strongly bound state like most other motors, but that, during the weakly bound state, they can freely slide along the microtubule (Okada and Hirokawa, 2000). KIF1A's diffusive interaction with microtubules was found to be mediated by the lysine-rich, positively charged K-loop unique to the KIF1A motor domain interacting with the glutamate-rich, negatively charged C-terminal region of the microtubule

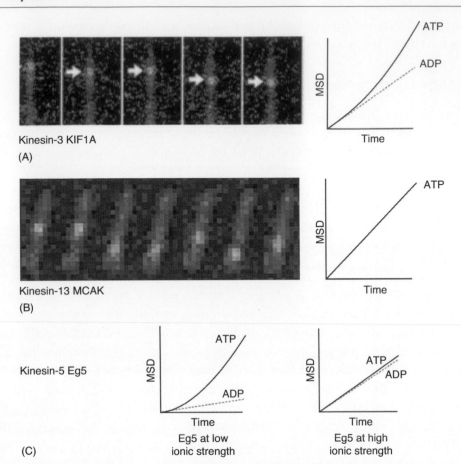

Figure 2.4: Lattice diffusion as an additional motility mode for various nonconventional kinesins. (A) Biased diffusion of Kinesin-3. Video frames of KIF1A motility along a microtubule. Adapted from Okada and Hirokawa (1999) by permission from AAAs. The mean-squared displacement (MSD) will have a quadratic term corresponding to its overall directionality ($v^2 t^2$), as well as a large linear term corresponding to its diffusive spreading ($2Dt$); with v the velocity, D the diffusion constant, and t the time interval. In the presence of ADP, the motility is only unbiased and the MSD is linear. (B) Unbiased diffusion for Kinesin-13. Video frames of MCAK motility along a microtubule. Adapted from Helenius et al. (2006) by permission from Macmillan Publishers Ltd: Nature, Copyright 2006. (C) Biased or unbiased diffusion of Kinesin-5, depending on ionic strength (Kapitein et al., 2008)

(E-hook). It has been hypothesized that this E-hook is disordered and mobile enough to generate an attractive landscape along the whole tubulin subunit, preventing the motor from moving away but allowing the motor to freely move along the microtubule lattice. In addition, the plus-end-directed power stroke during the strongly bound state is thought to impose a bias on the motor domain such that the probability is slightly increased to bind to the next binding site on the plus-end direction of the microtubule. This combination of a power stroke and delocalized interactions provide a simple mechanism for the diffusive, directional, and processive motility of KIF1A (Okada and Hirokawa, 1999, 2000).

It should, however, be noted that, while these experiments and models represent an important conceptual advance to our understanding of molecular motion, the biological relevance of this motion remains unclear in the case of Kinesin-3 (see also the next section). This typical motility behavior has only been observed for certain chimeric constructs at ionic strength considerably lower than physiological and has not been detected for other monomeric family members (Okada and Hirokawa, 1999; Tomishige et al., 2002).

Diffusion without Bias along Microtubules by Kinesin-13

Apart from Kinesin-3, several other types of kinesins have been found capable of lattice diffusion. Kinesin-13 family members have an unusual structure in which the motor domain is not at the C- or N-terminus, but located in the middle of the polypeptide chain. These motors have been shown to depolymerize microtubules (Desai et al., 1999). Immunostaining indicated that these motors interact specifically with the ends of microtubules and it had remained unclear whether and how these motors interact with the remaining microtubule lattice away from the ends (Desai et al., 1999). Single molecule fluorescence experiments using a recombinant GFP-tagged MCAK construct have shown that these motors diffuse along the microtubule axis without directional bias (Figure 2.4B) (Helenius et al., 2006). It has been argued that this lattice diffusion helps MCAK to target both microtubule plus and minus ends at enhanced rates. At ionic conditions close to physiological, MCAK lattice diffusion is less apparent and binding events are much shorter; hence, the importance of this motility mode for Kinesin-13's function in vivo is still unclear (Helenius et al., 2006).

The Motility of Kinesin-5 is a Mixture of Diffusion and Directional Motion

Kinesin-5 family members are tetrameric motor proteins with two motor domains at each end of a central stalk (Kashina et al., 1996). This structure allows the motor to cross-link two microtubules and drive their relative sliding by walking toward the plus end of each microtubule (Kapitein et al., 2005). This sliding mechanism resembles the force-generating system in muscle, where aggregates of nonprocessive myosin motors cooperate to drive sliding of actin filaments (Alberts et al., 2002, Chapter 1). Optical tweezers experiments examining the motility of single truncated dimeric constructs of human Kinesin-5 have shown that these motors can move processively, but only make very short runs of on average 8 steps (Valentine et al., 2006). In contrast, fluorescence experiments using GFP-tagged full-length Kinesin-5 from *X. laevis* revealed much longer runs (Kwok et al., 2006). These runs, however, were very irregular and appeared similar to those observed for KIF1A, representing a biased random walk (Figure. 2.4C). In the presence of ADP, unbiased one-dimensional diffusion was observed. These observations suggest that the motility of the full-length motor is a mixture of short processive runs and diffusive intervals. In contrast to KIF1A, the diffusive interaction is likely to be mediated by microtubule-binding regions outside the motor domain since no diffusive interaction was observed for the dimeric construct. Indeed, experiments examining a series of tetrameric mutants from the *Drosophila* Kinesin-5 KLP61F have revealed that tetrameric constructs lacking the motor domains are still capable of cross-linking microtubules, indicating the existence of such passive binding modes (Tao et al., 2006).

In addition to these examples, diffusive features corresponding to additional microtubule-binding modes have been found for various other motor enzymes for which they are believed to enhance processivity (Chandra et al., 1993; Culver-Hanlon et al., 2006; Vale et al., 1989). Further work is required to find out the precise mechanisms of such diffusional motility. In one picture, the interaction between an individual binding domain and the microtubule is nonlocalized, i.e., the binding domain feels a smeared-out potential, in such a way that the protein easily gets "kicked" along the lattice by thermal excitation. This mechanism resembles that of certain processivity factors of DNA-bound motors, which can prevent off-axis displacement by forming a (partial) loop around the DNA, while still allowing free displacement along it (Kong et al., 1992; Stukenberg et al., 1991). Alternatively, binding modes could have transient (localized) interactions with the microtubule, but conspire with other binding regions to maintain attachment.

Regulation of Kinesin Motors

In order to perform specific tasks in the cell, molecular motors need to be regulated, for example, to control the type of cargo or the activation of the motors during various stages of the cell cycle. In general, these mechanisms involve complicated pathways and a large variety of proteins participating in processes varying from expression regulation and posttranslational modifications (such as phosphorylation) to motor–cargo complex formation and possibly even the modification of the microtubule surface (Adio et al.,

2006b; Luduena, 1998; Reed et al., 2006). At the moment, the exact regulatory pathways are not known for most motors and are being studied mainly using biochemical and genetic approaches that are outside the scope of this chapter. Nevertheless, single molecule experiments have revealed several activation mechanisms of kinesin motor proteins.

Regulation by Cargo-Binding-Induced Conformational Changes

Early biochemical experiments exploring kinesin's kinetic cycle used tissue-purified Kinesin-1 to measure parameters such as microtubule affinity and ATP hydrolysis rate (Cross, 2004). It was found that the ATPase activity of Kinesin-1 is greatly enhanced in the presence of microtubules, revealing a basic regulatory mechanism that prevents futile hydrolysis when the motors are not bound to their track (Kuznetsov and Gelfand, 1986). However, these experiments often yielded conflicting and ambiguous results. Eventually, it was found that many of these problems could be overcome by measuring on truncated kinesins, consisting of either only the motor and dimerization domains or only the motor domain. These constructs exhibited higher and more consistent ATPase activity, suggesting an additional regulatory mechanism in which kinesin's tail can fold back on the motor domain inhibiting microtubule binding (Hackney and Stock, 2000; Hackney et al., 1992; Stock et al., 1999). Cargo binding to the tail would prevent such back-folding and hence facilitate microtubule binding and motility.

These ideas have been tested in vitro by single molecule fluorescence studies comparing the motility of a series of recombinant Kinesin-1 constructs (Friedman and Vale, 1999). A truncated kinesin construct lacking the tail domains moved rapidly and smoothly. In contrast, full-length constructs with and without associated light chains were found to interact with microtubules more than ten times less frequently and their runs were very irregular, showing frequent pausing interspersed with bursts of rapid motility (Figure 2.5A). Inhibition could be suppressed almost completely by introducing mutations stabilizing the coiled coils and thereby preventing back-folding (Figure 2.5A). In addition, mutations in the neck region also suppressed inhibition, indicating that the back-folded tail docks onto the neck region to prevent motility.

Recent in vivo experiments have revealed an additional conformational change associated with inhibition and activation (Cai et al., 2007). Complete inhibition of MT binding is established by the back-folding of the tail onto the neck region, in combination with an increased separation of the motor domains induced by the light chains. Activation requires the binding of a cargo-binding protein (JIP1) to the tail as well as the binding of another protein (FEZ1) to the light chains (Blasius et al., 2007).

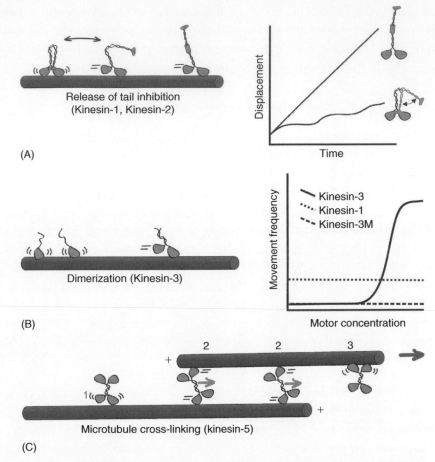

(A)

(B)

(C)

Figure 2.5: Regulatory mechanisms revealed by single molecule experiments.
(A) Binding of Kinesin-1's tail to its neck region disfavors microtubule binding and
inhibits motility. Replacement of the hinge region by a stable coiled coil suppressed this
inhibition. Data based on Friedman and Vale (1999). (B) Dimerization of monomeric
Kinesin-3 activates fast processive motility. Addition of increasing amounts
of nonfluorescent monomers of Unc104 triggers a switch toward processive motion.
Addition of nonfluorescent analogues does not change the motility of Kinesin-1 or
a Unc104 mutant that cannot dimerize (Kinesin-3M). Data based on Tomishige
et al. (2002). (C) Microtubule cross-linking of Kinesin-5 activates directional motility.
Eg5 molecules exhibit unbiased lattice diffusion on single microtubules, but become
directional upon binding to a second microtubule

A similar form of inhibition by back-folding and activation by cargo binding has been demonstrated recently for the Kinesin-2 OSM-3 from *C. elegans* (Imanishi et al., 2006). Neither motility nor microtubule binding of the full-length construct was observed in single molecule fluorescence experiments, whereas processive movement of beads coated with several of these motors was observed in an optical tweezers assay. As with Kinesin-1, replacing the hinge region by a stable coiled coil yielded motors that were also active in the fluorescence assay, without attached cargo. Amazingly, a single point mutation in the hinge region, corresponding to a chemosensory defect in *C. elegans*, was also sufficient to fully activate this motor.

Regulation by Cargo-Induced Dimer Formation

In the previous section, we have described experiments on the Kinesin-3 KIF1A that revealed its capability to move directionally as a monomer. Sequence analysis of the related Kinesin-3 Unc104 from *C. elegans* indicated that while these motors are predominantly monomeric in solution, they do have potential coiled-coil regions in the neck region (Tomishige et al., 2002). It was then speculated that these motors could dimerize at high enough concentrations, allowing them to move processively like Kinesin-1 (Figure 2.5B). However, the concentrations required to directly test this hypothesis using fluorescent Unc104 monomers were too high to still allow detection of single motors by fluorescence. Therefore, the amount of fluorescent monomers was kept low (7 nM), while the overall Unc104 concentration was increased by adding nonfluorescent monomers (up to 7 μM) (Tomishige et al., 2002). At overall concentrations above 1 μM, processive runs were observed, consistent with the idea of concentration-dependent multimerization (Figure 2.5B). Additional experiments using fluorescent lipid vesicles capable of binding GFP-tagged Unc104 monomers revealed that these vesicles could be transported with only very few monomers attached, suggesting that dimerization switches the motors to move processively. This was confirmed by the observation that mutants unable to dimerize could only drive vesicle motility at much higher concentrations (Tomishige et al., 2002).

Regulation by Microtubule Cross-Linking

For the Kinesin-5 Eg5, single molecule experiments have revealed that increasing the ionic strength to physiological levels induces a transition toward a nondirectional diffusive state, raising questions about how the motor gets activated in vivo (Figure 2.4C) (Kapitein et al., 2008). To examine how these motors drive relative sliding of microtubules at close to physiological salt conditions, the assay mix used for single molecule imaging was supplemented with additional microtubules (Kapitein et al., 2008). Occasionally, these microtubules became cross-linked onto an immobilized microtubule by Eg5 and then started sliding along it. While on single microtubules Eg5 motility was diffusive and unbiased, individual directional runs of Eg5-GFP could be observed in the overlap zone of a sliding microtubule pair. The speed of these runs was on average half that of the sliding microtubule, indicating that these motors were driving the microtubule sliding by moving directionally on both microtubules that they cross-linked. These results indicate that the motility of bipolar tetrameric Kinesin-5 can be regulated between a diffusive and a directional mode by microtubule cross-linking (Figures 2.4C and 2.5C) (Kapitein et al., 2008).

Final Thoughts

In this chapter, we have highlighted a variety of experiments that have explored the behavior of individual kinesin motor proteins using single molecule techniques. These experiments have helped to unravel the molecular mechanisms underlying the remarkable capabilities of these cellular workhorses. More and more single molecule studies currently explore nonconventional kinesins and report new features corresponding to additional binding modes and regulatory pathways. Additionally, the interaction between these motors and other proteins is increasingly becoming a subject of in vitro assays. Furthermore, experiments addressing the motility of individual motor proteins in living cells are now within reach. Since many members of the kinesin superfamily are still unexplored, new and exciting findings are likely to be revealed with the aid of these powerful techniques.

However, not only has our knowledge of kinesin benefited greatly from these kind of experiments, the reverse is also true. Experiments developed to study kinesin have resulted in new tools, insights, and concepts that have been successfully applied to other biological systems. For example, the development of optical tweezers technologies that allow detecting steps or power strokes as well as the application of well-controlled forces has been pioneered to a large extent for the study of kinesin, but the techniques have been applied to numerous other biological systems as well. More or less the same holds for the approach to assay protein dynamics using single molecule fluorescence in combination with site-specific mutagenesis, which allows us to conclude that single molecule experiments and the kinesin superfamily are walking hand in hand to progress our understanding of molecular mechanisms in the cell.

Acknowledgments

We thank Christoph Schmidt and our group members for many inspiring discussions on kinesins and single molecule methods. Our work is supported by a VIDI fellowship to

E.P. from the Research Council for Earth and Life Sciences (ALW), with financial aid from the Netherlands Organization of Scientific Research (NWO), and by the Foundation for Fundamental Research on Matter (FOM).

References

Adio, S., Bloemink, M., Hartel, M., Leier, S., Geeves, M. A., and Woehlke, G. (2006a). Kinetic and mechanistic basis of the nonprocessive Kinesin-3 motor NcKin3. *J Biol Chem* 281, 37782–37793.

Adio, S., Reth, J., Bathe, F., and Woehlke, G. (2006b). Review: regulation mechanisms of Kinesin-1. *J Muscle Res Cell Motil* 27, 153–160.

Alberts, B., Johnson, A., Lewis, J., Raff, M., Roberts, K., and Walter, P. (2002). *Molecular Biology of the Cell*. Garland Science, New York.

Allen, R. D., Allen, N. S., and Travis, J. L. (1981). Video-enhanced contrast, differential interference contrast (AVEC-DIC) microscopy: a new method capable of analyzing microtubule-related motility in the reticulopodial network of Allogromia laticollaris. *Cell Motil* 1, 291–302.

Amos, L. and Klug, A. (1974). Arrangement of subunits in flagellar microtubules. *J Cell Sci* 14, 523–549.

Asbury, C. L. (2005). Kinesin: world's tiniest biped. *Curr Opin Cell Biol* 17, 89–97.

Asbury, C. L., Fehr, A. N., and Block, S. M. (2003). Kinesin moves by an asymmetric hand-over-hand mechanism. *Science* 302, 2130–2134.

Asenjo, A. B., Weinberg, Y., and Sosa, H. (2006). Nucleotide binding and hydrolysis induces a disorder–order transition in the kinesin neck-linker region. *Nat Struct Mol Biol* 13, 648–654.

Blasius, T. L., Cai, D., Jih, G. T., Toret, C. P., and Verhey, K. J. (2007). Two binding partners cooperate to activate the molecular motor Kinesin-1. *J Cell Biol* 176, 11–17.

Block, S. M., Goldstein, L. S., and Schnapp, B. J. (1990). Bead movement by single kinesin molecules studied with optical tweezers. *Nature* 348, 348–352.

Bringmann, H., Skiniotis, G., Spilker, A., Kandels-Lewis, S., Vernos, I., and Surrey, T. (2004). A kinesin-like motor inhibits microtubule dynamic instability. *Science* 303, 1519–1522.

Cai, D. W., Hoppe, A. D., Swanson, J. A., and Verhey, K. J. (2007). Kinesin-1 structural organization and conformational changes revealed by FRET stoichiometry in live cells. *J Cell Biol* 176, 51–63.

Case, R. B., Pierce, D. W., Hom-Booher, N., Hart, C. L., and Vale, R. D. (1997). The directional preference of kinesin motors is specified by an element outside of the motor catalytic domain. *Cell* 90, 959–966.

Chandra, R., Endow, S. A., and Salmon, E. D. (1993). An N-terminal truncation of the Ncd motor protein supports diffusional movement of microtubules in motility assays. *J Cell Sci* 104 (Pt 3), 899–906.

Coy, D. L., Wagenbach, M., and Howard, J. (1999). Kinesin takes one 8-nm step for each ATP that it hydrolyzes. *J Biol Chem* 274, 3667–3671.

Cross, R. A. (2004). The kinetic mechanism of kinesin. *Trends Biochem Sci* 29, 301–309.

Culver-Hanlon, T. L., Lex, S. A., Stephens, A. D., Quintyne, N. J., and King, S. J. (2006). A microtubule-binding domain in dynactin increases dynein processivity by skating along microtubules. *Nat Cell Biol* 8, 264–270.

Dagenbach, E. M. and Endow, S. A. (2004). A new kinesin tree. *J Cell Sci* 117, 3–7.

deCastro, M. J., Fondecave, R. M., Clarke, L. A., Schmidt, C. F., and Stewart, R. J. (2000). Working strokes by single molecules of the kinesin-related microtubule motor Ncd. *Nat Cell Biol* 2, 724–729.

Desai, A., Verma, S., Mitchison, T. J., and Walczak, C. E. (1999). Kin I kinesins are microtubule-destabilizing enzymes. *Cell* 96, 69–78.

Endres, N. F., Yoshioka, C., Milligan, R. A., and Vale, R. D. (2006). A lever-arm rotation drives motility of the minus-end-directed kinesin Ncd. *Nature* 439, 875–878.

Finer, J. T., Simmons, R. M., and Spudich, J. A. (1994). Single myosin molecule mechanics: piconewton forces and nanometre steps. *Nature* 368, 113–119.

Friedman, D. S. and Vale, R. D. (1999). Single molecule analysis of kinesin motility reveals regulation by the cargo-binding tail domain. *Nat Cell Biol* 1, 293–297.

Gelles, J., Schnapp, B. J., and Sheetz, M. P. (1988). Tracking kinesin-driven movements with nanometre-scale precision. *Nature* 331, 450–453.

Gibbons, I. R. and Rowe, A. J. (1965). Dynein-A protein with adenosine triphosphatase activity from cilia. *Science* 149, 424–426.

Guydosh, N. R. and Block, S. M. (2006). Backsteps induced by nucleotide analogs suggest the front head of kinesin is gated by strain. *Proc Natl Acad Sci U S A* 103, 8054–8059.

Hackney, D. D., Levitt, J. D., and Suhan, J. (1992). Kinesin undergoes a 9-S to 6-S conformational transition. *J Biol Chem* 267, 8696–8701.

Hackney, D. D. and Stock, M. F. (2000). Kinesin's IAK tail domain inhibits initial microtubule-stimulated ADP release. *Nat Cell Biol* 2, 257–260.

Helenius, J., Brouhard, G., Kalaidzidis, Y., Diez, S., and Howard, J. (2006). The depolymerizing kinesin MCAK uses lattice diffusion to rapidly target microtubule ends. *Nature* 441, 115–119.

Howard, J. (2001). *Mechanics of Motor Proteins and the Cytoskeleton*. Sinauer Associates, Sunderland, MA.

Howard, J., Hudspeth, A. J., and Vale, R. D. (1989). Movement of microtubules by single kinesin molecules. *Nature* 342, 154–158.

Hua, W., Young, E. C., Fleming, M. L., and Gelles, J. (1997). Coupling of kinesin steps to ATP hydrolysis. *Nature* 388, 390–393.

Imanishi, M., Endres, N. F., Gennerich, A., and Vale, R. D. (2006). Autoinhibition regulates the motility of the *C. elegans* intraflagellar transport motor OSM-3. *J Cell Biol* 174, 931–937.

Inoue, S. (1981). Video image processing greatly enhances contrast, quality, and speed in polarization-based microscopy. *J Cell Biol* 89, 346–356.

Kapitein, L. C., Kwok, B. H., Weinger, J. S., Schmidt, C. F., Kapoor, T. M., and Peterman, E. J. G. (2008). *J Cell Biol* 182, 421–428.

Kapitein, L. C., Peterman, E. J. G., Kwok, B. H., Kim, J. H., Kapoor, T. M., and Schmidt, C. F. (2005). The bipolar mitotic kinesin Eg5 moves on both microtubules that it crosslinks. *Nature* 435, 114–118.

Kaseda, K., Higuchi, H., and Hirose, K. (2003). Alternate fast and slow stepping of a heterodimeric kinesin molecule. *Nat Cell Biol* 5, 1079–1082.

Kashina, A. S., Scholey, J. M., Leszyk, J. D., and Saxton, W. M. (1996). An essential bipolar mitotic motor. *Nature* 384, 225.

Kikkawa, M., Sablin, E. P., Okada, Y., Yajima, H., Fletterick, R. J., and Hirokawa, N. (2001). Switch-based mechanism of kinesin motors. *Nature* 411, 439–445.

Kong, X. P., Onrust, R., Odonnell, M., and Kuriyan, J. (1992). 3-Dimensional structure of the beta-subunit of *Escherichia coli* DNA polymerase-III holoenzyme – a sliding DNA clamp. *Cell* 69, 425–437.

Kuznetsov, S. A. and Gelfand, V. I. (1986). Bovine brain kinesin is a microtubule-activated ATPase. *Proc Natl Acad Sci U S A* 83, 8530–8534.

Kwok, B. H., Kapitein, L. C., Kim, J. H., Peterman, E. J. G., Schmidt, C. F., and Kapoor, T. M. (2006). Allosteric inhibition of kinesin-5 modulates its processive directional motility. *Nat Chem Biol* 2, 480–485.

Lakämper, S., Kallipolitou, A., Woehlke, G., Schliwa, M., and Meyhofer, E. (2003). Single fungal kinesin motor molecules move processively along microtubules. *Biophys J* 84, 1833–1843.

Luduena, R. F. (1998). Multiple forms of tubulin: different gene products and covalent modifications. *Int Rev Cytol* 178, 207–275.

McDonald, H. B., Stewart, R. J., and Goldstein, L. S. B. (1990). The kinesin-like Ncd protein of *Drosophila* is a minus end-directed microtubule motor. *Cell* 63, 1159–1165.

Molloy, J. E., Burns, J. E., Kendrick-Jones, J., Tregear, R. T., and White, D. C. (1995). Movement and force produced by a single myosin head. *Nature* 378, 209–212.

Okada, Y. and Hirokawa, N. (1999). A processive single-headed motor: kinesin superfamily protein KIF1A. *Science* 283, 1152–1157.

Okada, Y. and Hirokawa, N. (2000). Mechanism of the single-headed processivity: diffusional anchoring between the K-loop of kinesin and the C terminus of tubulin. *Proc Natl Acad Sci U S A* 97, 640–645.

Peterman, E. J. G., Sosa, H., and Moerner, W. E. (2004). Single molecule fluorescence spectroscopy and microscopy of biomolecular motors. *Annu Rev Phys Chem* 55, 79–96.

Pierce, D. W. and Vale, R. D. (1998). Assaying processive movement of kinesin by fluorescence microscopy. *Methods Enzymol* 298, 154–171.

Reed, N. A., Cai, D. W., Blasius, T. L., Jih, G. T., Meyhofer, E., Gaertig, J., and Verhey, K. J. (2006). Microtubule acetylation promotes kinesin-1 binding and transport. *Curr Biol* 16, 2166–2172.

Rice, S., Lin, A. W., Safer, D., Hart, C. L., Naber, N., Carragher, B. O., Cain, S. M., Pechatnikova, E., Wilson-Kubalek, E. M., Whittaker, M., Pate, E., Cooke, R., Taylor, E. W., Milligan, R. A., and Vale, R. D. (1999). A structural change in the kinesin motor protein that drives motility. *Nature* 402, 778–784.

Schnitzer, M. J. and Block, S. M. (1997). Kinesin hydrolyses one ATP per 8-nm step. *Nature* 388, 386–390.

Schnitzer, M. J., Visscher, K., and Block, S. M. (2000). Force production by single kinesin motors. *Nat Cell Biol* 2, 718–723.

Sheetz, M. P. and Spudich, J. A. (1983). Movement of myosin-coated fluorescent beads on actin cables in vitro. *Nature* 303, 31–35.

Skiniotis, G., Surrey, T., Altmann, S., Gross, H., Song, Y. H., Mandelkow, E., and Hoenger, A. (2003). Nucleotide-induced conformations in the neck region of dimeric kinesin. *EMBO J* 22, 1518–1528.

Stock, M. F., Guerrero, J., Cobb, B., Eggers, C. T., Huang, T. G., Li, X., and Hackney, D. D. (1999). Formation of the compact confomer of kinesin requires a COOH-terminal heavy chain domain and inhibits microtubule-stimulated ATPase activity. *J Biol Chem* 274, 14617–14623.

Stukenberg, P. T., Studwellvaughan, P. S., and Odonnell, M. (1991). Mechanism of the sliding beta-clamp of DNA polymerase-III holoenzyme. *J Biol Chem* 266, 11328–11334.

Svoboda, K. and Block, S. M. (1994). Force and velocity measured for single kinesin molecules. *Cell* 77, 773–784.

Svoboda, K., Schmidt, C. F., Schnapp, B. J., and Block, S. M. (1993). Direct observation of kinesin stepping by optical trapping interferometry. *Nature* 365, 721–727.

Szent-Györgyi, A. (1945). Studies on muscle. *Acta Physiol Scand* 9 (Suppl 25), 1–158.

Tao, L., Mogilner, A., Civelekogiu-Scholey, G., Wollman, R., Evans, J., Stahlberg, H., and Scholey, J. M. (2006). A homotetrameric kinesin-5, KLP61F, bundles microtubules and antagonizes Ncd in motility assays. *Curr Biol* 16, 2293–2302.

Tomishige, M., Klopfenstein, D. R., and Vale, R. D. (2002). Conversion of Unc104/KIF1A kinesin into a processive motor after dimerization. *Science* 297, 2263–2267.

Tomishige, M., Stuurman, N., and Vale, R. D. (2006). Single molecule observations of neck linker conformational changes in the kinesin motor protein. *Nat Struct Mol Biol* 13, 887–894.

Tomishige, M. and Vale, R. D. (2000). Controlling kinesin by reversible disulfide cross-linking: identifying the motility-producing conformational change. *J Cell Biol* 151, 1081–1092.

Vale, R. D., Funatsu, T., Pierce, D. W., Romberg, L., Harada, Y., and Yanagida, T. (1996). Direct observation of single kinesin molecules moving along microtubules. *Nature* 380, 451–453.

Vale, R. D. and Milligan, R. A. (2000). The way things move: looking under the hood of molecular motor proteins. *Science* 288, 88–95.

Vale, R. D., Reese, T. S., and Sheetz, M. P. (1985a). Identification of a novel force-generating protein, kinesin, involved in microtubule-based motility. *Cell* 42, 39–50.

Vale, R. D., Schnapp, B. J., Mitchison, T., Steuer, E., Reese, T. S., and Sheetz, M. P. (1985b). Different axoplasmic proteins generate movement in opposite directions along microtubules in vitro. *Cell* 43, 623–632.

Vale, R. D., Schnapp, B. J., Reese, T. S., and Sheetz, M. P. (1985c). Organelle, bead, and microtubule translocations promoted by soluble factors from the squid giant axon. *Cell* 40, 559–569.

Vale, R. D., Soll, D. R., and Gibbons, I. R. (1989). One-dimensional diffusion of microtubules bound to flagellar dynein. *Cell* 59, 915–925.

Valentine, M. T., Fordyce, P. M., Krzysiak, T. C., Gilbert, S. P., and Block, S. M. (2006). Individual dimers of the mitotic kinesin motor Eg5 step processively and support substantial loads in vitro. *Nat Cell Biol* 8, 470–476.

Varga, V., Helenius, J., Tanaka, K., Hyman, A. A., Tanaka, T. U., and Howard, J. (2006). Yeast kinesin-8 depolymerizes microtubules in a length-dependent manner. *Nat Cell Biol* 8, 957–962.

Walker, R. A., Salmon, E. D., and Endow, S. A. (1990). The *Drosophila* claret segregation protein is a minus-end directed motor molecule. *Nature* 347, 780–782.

Yildiz, A., Tomishige, M., Vale, R. D., and Selvin, P. R. (2004). Kinesin walks hand-over-hand. *Science* 303, 676–678.

Zhang, Y. and Hancock, W. O. (2004). The two motor domains of KIF3A/B coordinate for processive motility and move at different speeds. *Biophys J* 87, 1795–1804.

CHAPTER 3

Force-Generating Mechanisms of Dynein Revealed through Single Molecule Studies

Kazuhiro Oiwa

*Kobe Advanced ICT Research Center, National Institute of Information and Communications Technology,
588-2 Iwaoka, Nishi-ku, Kobe 6512492, Japan
Graduate School of Life Science, University of Hyogo, Harima Science Park City, Hyogo 6781297, Japan*

Hiroaki Kojima

*Kobe Advanced ICT Research Center, National Institute of Information and Communications Technology,
588-2 Iwaoka, Nishi-ku, Kobe 6512492, Japan*

Summary

Dynein, which is crucial to a range of cellular processes, is a minus-end-directed
microtubule motor and the largest and most complex of the three classes of linear motor
proteins present in eukaryotic cells. The mass of its motor domain is about 10 times that
of the other microtubule motor, kinesin. Its large size and the difficulty of expressing
and purifying mutants have hampered progress in dynein research. Recently, however,
electron microscopic observations and single molecule measurements have shed light on
several key unsolved questions concerning how the dynein molecule is organized, what
conformational changes in the molecule accompany ATP hydrolysis, and whether two or
three motor domains are coordinated in the movements of dynein. This chapter describes
our current knowledge of the force-generating mechanism of dynein, with emphasis on
findings from electron microscopy and single molecule nanometry.

Key Words

dynein; microtubule; cilia; flagella; axoneme; optical trap; step size; processivity;
single molecule nanometry

Introduction

A high-molecular-weight ATPase extracted from *Tetrahymena* cilia was the first microtubule-based force-generating ATPase to be discovered (Gibbons, 1963). It was named "dynein" after the CGS unit of force, the dyne (Gibbons and Rowe, 1965). Dynein is now known to consist of a functionally diverse family of proteins, the members of which are involved in a wide range of cellular functions in various cells. These functions have been reviewed by Hirokawa (1998), Di Bella and King (2001), Vale (2003), and Vallee et al. (2004). Dyneins are bulky molecules that contain one to three heavy chains, each with a molecular mass of more than 500 kDa and consisting of approximately 4500 amino acid residues. The number of heavy chains depends on the origin of the protein, and each heavy chain consists of a C-terminal toroidal-shaped head domain together with two elongated flexible structures called the stalk (microtubule-binding domain) and an N-terminal tail (cargo-binding domain, formerly known as the stem) to which the intermediate chains (40–120 kDa) and most light chains (8–30 kDa) bind. The N-terminal tail is thought to be involved in binding the dynein motor to its various cargoes in an ATP-insensitive manner.

Among other linear motors, at least 24 classes of myosin (Foth et al., 2006) and 14 classes of kinesin (Dagenbach and Endow, 2004; Lawrence et al., 2004) have been identified, but dyneins originally fell into only two major classes: the axonemal dyneins, which are located in cilia and flagella of eukaryotes, and cytoplasmic dyneins, which have been isolated from a variety of sources. Although discrimination into these classes was originally based on the function and localization of dyneins, phylogenetic analyses of full-length dynein sequences have confirmed the existence of differences among various dyneins (Höök and Vallee, 2006), and recently, nine classes of dyneins (two cytoplasmic, two outer-arm, and five inner-arm) have been identified (Wickstead and Gull, 2007).

The axonemal dyneins are localized on peripheral doublet microtubules in axonemes (the cytoskeletal structures within cilia and flagella) and form projections called dynein arms, each of which is composed of several dynein molecules. The axonemal dyneins are further classified into two subclasses, outer-arm dyneins and inner-arm dyneins, on the basis of their localization. These dynein molecules are organized with a few heavy chains that form heterotrimers, heterodimers, or monomers, together with intermediate, light intermediate, and light chains. The number of heavy chains in outer-arm dyneins depends on the species of origin. Outer-arm dyneins from most sources consist of two distinct heavy chains (Hastie et al., 1986; King et al., 1990), whereas those from *Tetrahymena* and *Chlamydomonas* each contain three distinct heavy chains. Inner-arm dyneins contain one or two heavy chains (Goodenough et al., 1987; Piperno and Luck, 1979; Pfister et al.,

1982; Piperno et al., 1990), and at least seven subspecies are identified in *Chlamydomonas* axonemes (Kagami and Kamiya, 1992).

The various heavy chains of axonemal dyneins so far studied have distinct properties and specific functions in flagellar motility (Asai, 1995; Kagami and Kamiya, 1992; Yagi et al., 2005; reviewed in Kamiya, 2002). Not all dyneins are necessary, but the presence of certain combinations of dyneins seems to be crucial for the correct functioning of the axoneme (Kamiya, 2002). Furthermore, some heavy chains of axonemal dyneins permit the generation of torque (Kagami and Kamiya, 1992; Vale and Toyoshima, 1988) and/or oscillations (Shingyoji et al., 1998). Coordinated beating and bend propagation of cilia and flagella are generated by active sliding (shear) of peripheral doublet microtubules driven by ensembles of these various types of axonemal dyneins.

Cytoplasmic dynein was reported in the early 1980s (Hisanaga and Pratt, 1984; Pallini et al., 1982), but its presence was not widely accepted until Vallee and colleagues showed that the microtubule-associated protein MAP-1C has many characteristics in common with axonemal dyneins (Paschal et al., 1987; Vallee et al., 1988). It is now clear that cytoplasmic dynein is present in many eukaryotic cells and drives a variety of fundamental cellular processes, including nuclear migration, organization of the mitotic spindle, chromosome separation during mitosis, and the positioning and function of many intracellular organelles (reviewed in Hirokawa and Takemura, 2004; Vallee and Stehman, 2005; Vallee et al., 2004). Cytoplasmic dynein is thought to be a heavy-chain homodimer, in which each heavy chain has a stalk–head–tail organization like axonemal dyneins (Gee et al., 1997; Koonce and Samsó, 1996).

The bulkiness of the molecule and consequent difficulties in expressing and purifying mutants in large quantity has hampered progress in structural and mechanistic studies on dyneins. To date, no atomic resolution structures of dynein have been solved. As a result, the force-generating mechanism of dyneins remains poorly understood. Early progress in electron microscopic observations provided many insights into the force-generating action of dyneins, especially outer-arm dyneins in flagellar axonemes (Avolio et al., 1984, 1986; Burgess, 1995; Goodenough and Heuser, 1982, 1984; Johnson and Wall, 1983; Sale et al., 1985). However, for a breakthrough in structure-based studies on dynein motility, it was essential to develop a system in which dynein heavy-chain genes could be readily mutated or engineered and produced in biochemical quantities suitable for biochemical and biophysical analyses. Recent success in expressing active cytoplasmic dyneins in *Dictyostelium discoideum* (Nishiura et al., 2004), yeast (Reck-Peterson and Vale, 2004), or insect cells (Höök et al., 2005) has therefore ushered in a new era of dynein research.

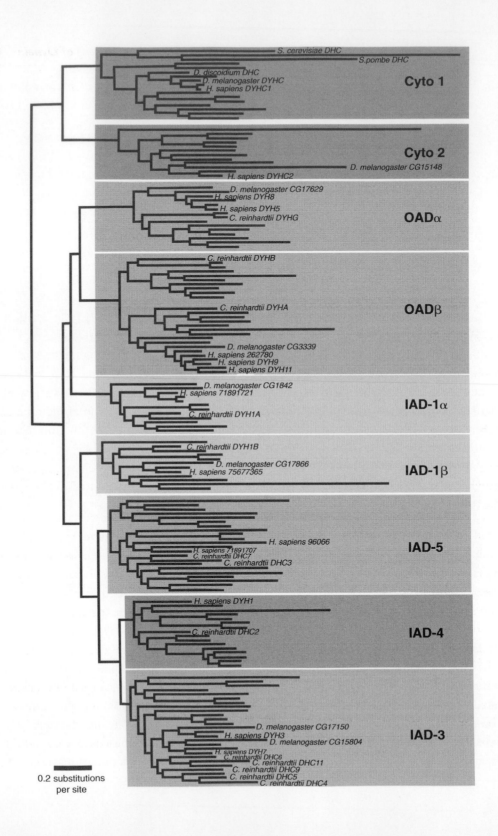

Cyto 1

S. cerevisiae DHC
S.pombe DHC
D. discoidium DHC
D. melanogaster DYHC
H. sapiens DYHC1

Cyto 2

D. melanogaster CG15148
H. sapiens DYHC2

OADα

D. melanogaster CG17629
H. sapiens DYH8
H. sapiens DYH5
C. reinhardtii DYHG

OADβ

C. reinhardtii DYHB
C. reinhardtii DYHA
D. melanogaster CG3339
H. sapiens 262780
H. sapiens DYH9
H. sapiens DYH11

IAD-1α

D. melanogaster CG1842
H. sapiens 71891721
C. reinhardtii DYH1A

IAD-1β

C. reinhardtii DYH1B
D. melanogaster CG17866
H. sapiens 75677365

IAD-5

H. sapiens 96066
H. sapiens 71891707
C. reinhardtii DHC7
C. reinhardtii DHC3

IAD-4

H. sapiens DYH1
C. reinhardtii DHC2

IAD-3

D. melanogaster CG17150
H. sapiens DYH3
D. melanogaster CG15804
H. sapiens DYH7
C. reinhardtii DHC6
C. reinhardtii DHC11
C. reinhardtii DHC9
C. reinhardtii DHC5
C. reinhardtii DHC4

0.2 substitutions
per site

Single molecule measurements combined with protein engineering of dyneins and advanced electron microcopy have now shed light on key unsolved questions concerning how the molecule is organized, what conformational changes accompany ATP hydrolysis, and whether two or three motor domains are coordinated in their motions (reviewed in Koonce, 2006; Oiwa and Sakakibara, 2005; Spudich, 2006). In this chapter, we will discuss the structure of dynein molecules and also consider their mechanochemistry and force-generating mechanism.

Molecular Organization of Dynein

Phylogeny of Dynein Heavy-Chain Sequences

The amino acid sequence of the dynein heavy chain was determined independently by two groups, Gibbons et al. (1991) and Ogawa (1991), who cloned and sequenced the cDNA of the β dynein heavy chain of sea urchin sperm flagella. The motor domain of these dynein heavy chains is well conserved and has four nucleotide-binding sites ("P-loop" motifs) and a pair of sequences that has a high probability of forming an antiparallel coiled-coil structure known as the stalk (see below). Many full-length sequences of axonemal dynein and cytoplasmic dynein heavy chains are now available as a result of genome sequencing of a broad range of eukaryotes; these analyses have revealed the existence of a high level of sequence conservation among motor domains.

Phylogenetic analysis of the full-length sequences of dynein heavy chains provides insights into relationships among the family of dyneins. The boundaries of each dynein

Figure 3.1: A Bayesian phylogeny for the dynein heavy-chain sequences from 24 diverse eukaryotes. Reproduced and modified from Wickstead and Gull (2007) with permission. Only dynein heavy chains described in this article are indicated in this phylogeny for clarity, although the original phylogeny (Wickstead and Gull, 2007) includes sequences of dynein heavy chains of *Homo sapiens, Takifugu rubripes, Drosophila melanogaster, Caenorhabditis elegans, Cryptosporidium parvum, Dictyostelium discoideum, Giardia lamblia, Tetrahymena thermophila, Leishmania major, Osteococcus lucimarinus, Phaeodactylum tricornutum, Phytophthora sojae, Plasmodium falciparum, Chlamydomonas reinhardtii, Saccharomyces cerevisiae, Schizosaccharomyces pombe, Thalassiosira psuedonana, Toxoplasma gondii,* and *Trypanosoma brucei.* No dynein heavy chains were found in *Arabidopsis thaliana, Cyanidioschyzon merolae, Entamoeba histolytica, Oryza sativa,* or *Populus trichocarpa* (Wickstead and Gull, 2007). Cytoplasmic dynein is composed of two families. Axonemal dynein heavy chains were divided into seven families: OADα, OADβ, IAD-1α, IAD-1β, IAD-5, IAD-4, and IAD-3

family were originally drawn on the basis of differences in physiological functions and localization in cells and organelles. Four functional groups are identified in two major groups: cytoplasmic dyneins 1 and 2, axonemal outer-arm dyneins and inner-arm dyneins. By using available genome sequences from 24 diverse eukaryotes, a Bayesian phylogeny for the complete repertories of dynein heavy chains recently identified nine dynein heavy-chain families: two cytoplasmic and seven axonemal (Wickstead and Gull, 2007) (Figure 3.1). This phylogeny is consistent with the classification of dynein heavy chains into two groups: cytoplasmic and axonemal. It also indicates that outer-arm dyneins and inner-arm dyneins constitute discrete groups and their divergence must have occurred in the early stages of evolution, after which they have evolved independently.

As a result of previous research, axonemal dyneins have been divided into six families: OADα, OADβ, OADγ, IAD-1α, IAD-1β, and the single-headed dyneins (Asai and Wilkes, 2004). More-recent work based on the sea urchin genome assembly has suggested that there may be more families within the single-headed category (Morris et al., 2006). Most recently, Wickstead and Gull (2007) have suggested that the OADα, OADβ, and OADγ groups encompass two well-supported ancestral families, and the single-headed dyneins can be classified into three distinct families: IAD-3, IAD-4, and IAD-5. The heavy chains of double-headed inner-arm dyneins are classified into IAD-1α and IAD-1β (Figure 3.1). Note that it has also been confirmed that dyneins have been lost from higher plants (Wickstead and Gull, 2007).

Sequence analysis has also shown that dyneins are members of the AAA^+ protein superfamily, sharing similarities both in sequence and characteristics with the ATPase domain secondary structure (Neuwald et al., 1999; reviewed in Asai and Koonce, 2001; King, 2000a; Ogura and Wilkinson, 2001; Vale, 2000; Vallee and Höök, 2006). As is the case with many other AAA^+ ATPases, the motor domain of dynein is composed of six AAA^+ ATPase-like domains (AAA modules), each of 35–40 kDa. These, however, are concatenated into a single large polypeptide (Figure 3.2A). Hereafter, we refer to the most N-terminal AAA module as AAA1, the second as AAA2, and so on. The incorporation of all six AAA modules within a single polypeptide is characteristic of this branch of the AAA^+ protein superfamily, which also includes midasin (Iyer et al., 2004). Dynein is thus classified within the P-loop NTPases and is on the branch of the RecA fold, which is different from the G-protein fold that includes kinesin and myosin (Lupas and Martin, 2002).

A high level of sequence conservation among motor domains supports the idea that findings for one experimental model system can be generalized to other dyneins.

Therefore, any system in which one can readily mutate or engineer dynein heavy-chain genes and produce biochemical quantities of dynein for analysis is valuable for the elucidation of the structural basis of dynein motility in general. For cytoplasmic dyneins, few expression systems have been established; Samsó et al. (1998) expressed cytoplasmic dynein constructs in *Dictyostelium* cells. Nishiura et al. (2004) succeeded in expression of motile constructs of cytoplasmic dynein in large quantities from *Dictyostelium* cells. Reck-Peterson and Vale (2004) showed that *Saccharomyces cerevisiae* affords an attractive system for manipulating a nonessential genomic copy of the dynein heavy chain by homologous recombination. Truncated dynein constructs of rat cytoplasmic dynein can be expressed in the baculovirus expression system (Mazumdar et al., 1996).

In contrast, research on axonemal dyneins awaits the establishment of an expression system for these proteins. In the absence of a suitable expression system, studies using either sea urchin sperm flagella or *Tetrahymena* cilia have provided most of the important information available on the biochemical properties of axonemal dyneins. *Tetrahymena* is easy to grow in large quantities, and a *Tetrahymena* cell has 500–700 cilia. Therefore, it provides the best source for biochemical studies such as kinetic analysis of dynein's ATPase (reviewed in Johnson, 1985).

The unicellular green alga *Chlamydomonas reinhardtii* has also been used for research on axonemal dyneins (reviewed by Di Bella and King, 2001; Kamiya, 2002). A *Chlamydomonas* cell has two flagella extending from its anterior end, which it uses to swim in a "breaststroke" fashion. Fortunately, the basic structure of these flagella is indistinguishable from that of the cilia lining the human airway. Many outer-arm subunits have one or more orthologues in humans (Pazour et al., 2006), and many useful mutants of dyneins (and several axonemal components) have also been isolated and characterized, making *Chlamydomonas* a useful model for the study of dyneins.

Molecular Configurations of Dyneins

The absence of any crystal structure makes dynein the poor cousin among the linear motor proteins (kinesin and myosin). Biochemical and molecular biological studies on cytoplasmic and axonemal dyneins have shown that the dynein heavy chain, which has a molecular mass of more than 500 kDa, contains a large fundamental motor domain in the C-terminal 320–380 kDa fragment. This domain incorporates sites for both ATP hydrolysis and microtubule binding (Gee et al., 1997; Koonce and Tikhonenko, 2000; Nishiura et al., 2004; Reck-Peterson and Vale, 2004). The motor domain is composed of six AAA modules each of 35–40 kDa, the microtubule-binding stalk and the linker, which connects

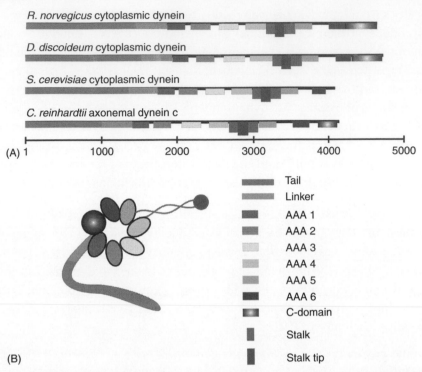

Figure 3.2: (A) Schematic diagrams of the heavy chains of rat cytoplasmic (AAA41103, *Rattus norvegicus*), *Dictyostelium* cytoplasmic (XP643185, *D. discoideum*), the budding yeast cytoplasmic (CAA79923, *S. cerevisiae*), and *Chlamydomonas* axonemal (BAE19786, *C. reinhardtii*) dyneins, showing the domain structure: tail, putative linker, AAA domains, and microtubule binding region. Amino acid numbers are shown at the bottom. Note that the C-terminal domain found in *Dictyostelium* and rat cytoplasmic dyneins and in flagellar dyneins is missing in yeast cytoplasmic dynein. (B) Model for the organization of the dynein heavy chain. The six AAA domains and the C-terminal domain are arranged in a ring. The linker is drawn as undocked from the head ring since two different models can be constructed depending upon a clockwise or counterclockwise arrangement of AAA domains with respect to the docked linker (Burgess et al., 2004). At present we cannot discriminate decisively between these two alternatives

the head ring and the tail and is normally docked onto the head ring (Figure 3.2). Electron microscopic observations show that dynein has a complex morphology, in which the six subdomains (probably corresponding to AAA modules) are arranged in a ring to form a head domain (Burgess et al., 2003, 2004; Kotani et al., 2007; Samsó and Koonce, 2004; Samsó et al., 1998) (Figures 3.2 and 3.3). The external diameter of the ring is

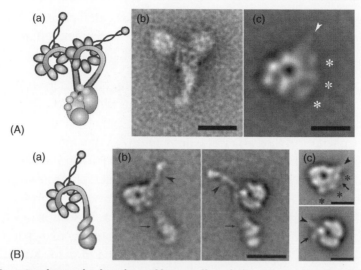

Figure 3.3: (A) a. A schematic drawing of heterodimeric inner-arm dynein I1 (dynein-f) from *Chlamydomonas reinhardtii* flagella. b. Electron microscopic images of the negative-stained molecule. Single-particle image analysis confirms the organization of the molecule. Despite considerable variability in the spatial arrangement of the two heads among all individual molecules examined, several classes show a surprisingly characteristic structure in which the two heads are relatively closely apposed at one end of a tail structure that itself has a characteristic asymmetric structure. c. Alignment and averaging of the head domains clearly shows them to comprise an asymmetric ring-like morphology very similar to that of another *Chlamydomonas* flagellar inner-arm subspecies, dynein-c (Figure 3.3B(c)). Subdomains (asterisks) and the position of stalk (arrowhead) are indicated. Scale bar: b, 20 nm; c, 10 nm. (B) a. A schematic drawing of the single-headed inner-arm dynein-c. Dynein-c contains one actin monomer and two copies of a 28 kDa protein as light chains, which all bind to the N-terminal region of the heavy chain (Yanagisawa and Kamiya, 2001). b. Image averages of dynein-c molecules; right and left views in the apo state. Stalk (arrowhead) and tail (arrow) are clearly seen. c. Image average of the head of right views (upper panel) and left views (lower panel) of dynein-c in the absence of nucleotide, showing subdomains in the head ring. Seven subdomains are seen, three of which are clearly identified (asterisks) in the right view. The positions of stalk (arrowhead) and tail (arrow) domains are indicated. Dynein clearly has two different faces. Scale bar: b, 20 nm; c, 10 nm

10–14 nm, the thickness 4–6 nm, and the diameter of the central channel 2.5 nm (Burgess et al., 2003; Samsó et al., 1998). The two faces of the head ring are different (Figure 3.3B(b) and (c)), suggesting that the head is not a simple planar ring (Burgess et al., 2003).

The first four AAA modules (AAA1–AAA4) contain a highly conserved Walker A motif (GXXXXGKT, a so-called P-loop) and a Walker B motif (DEXX) (Gibbons et al., 1991; Koonce et al., 1992; Mikami et al., 1993; Ogawa, 1991) and are thought to bind nucleotide. In contrast, the sequences of the two most C-terminal AAA domains (AAA5 and AAA6) have highly degraded Walker motifs. A principal site of ATP hydrolysis had been mapped to the Walker A and B motifs of AAA1 by vanadate-mediated photocleavage of the heavy chain of axonemal dynein (Gibbons et al., 1987). Further strong support for a functional role for the Walker motif of AAA1 is provided by molecular dissection of cytoplasmic dyneins, in which mutation of the P-loop eliminates their motor activities in vivo (Reck-Peterson and Vale, 2004; Silvanovich et al., 2003) and in vitro (Kon et al., 2004).

It seems likely that the additional Walker motifs (in AAA2–AAA4) act in a regulatory manner by binding either ADP or ATP. For example, in some dynein isoforms, the presence of ADP is known to be essential for motile activity in vitro (Shiroguchi and Toyoshima, 2001; Yagi, 2000), and in others, ADP increases the velocity of microtubule movement, indicating that ADP binds to at least one of these AAA modules (Kikushima et al., 2004; Yagi, 2000). A hypothetical atomic structure produced by homology modeling of the dynein AAA modules suggests that the nucleotide-binding Walker A motifs lie close to the interface between adjacent modules (Mocz and Gibbons, 2001). Interactions between adjacent AAA modules through their nucleotide pockets support the idea that they may act in concert to produce a functional motor.

The function of these additional Walker motifs has now been further elucidated by molecular dissection in expressed cytoplasmic dyneins (Kon et al., 2004; Reck-Peterson and Vale, 2004) and by ATPase measurements of truncated cytoplasmic dyneins expressed in *Escherichia coli* (Takahashi et al., 2004). Manipulation of the Walker A motif of AAA3 in cytoplasmic dynein showed that the resulting construct failed to be released from microtubules by ATP, and that it lost most of its microtubule-activated ATPase activity. Motor activity of the construct was retained, although the velocity of microtubule sliding was less than 1/20th of that of the wild type. Mutations in Walker A motifs in AAA2 and AAA4 did not affect the apparent binding affinity for microtubules in the presence of ATP but reduced both ATPase and motor activity (Kon et al., 2004). These results clearly indicate that additional Walker motifs regulate the motor activities through nucleotide binding and/or hydrolysis.

The origin of the seventh subdomain of the head ring remains unknown. The C-terminal sequence has been proposed to form the seventh subdomain of the head ring (King, 2000b). Heavy-chain constructs devoid of this sequence lack ATPase activity (Gee et al., 1997; Koonce and Samsó, 1996), and proteolytic cleavage near the C-terminal domain abolished sensitivity of vanadate-mediated photocleavage of the first ATPase site (Höök et al., 2005). If we assume that the C-terminal domain participates in forming the head ring as the seventh subdomain, it lies relatively close to AAA1 within the folded molecule and could interact with AAA1. The C-terminal domain may play a role in controlling entry and/or exit of substrate from the major ATPase sites (in AAA1–AAA4) within the motor domain (Höök et al., 2005). Höök et al. (2005) proposed a model in which elements within the C-terminal half of the dynein motor domains control catalytic activity by inducing conformational changes at the active sites within the head ring. However, yeast cytoplasmic dynein lacking sequence at its C-terminal domain still has motile activity (Reck-Peterson et al., 2006). The absence of the C-terminal domain in some organisms suggests that it is not necessary for dynein function but may be needed for regulation (Vallee and Höök, 2006).

Structure and Mechanical Properties of the Stalk

The stalk is up to 15 nm long and comprises an antiparallel coiled coil, which lies between AAA4 and AAA5, and a distal microtubule-binding domain (Figure 3.2). An expressed fragment of dynein containing this coiled-coil region forms a rodlike structure similar to the stalk (Gee et al., 1997). A small globular domain at a tip of the stalk has been identified as the ATP-sensitive microtubule-binding domain (Gee et al., 1997). Although poor sequence conservation exists in this globular domain at the stalk tip (Asai and Koonce, 2001), mutagenesis of conserved residues clearly interferes with microtubule binding (Koonce and Tikhonenko, 2000). Microtubule binding of the stalk tip was also examined with a recombinant stalk-tip peptide (Mizuno et al., 2004). The stalk-tip peptide was observed to bind to a microtubule with a periodicity of 8 nm and to share the binding region on the microtubule with kinesin (Mizuno et al., 2004).

Binding of ATP to AAA1 causes dissociation of dynein from the microtubule, and binding of the stalk tip to the microtubule enhances the dissociation of hydrolysis products from the motor domain. Because the sites of microtubule binding and primary ATP hydrolysis are spatially segregated (~25 nm), elucidation of the signaling pathway between them is an important problem in understanding dynein function. Although the structure of the stalk is not fully understood, it is predicted to transmit conformational changes along its length, implying a requirement for dynamic changes in its helix–helix interactions.

Sequences of left-handed coiled coils are characterized by a seven-residue periodicity (heptad repeat) represented by the expression (a–b–c–d–e–f–g) (reviewed by Burkhard et al., 2001), and sequences of the stalk region have this periodicity (Gibbons et al., 2005). Apolar residues appear preferentially in the first (a) and fourth (d) positions of the heptad. A "knobs-into-holes" packing of the apolar side chains into a hydrophobic core is mainly responsible for stabilizing coiled coils. On the basis of the amino acid sequences, a discontinuity of a half repeat shift in the heptad repeat is predicted to occur halfway along the outward α-helix of the stalk of cytoplasmic dyneins (see Figure 2b in Asai and Koonce, 2001; Gibbons et al., 2005). In contrast, the heptad repeat is highly ordered throughout the inward α-helix (Gibbons et al., 2005). This discontinuity probably modulates the stability of the coiled coil of the stalk, although large-scale melting of a coiled coil in a nucleotide-bound state seems unlikely. In nucleotide-bound states, electron microscopy of axonemes suggested that the stalk remains intact and maintains contact with the microtubule, even when the axonemes are fixed in the presence of ATP (Goodenough and Heuser, 1984, 1985), and it remains the same length in isolated dynein (Burgess et al., 2003). Stalk flexibility is reduced in apo molecules (those without bound nucleotide), indicating a greater stiffness of this domain (Burgess et al., 2003, 2004). Thus, some mechanism must exist that triggers changes in the stability of the coiled coil, and hence in the stiffness of the stalk.

To investigate the structure of the stalk, Gibbons et al. (2005) designed a series of chimeric constructs in which the microtubule-binding domain, along with a portion of its predicted coiled-coil stalk, is fused onto a stable antiparallel coiled-coil base found in the native structure of seryl-tRNA synthetase. They attempted to identify the optimal alignment between the hydrophobic heptad repeats in the two strands of the coiled-coil stalk. Alterations in the phase of the heptad repeats in the outward helix changed the affinity of the stalk tip toward the microtubules. On the basis of these results, Gibbons et al. (2005) hypothesized that, during the mechanochemical cycle, the two strands of its coiled-coil stalk undergo a small amount of sliding displacement as a means of communicating between the AAA core of the motor and the microtubule-binding domain.

A coiled-coil peptide has a flexural rigidity on the order of $400\,\mathrm{pN\,nm^2}$ (Howard and Spudich, 1996). When the coiled coil is clamped at one end and has a transverse force F applied to its free end, it will deflect by a distance y like a rigid beam of length L, where $y = FL^3/3EI$. E is the Young's modulus and I the moment of inertia. The product of these, $E \times I$, is the flexural rigidity. Thus a $15\,\mathrm{nm}$ length of coiled-coil peptide clamped at one end has a stiffness of about $0.36\,\mathrm{pN\,nm^{-1}}$. Since the range of conformations for an axonemal dynein molecule observed in negatively stained samples (Burgess et al.,

2003, 2004) depends on the distribution of thermal energy, it gives an estimation of the stiffness of the stalk. The stiffness was estimated to be $0.5 \, pN \, nm^{-1}$ in apo molecules and $0.14 \, pN \, nm^{-1}$ in ATP–vanadate molecules (Lindemann and Hunt, 2003). A transverse force of 1–8 pN, which corresponds to the maximal force of dynein (Gennerich et al., 2007; Hirakawa et al., 2000; Mallik et al., 2004; Sakakibara et al., 1999; Shingyoji et al., 1998; Toba et al., 2006), when applied to the free end of the coiled coil deflects it through a distance of about 3–22 nm. The negative-stained electron microscope images also indicated that the tail of the dynein molecule was flexible with a stiffness of $0.1 \, pN \, nm^{-1}$ (Lindemann and Hunt, 2003). Because the stiffnesses of the stalk and the tail are in series, this gives a total stiffness between 0.05 and $0.08 \, pN \, nm^{-1}$, which is similar to that estimated by optical-trap experiments (Sakakibara et al., 1999). Thus dynein is a flexible molecule, suggesting that under low- or zero-load conditions, it might be able to search for binding sites on the microtubule track by diffusion, as does myosin VI (Rock et al., 2005), and to take steps much larger than the mean stroke size of 15 nm predicted by electron microscopy (Burgess et al., 2003). The flexibility of the molecule might explain the variable step size of dyneins. For details, see discussion below.

The Linker

The linker is a structure located in the portion of the tail proximal to AAA1 that serves as a connection between AAA1 and the main part of the tail. The existence of the linker was first indicated in images of negative-stained monomeric axonemal dynein (dynein-c). It has been suggested that the linker, which is normally docked onto the head ring in both apo and ADP-V_i (ADP–vanadate) states, is involved in generation of force through its interaction with the head ring (Burgess et al., 2003) (Figure 3.4). When the linker is undocked from the head ring, it is revealed as a relatively large structure about 2 nm wide and 10 nm long (Burgess et al., 2003).

If this mechanism is common in all dyneins, the linker domain sequences are likely to be conserved among them. Conserved residues are more readily detected along the ~600-residue portion of the tail next to AAA1. Vanadate photocleavage of recombinant dynein heavy chains showed essential sequences are contained in residues from 1137 to 1455 in rat cytoplasmic dynein, in which the conserved element R[K]-X-R-H-W-X-X-I[L] is found (Gee et al., 1997).

However, the minimal motor domain of cytoplasmic dynein was defined by Reck-Peterson et al. (2006), who prepared N-terminal truncated cytoplasmic dynein constructs of *S. cerevisiae* and examined their motility in vitro. They established that the dynein construct ($Dyn1_{314kDa}$) is reasonably close to the minimal size of motor required for

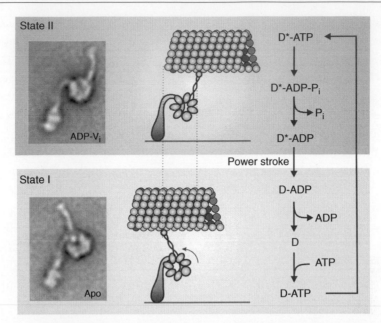

Figure 3.4: Proposed mechanism of dynein's power stroke. Negative-stain electron microscopy followed by single-particle analysis succeeded in capturing two distinct conformations of dynein-c molecules isolated from *Chlamydomonas* flagella in the ADP-V_i (state II) and apo (state I) states (Burgess et al., 2003). On ATP binding to AAA1, the tail emerges far from the stalk (state II, top panel). Upon product release, the tail emerges closer to the stalk (state I, bottom panel), suggesting movement of the unseen linker domain (and tail relative to head ring and stalk). This conformational change swings the stalk by ca. 15 nm. This model is based upon Burgess et al. (2003) and Kon et al. (2005)

dynein motility; this construct drives microtubules and consists of 394 amino acid residues before the start of the AAA1. This sequence probably forms the linker domain (Reck-Peterson et al., 2006). However, the conserved element described above is not included in these residues.

Dynamic measurements of linker movement were performed by measuring the Förster resonance energy transfer (FRET) between a green fluorescent protein (GFP) and a blue fluorescent protein (BFP) both fused into a dynein construct molecule (Kon et al., 2005). A series of 380 kDa dynein constructs (which include the monomeric motor domain containing the more highly conserved portion of the tail domain just upstream of AAA1 and correspond to $Dyn1_{331\,kDa}$ of *S. cerevisiae*) from *Dictyostelium* were prepared. These constructs had a GFP attached at the N-terminus and a BFP inserted into various sites on the dynein head ring. The efficiency of FRET was measured in each construct at various nucleotide states under steady-state conditions. The results showed two distinct values: a high FRET efficiency and a low FRET efficiency, suggesting movement of the N-terminus relative to the head ring. Using mutants that were trapped in specific intermediate states, it was shown that this movement is coupled to ATPase steps (Kon et al., 2005).

These observations suggest a model of the linker motion: On binding of ATP to AAA1, the orientation of the docked linker on the head ring causes the tail to emerge far from the stalk (state II, top panel in Figure 3.4). Upon release of products, the linker orientation on the ring changes and brings the tail closer to the stalk (state I, bottom panel in Figure 3.4). Mogami et al. (2007) performed pre-steady-state kinetic studies by FRET on the above constructs and showed that, in the absence of microtubules, the recovery stroke proceeds rapidly (ca. $180 \, s^{-1}$), whereas the power stroke is very slow (ca. $0.2 \, s^{-1}$). In the presence of microtubules, the power stroke is accelerated, but not the recovery stroke.

The thermodynamic properties of the interaction between the linker and AAA domains remain unknown. However, the series of dynein constructs in which GFP and BFP are both fused (Kon et al., 2005) will provide a useful tool for studies at the single molecule level of how force affects the interaction between the linker and the ring. Applying a load to the linker perturbs mechanical transitions between the high-FRET state and the low-FRET state and should result in a load-dependent shift in the dwell time distribution. From the shift of the dwell time distribution, we should be able to draw a free-energy diagram of the linker–ring interaction as has been shown in kinesin (Schnitzer et al., 2000). Gennerich et al. (2007) recently showed that dynein walks processively toward either the minus or plus ends of microtubules under an applied force in the absence of nucleotide hydrolysis and that a small assisting force causes dynein to step forward, while a much larger force is required to induce dynein to step backward. The directional asymmetry of this force-induced, nucleotide-independent stepping may be intimately related with the linker–head interaction.

Mechanism of Force Generation by Dyneins

Although the force-generating mechanism of dynein is not yet fully understood, at least three models have been proposed for this process: rotation about the junction between head ring and stalk owing to the rearrangement of the AAA modules in the head ring coupled with ATPase, causing the microtubule-binding stalk to swing relative to the tail (Mocz and Gibbons, 2001; Samsó and Koonce, 2004; Vale, 2000); rotation about the junction between head ring and tail coupled to ATPase activity, again causing the head and the microtubule-binding stalk to swing (Asai and Koonce, 2001); and change in its orientation in a previously unseen part of the tail (the linker) by switching between two differently docked positions on the head ring (Burgess et al., 2003), thus producing a rotation of the head ring that causes the stalk to swing ("linker" model) (Figure 3.4).

Negative-staining electron microscopy followed by single-particle analysis succeeded in capturing two distinct conformations of *Chlamydomonas* flagellar dynein-c molecules in the ADP-V_i and apo states. On the basis of these observations, the linker model was proposed (Burgess et al., 2003). This static view was supported by FRET measurements that showed ATP-dependent movements of the linker relative to the head, as described in the previous section (Kon et al., 2005).

Dynein constructs (*Dictyostelium* cytoplasmic dynein) with site-directed biotin tags at various locations permit the selective attachment of the motor to a substrate in motility assays conducted in vitro and, consequently, the assignment of the parts of the dynein molecule that are involved in producing movement (Shima et al., 2006). Tethering a dynein molecule at the most distal position of the linker permits it to drive microtubules at the fastest rate. On the other hand, when the dynein molecule is tethered at more proximal positions within the linker or directly at the head ring, the sliding velocity of microtubules is significantly reduced. These results show that the motion of the linker (and therefore the tail) is directly coupled with the generation of force. Movement of the microtubules was, however, not eliminated completely when the dynein molecule was held at the head ring. Under these circumstances, movement of the linker cannot contribute directly to microtubule sliding. Although this slow microtubule sliding is mechanically distinct from the sliding driven by the power stroke of the linker, its dependence upon ATP concentrations and minus-end-directed motion of microtubules suggest that it is also generated by an active energy transduction process (Shima et al., 2006).

Although experimental support has accumulated for the linker-docking model, single molecule measurements on force generation of dyneins have raised some questions: The stall force generated by single dynein molecules varies from measurement to measurement. For axonemal dyneins, a value of ~6 pN was reported by Shingyoji et al. (1998), a value of 1–2 pN was reported by Sakakibara et al. (1999), and a value of 4.7 pN was reported by Hirakawa et al. (2000). For cytoplasmic dyneins, a value of ~1 pN was reported by Mallik et al. (2004, 2005), and a value of 7–8 pN was reported by Toba et al. (2006) and Gennerich et al. (2007). This variation in force may depend upon the type of dynein used and its role in vivo.

In addition, variations in the mode of movements are not well understood. Several research groups have shown that dynein can display a diffusive mode of bidirectional motion along the microtubule as well as processive stepwise motion (Mallik et al., 2004; Reck-Peterson et al., 2006; Ross et al., 2006). Regulation mechanisms may exist that switch from the stepping mode to the diffusive mode and vice versa. Variation of the modes may also reflect physiological roles of dyneins in vivo. Further work will be required to resolve these reported differences in dynein behavior, with particular attention paid to possible species variation, protein preparation, and assay conditions, as stated by Reck-Peterson et al. (2006).

Furthermore, in vitro motility assays have shown that some inner-arm dyneins have the ability to generate a torque that causes rotation of microtubules around their longitudinal axis during forward translocation (Kagami and Kamiya, 1992; Vale and Toyoshima, 1988). The direction of microtubule rotation observed in *Chlamydomonas* inner-arm dyneins was always clockwise when viewed from the minus end (Kagami and Kamiya, 1992). The direction of rotation is the same as that observed with *Tetrahymena* 14S dynein (Vale and Toyoshima, 1988). Torque generation suggests that dynein performs its power stroke in three dimensions rather than within a plane. Considerable work remains to be done before we fully understand the force-generating mechanism of dynein.

Mechanical Properties of Dyneins Studied by Single Molecule Methods

In recent decades, the development of a number of technologies such as atomic-force microscopy, optical-trap nanometry, and fluorescence microscopy have provided tools for studying the dynamics of single molecules in situ over timescales from milliseconds to seconds. The single molecule sensitivities of these methods permit studies to be made on conformational changes and functions of protein motors that are masked in ensemble-averaged experiments. Processivity, step size, and dwell time distributions are among the properties that can be directly measured by single molecule techniques. Our understanding of the functions of dyneins has benefited considerably from the application of single molecule techniques.

The optical trap is now a popular and versatile technique for the manipulation of objects with micrometer dimensions. In this technique, refractive particles are captured by means of photon pressure from a laser beam focused to a diffraction-limited spot by a large-numerical-aperture objective lens (Ashkin, 1992; Svoboda and Block, 1994) (see techniques chapter by Knight, this volume). For spherical objects, the restoring force increases linearly with displacement from the center of the trap, provided the particle does not move too far from the center of the trap, that is, the optical trap behaves like a Hookean spring.

Using an optical detector based on a quadrant photodiode, the position of the particle held in the optical trap can be measured with nanometer precision and millisecond temporal resolution. The optical trap can be used in combination with these precise optical detectors as a force transducer to detect mechanical interactions between individual protein-motor molecules. Using optical-trap nanometry on various types of dynein, forces in the piconewton range and displacements in the nanometer range have been measured (Gennerich et al., 2007; Hirakawa et al., 2000; Mallik et al., 2004, 2005; Sakakibara et al.,1999; Shingyoji et al., 1998; Toba et al., 2006). For some protein motors, a combination of optical-trap techniques and single molecule fluorescence-imaging techniques has provided a hint of the existence of a coupling between biochemical and mechanical events during ATPase cycles (Adachi et al., 2007; Ishijima et al., 1998; Nishizaka et al., 2004). However, this technique has been applied to dyneins in rather few studies (Inoue and Shingyoji, 2007) because many of the fluorescent nucleotides that are suitable for single molecule imaging (e.g., Cy3–EDA–ATP; Oiwa et al., 2000) are poor substrates for dyneins.

Optical-Trap Nanometry Studies on Dynein Motility

The mechanical properties of several types of dynein have been measured at nanometer and millisecond spatiotemporal resolutions by means of the optical-trap technique. However, the step sizes and modes of movement of these dyneins are still a matter of controversy (see previous section). For axonemal dyneins, the single-headed inner-arm dynein-c shows processive movement and has a step size of 8.2 nm (Sakakibara et al., 1999). The maximal force generated by a dynein-c molecule was at 1.6 pN, which is smaller than the corresponding value for kinesin. The outer-arm dynein of *Tetrahymena* cilia, a 22S dynein that is a heterotrimer, also showed processive movement with an 8 nm step size at low concentrations of ATP ($<20\,\mu M$), whereas at higher concentrations ($\geq 20\,\mu M$), it does not move processively, but instead shows pulse-like force generation similar to that observed in skeletal muscle myosin II (Hirakawa et al., 2000). Under high loads, both these dyneins often fail to step forward and can slip backward (Hirakawa et al., 2000; Sakakibara et al., 1999).

Mallik et al. (2004) reported that single cytoplasmic dynein molecules purified from bovine brains primarily took large steps (24–32 nm) at low loads, but the step size decreased from 32 to 8 nm as the load increased to the stall force. The stall force generated by the cytoplasmic dynein was reported to be ~ 1 pN, which was smaller than that generated by kinesin. In addition, where multiple dynein molecules interact with a microtubule and contribute to movement, the dynein molecules move predominantly in 8 nm steps (Mallik et al., 2005).

Toba et al. (2006) performed optical-trap nanometric studies on cytoplasmic dynein purified from porcine brain. They observed processive movement of dynein with regular 8 nm steps, irrespective of the load, and measured that the stall force generated by a single dynein molecule was 7–8 pN, a value that is comparable to that generated by a kinesin-1 molecule (Toba et al., 2006). A large stall force of ~7 pN was also measured for yeast cytoplasmic dynein (Gennerich et al., 2007). These results contrast with a previous report on mammalian cytoplasmic dynein (Mallik et al., 2004).

Gennerich et al. (2007) applied a force-feedback optical trap to native and artificially dimerized yeast cytoplasmic dyneins and analyzed their stepping behavior as a function of load. Surprisingly, they found that dynein moves processively toward either the minus or plus ends of microtubules under an applied force in the absence of ATP hydrolysis. This force-induced, nucleotide-independent stepping shows directional asymmetry: A small assisting force (~3 pN) causes dynein to step forward, while a much larger force (7–10 pN) is required to induce dynein stepping backward (Gennerich et al., 2007). In addition, by applying near-stall force to dynein in the presence of ATP, they observed no net movement, but dynein continued forward and backward stepping. They referred to this behavior of dynein as non-advancing stepping and analyzed the dwell times between steps. The rate of non-advancing backward stepping was increased by load but not affected by ATP, while the rate of non-advancing forward stepping was unaffected by load and dependent on ATP.

In optical-trap nanometry, a dynein molecule carries a relatively large bead (200 nm to 1 μm in diameter), so that the measured position of the bead does not directly represent the position of the dynein molecule. Without a force-feedback system, compliance between the bead and the dynein molecule attenuates displacement performed by the dynein molecules. Moreover, movement of the bead measured in an optical trap reflects movement of the center of mass of the entire dynein molecule. To study the mechanism of processivity in more detail, the movement of individual head domains should be monitored with high spatial and temporal resolutions. This can be achieved by a technique called "fluorescence imaging with one-nanometer accuracy" (FIONA) (Toprak and Selvin, 2007; Yildiz et al., 2003).

Studies on Dynein Motility by Fluorescence Imaging with One-Nanometer Accuracy

The image of a point source of light (of wavelength λ) that is produced by using the objective lens of a conventional microscope is a diffraction-limited spot with an Airy disc pattern (bright disk of light surrounded by alternating dark and bright diffraction rings). The spot approximates well to a Gaussian distribution with a standard deviation σ given by $0.3\lambda/\text{N.A.} \approx 125$ nm for $\lambda = 540$ nm and N.A. $= 1.3$, where N.A. is the numerical aperture of the objective lens. The size and shape of the image in three dimensions is known as the point-spread function, which can be measured directly for any given imaging system. The error in determining the central position of a two-dimensional projection at the image plane is governed by system noise and photon-counting statistics. With proper adjustments of the experimental system, precision is limited simply by the photon count (n), where the standard error of the mean (SEM) is equal to $\sigma n^{-0.5}$, where σ is the standard deviation of the spot image (Yildiz and Selvin, 2005; Yildiz et al., 2003). The mean position can be determined either by a center-of-mass calculation or by least-squares fitting to a Gaussian distribution. For real systems where the signal-to-noise ratio is limited, the optimal determination of position is provided by the Gaussian fitting technique (Cheezum et al., 2001).

Yildiz et al. (2003) used total-internal-reflection fluorescence microscopy combined with long exposure times (0.5 s) on a cooled charge-coupled device (CCD) camera system to observe nanometer-scale stepping motion of individual myosin V molecules as they walked along a single actin filament. They manipulated the experimental conditions by reducing the concentration of ATP to prolong the dwell time of each myosin step so that the 70 nm steps could be easily resolved above the noise.

Toba et al. (2006) adopted the FIONA technique to analyze the movement of single cytoplasmic dynein molecules purified from porcine brain. To improve the temporal resolution and tolerance to photobleaching, they used quantum dots, which are photostable and have an intense fluorescence. They finally achieved a temporal resolution of 2 ms and showed that the dynein undergoes stepwise movement with a regular 8 nm step size. A distribution of dwell times between successive steps was well fitted by a single exponential function, suggesting one transition is rate limiting for the stepwise movement. However, Toba et al. (2006) used a nonspecific cross-linker to label dynein molecules with quantum dots, so the precise interpretation of their FIONA data was limited by the unknown position of the fluorescent labels in the molecule.

Reck-Peterson et al. (2006), on the other hand, using a functional recombinant dimeric dynein of the budding yeast *S. cerevisiae*, labeled dynein in specific locations with fluorescent dyes or quantum dots and then tracked single molecules by FIONA. The trajectories of dynein movements showed a stepping behavior. When a dynein molecule was labeled at its tail domain, the step size measured was primarily 8 nm. In contrast, when the molecule was labeled at its head domain, the step size was 16 nm.

Reck-Peterson et al. (2006) also showed that the cytoplasmic dynein needs two motor domains to undergo processive movement. This was demonstrated through the creation of a dynein that can be converted between monomeric and dimeric states by a small molecule (rapamycin). Processive motion was observed only in the presence of rapamycin, which induced dimer formation. In addition, they showed that processivity of dynein does not require any of the known dynein-associated subunits in the yeast genome, despite an earlier report that the dynactin complex enhances the processivity of brain cytoplasmic dynein (King and Schroer, 2000)

In early studies of dynein motility, cytoplasmic dynein-coated beads exhibited greater lateral movements among microtubule protofilaments than did kinesin (Wang et al., 1995). Hence, dynein apparently does not have to walk along a single protofilament. However, close examination using FIONA showed that fluorescently labeled dynein also displayed lateral stepwise movements, which usually occurred simultaneously with forward stepping. This shows that dynein has the reach or flexibility to occasionally land on an adjacent protofilament (Reck-Peterson et al., 2006).

There are at least two models for processive movement of dimeric protein motors: the "inchworm" and "hand-over-hand" models (Figure 3.5). The inchworm model is named after the movement of caterpillars that progress with an inching gait by extending the front part of the body forward and bringing the rear up to meet it. In the inchworm model, one head of the motor always leads and the other always trails behind. This mechanism was considered to account for the lack of rotation of a microtubule moved by a single kinesin molecule (Hua et al., 2002). These authors investigated the stepping mechanism of kinesin by immobilizing kinesin molecules on a glass surface and measuring the orientations of microtubules moved by single kinesin molecules at sub-micromolar ATP concentrations. A symmetric hand-over-hand mechanism was hypothesized to produce 180° rotations of the microtubule relative to the immobilized kinesin neck. In fact, however, no such rotations were observed (Hua et al., 2002).

Other studies of processive motors, such as myosin V (reviewed in Vale and Milligan, 2000) and kinesin-1 (reviewed in Schief and Howard, 2001), have established the hand-over-hand mechanism. In this mechanism, one of two motor domains swings forward at each step while the other motor domain remains attached to the track. Each motor domain alternates between performing moving and anchoring roles. Experimental support for this hand-over-hand mechanism has accumulated steadily. Dynein, however, has some important differences from kinesins and myosins, such as dynein's larger size, greater flexibility, and apparent ability to walk along multiple (protofilament) tracks.

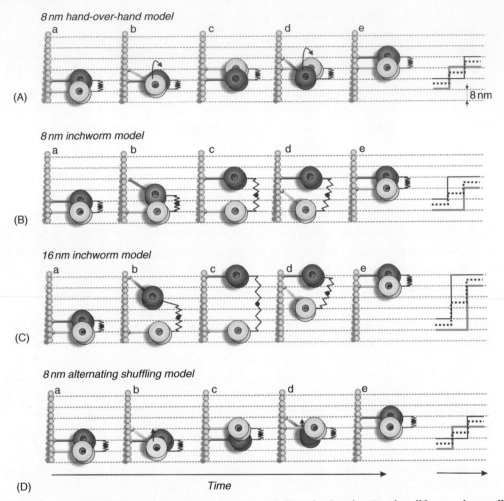

Figure 3.5: Models for processive movement of dimeric dynein. To simplify, we draw all presumed elastic elements in the tail domain as a simple spring that connects two head rings. Traces in the rightmost panels show the time course of positions of the heads (grey and black lines) and the center of mass (black dotted line) of the dimeric dynein, which would be expected in FIONA experiments. (A) Hand-over-hand model for dimeric dynein. Binding of ATP to the trailing head (white) releases the head from the microtubule (a) and then the head is swung forward by the strain stored in the connecting spring while the leading head (black) stays bound to the microtubule (b). Upon binding to the microtubule, the head changes its conformation coupled with products release (c). The trailing head is depicted as going across in front of the leading head and standing on the right of the leading head. (B) Inchworm model for dimeric dynein, in which both heads move

Given the apparently large size and flexibility of a dynein molecule, and the magnitude of the suggested power stroke (Burgess and Knight, 2004; Burgess et al., 2003) in relation to the tubulin lattice, the 8 nm step size displayed by dimeric dyneins is surprisingly small. This is in contrast to other linear motors since the step size of a motor, except myosin VI (Okten et al., 2004), can be predicted to be proportional to the length of its power stroke (Purcell et al., 2002; Sakamoto et al., 2003). One possible explanation is that when a dimeric dynein moves on a microtubule taking 8 nm steps, the molecule could be compact and stiff with its two head rings in close, intimate association as is seen in axonemal dyneins in situ (see below).

Based upon findings from yeast cytoplasmic dynein, Reck-Peterson et al. (2006) proposed a molecular model (which they called the "alternating shuffling model") to explain how processive motion is achieved by cytoplasmic dynein. In the model, two dynein heads alternate taking 16 nm steps, while the centroid position of the molecule moves by 8 nm for each step. The large dimensions of the head ring do not allow the heads to alternately "swing" forward at each step but overlap part of the head ring during the stepping motion. This is possibly supported by electron microscopic observations of the phi (Φ)-shaped structure of cytoplasmic dynein, in which the tails of the heavy chains are close to each other and the head rings are partially overlapped (Amos, 1989).

16 nm per ATP hydrolysis. a. Binding of ATP to the leading head (black) releases the head from the microtubule. b. The leading head undergoes diffusion search and rebinds to the microtubule. c. The leading head changes its conformation coupled with product release and produces tension in the connecting spring. d. Upon binding ATP, the trailing head (white) dissociates from the microtubule and performs diffusion search for the next binding site. Owing to tension produced by the leading head, the trailing head preferentially binds on forward sites. e. The trailing head rebinds to the microtubule and changes its conformation while releasing tension in the connecting spring. In this model, each head performs 16 nm stepping while the center of mass of the molecule shows 8 nm steps. (C) Inchworm model with a large step. In this model, each head should perform 32 nm stepping to obtain a 16 nm step of the center of mass, so that the intramolecular strain highly increases at step c. (D) Alternating shuffling model proposed by Reck-Peterson et al. (2006). Two dynein heads alternate taking 16 nm steps, whereas the position of the center of mass of the molecule moves by 8 nm for each step. Note that due to the large size of the head, two heads are partially overlapped during stepping without changing the relation of their lateral positions

However, in dynein heteromers, as found in axonemal dyneins, the motor domains need not behave in an identical manner. Hence, processive movement of dynein could be described simply in terms of an inchworm model coupled with a diffusion search if we assume that two heads can separate by up to 32 nm when the dynein shows processive movements with 16 nm steps (Figure 3.5C).

These single molecule experiments raise questions: For instance, how does the monomer move processively? Processive motion of a single-headed motor on microtubules has previously been reported for the kinesin-like motor KIF1A (Okada and Hirokawa, 1999; Okada et al., 2003). In this case, processive motion is achieved through electrostatic interactions between the negatively charged so-called E-hook of tubulin and the positively charged K-loop of KIF1A. A similar mechanism may apply to single-headed dyneins or other dyneins showing one-dimensional diffusion. Indeed, an outer-arm dynein and cytoplasmic dyneins were reported to interact with microtubules in a way that permits the latter to diffuse only along their longitudinal axes, as described above. Although the equivalent of the K-loop sequence has not been reported in dynein, Shima et al. (2006) suggest that motor-domain-anchored dynein could drive microtubules by a biased Brownian motion. This motility may be important in vivo for keeping dynein molecules on microtubules as a mechanism to maintain tension on the cytoskeleton.

Dyneins in Axonemes

Axonemes are highly ordered and precisely assembled superstructures. The most widespread form has a 9 + 2 arrangement of microtubules: nine doublets surrounding a pair of singlets (the central-pair microtubules) with radial spokes extending from each of the peripheral doublets toward the central pair (Figure 3.6A). Coordinated beating and bend propagation of cilia and flagella are generated by active sliding of peripheral doublet microtubules driven by ensembles of various types of dyneins. In an axoneme, the activity of dynein molecules propagates through linear or two-dimensional arrays of dynein arms closely packed on the peripheral doublet microtubules. Axonemal dyneins thus show large-scale integrated behavior that is responsible for the beating of flagella and for wave propagation.

Several important research findings on the concerted operation of multiple dyneins, or mixtures of different types of dynein, have been reported. Functionally intact dyncin arms have been prepared as follows: A demembranated axoneme is gently treated with the proteolytic enzyme trypsin (Summers and Gibbons, 1971) or elastase (Brokaw, 1980) to loosen the protein ties that normally hold it together. Addition of ATP to the axoneme activates the sliding of the individual doublets, and a group of the doublets protrudes from the axoneme. This activation causes disintegration of the axoneme into individual doublets and provides arrays of functional dynein arms on the doublet. Our group examined movements of fluorescently labeled singlet microtubules on dynein rows on axonemes exposed by sliding disintegration and showed that microtubules moved only toward the tip of the flagella but not in the opposite direction (Yamada et al., 1998). The results strongly suggest that dynein molecules in flagella are so oriented that they can efficiently move adjacent outer doublet microtubules in the direction away from the base of the flagellum (Sale and Satir, 1977; Yamada et al., 1998). A solitary dynein arm on an extruded doublet microtubule is, however, capable of oscillatory movement (Shingyoji et al., 1998). Axonemes immobilized on a glass surface along their entire length show a back-and-forth oscillation of a bundle of doublets at high frequencies driven by dyneins (Kamimura and Kamiya, 1989). This longitudinal oscillation of the axoneme is coupled to a transverse oscillation (Sakakibara et al., 2004). For coordinated activities of dynein arms, there must be a structural origin within the axoneme and the dynein ensemble must possess special mechanical properties.

Arrangement of Dynein Heavy Chains within a Dynein Arm

The axonemal dyneins are organized so that a few heavy chains form heterodimers, heterotrimers, or monomers. Nevertheless, precisely how dynein heavy chains are organized into the dynein arm complexes of an axoneme remains unknown. Studies on an isolated outer-arm dynein show that its heavy chains are tied at the ends of their tails, whereas their globular heads spread apart to form a "bouquet" structure (Goodenough and Heuser, 1984; Johnson and Wall, 1983). In contrast, recent three-dimensional reconstructions of in situ outer-arm dyneins have demonstrated that their head rings are intimately associated with one another on the microtubule. Electron tomography of metal replicas of rapidly frozen and cryo-fractured sperm axonemes from the dipteran *Monarthropalpus flavus* (Lupetti et al., 2005), cryo-electron tomography of sea urchin sperm (Nicastro et al., 2005) and *Chlamydomonas* flagella (Ishikawa et al., 2007; Nicastro et al., 2005, 2006) and in vitro cryo-electron microscopy studies of reconstituted outer-arm dyneins (Oda et al., 2007) have shown that an outer-arm dynein is composed of two or three plates corresponding to the head rings. These plates are stacked up in parallel and are oriented obliquely or parallel to the longitudinal axis of the microtubules in the axoneme (dotted lines in Figure 3.6B and C). The plates are connected to their tilted tail domains (arrows in Figure 3.6D). The molecular organization of the outer dynein arm has been predicted on the basis of three-dimensional reconstructions of *Chlamydomonas*

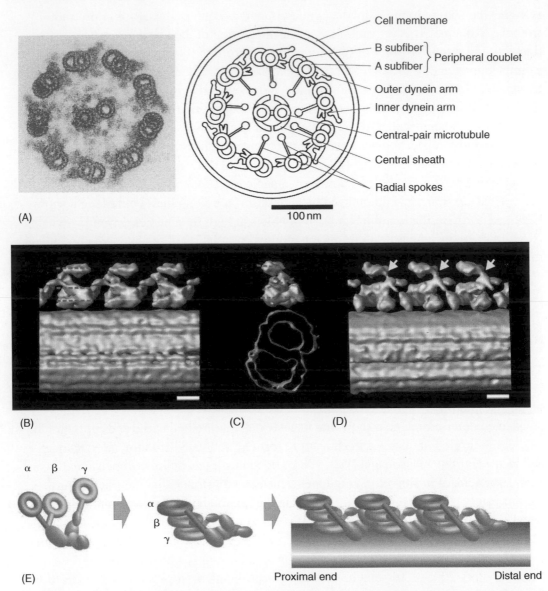

Figure 3.6: (A) Electron micrograph showing axonemal structures of a wild-type flagellum of *Chlamydomonas reinhardtii* in transverse section. The axoneme is demembranated and viewed from the base of the flagellum. Nine peripheral doublet microtubules surround two central singlet microtubules. The central microtubules are enclosed by the central sheath. Two rows of dynein arms are attached to the surface of the peripheral double microtubule and extended toward the adjacent doublet. The schematic drawing and this micrograph

axonemes of the α-heavy-chain-lacking mutant (*oda11*) or a β-heavy-chain-truncated mutant (*oda4-s7*). Although the stalk cannot be identified in these tomograms, the predicted arrangement implies that the stalk and tail domains are not in the same plane as the head rings. If this is the case, the swing motion between the stalk and the tail proposed for isolated dynein molecules can be interpreted as a two-dimensional projection of a three-dimensional movement (Ishikawa et al., 2007; Lupetti et al., 2005; Nicastro et al., 2006).

Tomographic studies of the axonemes further indicated the presence of three inter-arm connections between the outer arms: two outer arm–outer arm linkages and one outer arm–inner arm linkage. Although the origins of these linkages are not yet known, they may play important roles in the coordinating dynein arm activity. For example, dynein arms along an individual doublet in actively beating flagella have a "relaxed" conformation, but small groups of them occur with the "rigor" conformation (no-nucleotide conformation), suggesting localized coordination of their power strokes (Burgess, 1995).

Furthermore, the molecular organization of the outer dynein arm raises the question of whether the three dynein heads contribute equally to force generation and movement. The outer dynein arms of *Chlamydomonas* flagella are attached to a precise site on the outer doublet microtubules in vivo. This binding is mediated by the outer-dynein-arm docking complex, which is composed of three protein subunits (Takada et al., 2002). In the presence of this docking complex, outer-arm dyneins mixed with microtubules in vitro become organized into an armlike configuration on the microtubule with the axial

were provided by Dr Sakakibara (National Institute of Information and Communications Technology). (B–D) Surface rendered representations of averaged outer dynein arms arrayed on the microtubule doublets from cryo-tomograms. Modified from Ishikawa et al. (2007). (B) View from the external side of the axoneme. Three plates corresponding to the head rings are indicated by dotted lines. (C) Transverse section seen from the base of the flagellum. (D) View from the internal side of the axoneme. Arrows indicate the tilted tail domains. In (B), the base of the flagellum is on the right side and the tip on the left. In (D), the base is on the left and the tip on the right. (E) Schematic diagram to describe how the three heavy chains of the outer-arm dynein form the outer-arm complex. All three head rings are stacked. The γ-tail would be branched from the blob at the end of the tail. It then runs through the connection, which connects adjacent outer arms past the β-ring and down to the γ-ring underneath the β-ring. In this model, all rings in an arm belong to the same triple-headed "bouquet"

periodicity of 24 nm (Haimo et al., 1979; Oda et al., 2007). Reconstituted arms are still motile and can drive microtubules at faster velocities than outer-arm dyneins randomly distributed on a glass surface in in vitro motility assays (Aoyama and Kamiya, 2006). The organized structure and the ordered array probably improve the efficiency of the motility through the intermolecular and intramolecular coordination.

Minimal Components Required for Axonemal Beating

To understand the coordination mechanism of dynein arms, realistic computations of the coordinated motion of cilia and flagella were performed on the basis of solid evidence relating to the mechanics and biochemistry of dyneins and axonemes (Brokaw, 2002; Brokaw, 2005; Lindemann and Kanous, 1995). In these computations, the force exerted by dyneins on elastic components of the axoneme was evaluated and the motions of axonemes were computed. These computations provide us with new predictions that cannot be obtained by a simple description of mechanical parts and their relations. From an experimental point of view, however, the minimal functional unit of the axoneme should be defined, and its behavior needs to be examined to extract essential features of the interaction between outer doublet microtubules.

Studies on flagellar mutants of *Chlamydomonas* showed that mutant axonemes lacking the central-pair apparatus or radial spokes, which are usually nonmotile, beat in certain non-physiological solutions (Omoto et al., 1996; Wakabayashi et al., 1997; Yagi and Kamiya, 2000; reviewed in Kamiya, 2002). These observations indicate that the central-pair microtubules or radial spokes are not essential for beating of *Chlamydomonas* flagella. Flagella of mutants lacking outer arms beat with an almost normal waveform, although the beat frequency is decreased. Hence, doublet microtubules and inner dynein arms are considered to comprise the essential set of components for axonemal beating.

Aoyama and Kamiya (2005) have recently succeeded in establishing that a pair of microtubules in a frayed *Chlamydomonas* axoneme frequently associates with and dissociates from each other and bends cyclically in the presence of ATP. Through quantitative examinations, they showed that the dissociation of two microtubules is not accompanied by noticeable bending (Aoyama and Kamiya, 2005). This observation may imply that dynein–microtubule interaction is not necessarily regulated by the curvature of the axoneme.

Measurement of Force Generated by an Axoneme

A fine glass needle can be used as a cantilever spring to estimate the force generated by small organelles or cells attached to its tip. When a fine glass fiber of length L and radius r with a circular cross section is clamped at one end, the stiffness of the glass fiber, k, is given as follows:

$$k = \frac{3EI}{L^3} = \frac{3\pi E r^4}{4L^3} \tag{3.1}$$

where E is the Young's modulus and I the moment of inertia (Howard, 2001). To permit measurements to be made, the deflection of the fine needle for static measurements should be at least 10 nm since the spatial resolution of the detector is about 0.1–1 nm. The force generated by a single protein motor is expected to be less than about 10 pN. Hence, the stiffness of the glass needle should be less than 1 pN nm^{-1} for single protein motors. By using glass threads (Pyrex glass, $E = 70$ GPa) of lengths 100–200 μm and radii 0.1–1 μm, we can achieve a stiffness ranging from 0.001 to 10 pN nm^{-1}. The mechanical time constant of a cantilevered spring is given as follows:

$$\tau = 0.2 \frac{\eta L^4}{E r^4} \tag{3.2}$$

where η is the viscosity of the solution (Howard, 2001). The glass needles can have time constants of about 1 ms in aqueous solution, which makes them fast enough to resolve the individual mechanical events within a hydrolysis cycle.

Mechanical properties of dynein ensembles and a flagellar axoneme have been measured by using such fine glass needles (Kamimura and Takahashi, 1981; Oiwa and Takahashi, 1988; Okuno and Hiramoto, 1979; Yoneda, 1960; early studies in Japan were reviewed in Takahashi, 1995). Yoneda (1960) measured the torque generated by a giant cilium of *Mytilus edulis* by immobilizing the cilium with a calibrated fine glass needle. The measured bending moment generated by the cilium during its effective stroke averaged 0.04 pN m. Assumptions that this compound cilium contains 20 axonemes, and that 15–20 μm of its length is actively generating shear moment, lead to an approximate value for the shear moment per unit length in the active region of the axoneme of 117 pN m/m. The maximal sliding force per dynein arm was estimated from these assumptions as 8.8 pN (Brokaw, 1975).

Direct measurement of the force generated by sliding of microtubules in a flagellum or cilium is hampered by the small size of the organelles and the small force they develop. However, Kamimura and Takahashi (1981) showed that the sliding force could be measured by means of a pair of glass microneedles attached to a flagellar axoneme that had been demembranated with the detergent Triton X-100 and exposed to elastase (a protease) in a reactivating solution containing Mg-ATP. By this technique, it became possible to

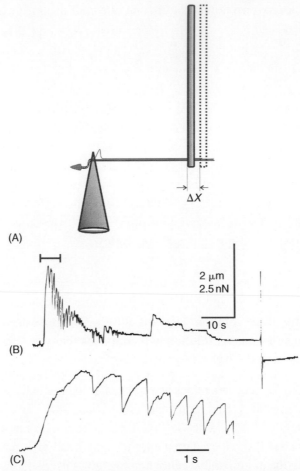

Figure 3.7: (A) The sliding force generated by doublet microtubules in a sea urchin sperm flagellum measured by a pair of fine glass needles. The needles were attached to a flagellar axoneme that had been demembranated with Triton X-100 and exposed to elastase in a reactivating solution containing ATP. One needle is stiffer with a steeply tapered tip and the other more flexible with a long tapered tip. The elastic coefficient of the flexible needle was predetermined. (B) A record showing the time course of force development of sliding microtubules in a demembranated axoneme. (C) An example of repetitive force transients commonly observed. Reproduced from Kamimura and Takahashi (1981) and Oiwa and Takahashi (1988) with permission

perform a direct measurement of the mechanical properties associated with microtubule sliding (Figure 3.7). Oiwa and Takahashi (1988) refined the method of Kamimura and Takahashi (1981) and determined the force–velocity relationship for sliding microtubules in a sea urchin sperm flagellum. In force measurements made on sliding microtubules in demembranated axonemes, the generation of oscillatory forces was often observed, which may be due to coordinated activity of dynein arms (Figure 3.7C). Note that these fine-needle techniques have been extended to studies on protein-motor mechanics at the single molecule level; for example, the forces generated by single myosin motors on actin filaments have been determined (Kishino and Yanagida, 1988; Kitamura et al., 1999; Kojima et al., 1994).

Among the major challenges in understanding the mechanism of ciliary motility, one is to determine how dynein-driven microtubule sliding is regulated. Mathematical models indicate that oscillatory bending and bend propagation require mechanical feedback mechanisms that control the timing and location of active sliding (Brokaw, 1985, 2000; Lindemann and Kanous, 1995). However, experimental evidence for the identity and nature of such a mechanical feedback mechanism is still limited, although it has been reported recently that changes in axoneme diameter may be responsible (Lindemann and Mitchell, 2007). Although earlier studies involving mechanical measurements of cilia and flagella or dynein arms provided insights into the operation of dynein arms in ensembles, recent advances in single molecule techniques with high temporal and spatial resolutions permit quantitative, reproducible, and precise experiments to be performed on the mechanism of coordinated behavior in dynein ensembles. Therefore, dynein ensembles on microtubules have become the target of characterization through single molecule techniques.

Perspectives: From Single Molecules to Ensembles

Mechanical experiments at the single molecule level raise questions such as whether dynein molecules are really organized into intact configurations on artificial substrates and how the dimer or the trimer coordinates its multiple motor domains during stepping. If we take into consideration the large size and flexibility of dynein molecules, the 8 nm step size displayed by dimeric dyneins is puzzling. When a dimeric dynein moves on a microtubule, the molecule could be compact and stiff with two head rings in intimate association, as seen in axonemal outer-arm dyneins in situ.

In studies of axonemal dyneins, we need to take into consideration three layers of coordination: intra-domain coordination within a heavy chain, intramolecular coordination among multiple motor domains within each arm, and intermolecular coordination among multiple dynein arm complexes.

In the first layer, the coordination and communication between domains needs to be studied. As shown in the previous sections, the microtubule-binding site of dynein is at the stalk tip and is segregated from the principal ATPase site in AAA1. Microtubule binding to dynein accelerates product release from the AAA1 site, and loss of nucleotide from AAA1 increases the affinity of the stalk tip for microtubules, suggesting that the coiled-coil stalk plays an important role in transducing conformational information between the stalk tip and AAA1.

Furthermore, studies on P-loop mutants have shown that at least two AAA domains (AAA1 and AAA3 for cytoplasmic dyneins) can hydrolyze ATP at distinct rates, and binding of nucleotides at both sites is required for ATP-induced release of microtubules and for microtubule activation of ATPase. These observations support the existence of long-range allosteric communication through conformational changes within the motor domain (Höök et al., 2005).

A remarkable feature of dynein activity in flagella is the presence of oscillations. Single molecule studies showed that forces oscillate with an amplitude of 2 pN, and the maximum frequency of the oscillation (at 0.75 mM Mg-ATP) is 70 Hz (Shingyoji et al., 1998). AAA1 and AAA3 are oscillators in the sense that they hydrolyze ATP in cycles but at distinct rates (Kon et al., 2004). These potential oscillators are connected through damping and elastic elements to establish long-range allosteric communication, and their kinetics could show modulation through mechanical signals. The swinging motion of the coiled-coil stalk can produce significant displacement when it pulls a microtubule like the rope of a winch (Burgess and Knight, 2004), but not when it pushes the microtubule like a lever arm. The flexibility of the stalk may provide anisotropy of force transduction through the stalk. In vitro motility assays on dynein-f have implied this nonlinearity (Kotani et al., 2007). Such anisotropic properties of the stalk and coupled oscillators may play important roles in the function of dyneins. Intra-domain communication is an attractive research target, and a combination of molecular dissections using mutants and single molecule studies should improve our understanding of the communication mechanisms.

In the second level of coordination, important questions remain concerning the possible coordination between two or three motor domains connected by a presumably flexible tail. In other words, force generation and kinetics of motor domains, which are intimately associated with each other, could be modulated by stress and strain in a multimeric dynein molecule. Such modulation could play important roles, especially in the processive movement of cytoplasmic dyneins. In other processive motors, such as conventional kinesin and myosin V, changes in the rates of nucleotide binding or release to either the leading head or trailing head occur as a result of intramolecular strain. Single molecule studies should show whether applying a load to a dynein molecule specifically perturbs mechanical transitions in the stepping cycle, resulting in a force-dependent shift in the dwell time distribution (Gennerich et al., 2007).

For the third level of coordination, thanks to technological advances of the past decade, electron tomography has enabled the analysis of complex and heterogeneous systems, such as the three-dimensional structure of dynein arms in situ in axonemes (Ishikawa et al., 2007; Lupetti et al., 2005; Nicastro et al., 2005, 2006). This technique, combined with mutant studies, should provide details of the structural basis for coordination of dynein arm complexes. Dynein arms may influence one another through stress and strain during coordinated beating and bend propagation of cilia and flagella. We are now able to perform quantitative and extremely reproducible and precise experiments on individual dynein molecules and ensembles in a manner that is very attractive to physicists.

Acknowledgments

We thank Dr Stan Burgess (University of Leeds), Dr Justin Molloy (National Institute for Medical Research) and Hitoshi Sakakibara (National Institute of Information and Communications Technology) for helpful discussions and critical comments on the manuscript. This work is supported by the Special Coordination Funds for Promoting Science and Technology (KO), Strategic International Cooperative Program, Japan Science and Technology Agency (KO), and Grant-in-Aid for Scientific Research on the Priority Area "Regulation of Nano-systems in Cells" by the Ministry of Education, Culture, Sports, Science and Technology (KO).

References

Adachi, K., Oiwa, K., Nishizaka, T., Furuike, S., Noji, H., Itoh, H., Yoshida, M., and Kinosita Jr., K. (2007). Coupling of rotation and catalysis in F_1-ATPase revealed by single molecule imaging and manipulation. *Cell* 130, 309–321.

Amos, L. A. (1989). Brain dyncin crossbridges microtubules into bundles. *J Cell Sci* 93, 19–28.

Aoyama, S. and Kamiya, R. (2005). Cyclical interactions between two outer doublet microtubules in split flagellar axonemes. *Biophys J* 89, 3261–3268.

Aoyama, S. and Kamiya, R. (2006). A dynein/microtubule system that can partially mimic the motile properties of flagellar axonemes. *Seibutsu Butsuri* 46, S215.

Asai, D. J. (1995). Multi-dynein hypothesis. *Cell Motil Cytoskeleton* 32, 129–132.

Asai, D. J. and Koonce, M. P. (2001). The dynein heavy chain: structure, mechanics and evolution. *Trends Cell Biol* 11, 196–202.

Asai, D. J. and Wilkes, D. E. (2004). The dynein heavy chain family. *J Eukaryot Microbiol* 51, 23–29.

Ashkin, A. (1992). Forces of a single-beam gradient laser trap on a dielectric sphere in the ray optics regime. *Biophys J* 61, 569–582.

Avolio, J., Glazzard, A. N., Holwill, M. E., and Satir, P. (1986). Structures attached to doublet microtubules of cilia: computer modeling of thin-section and negative-stain stereo images. *Proc Natl Acad Sci U S A* 83, 4804–4808.

Avolio, J., Lebduska, S., and Satir, P. (1984). Dynein arm substructure and the orientation of arm-microtubule attachments. *J Mol Biol* 173, 389–401.

Brokaw, C. J. (1975). Cross-bridge behavior in a sliding filament model for flagella. In: *Molecules and Cell Movement* (Inoue, S. and Stevens, R., Eds.), pp. 165–179. Raven Press, New York.

Brokaw, C. J. (1980). Elastase digestion of demembranated sperm flagella. *Science* 207, 1365–1367.

Brokaw, C. J. (1985). Computer simulation of flagellar movement. VI. Simple curvature-controlled models are incompletely specified. *Biophys J* 48, 633–642.

Brokaw, C. J. (2000). Stochastic simulation of processive and oscillatory sliding using a two-headed model for axonemal dynein. *Cell Motil Cytoskeleton* 47, 108–119.

Brokaw, C. J. (2002). Computer simulation of flagellar movement VIII: coordination of dynein by local curvature control can generate helical bending waves. *Cell Motil Cytoskeleton* 53, 103–124.

Brokaw, C. J. (2005). Computer simulation of flagellar movement IX. Oscillation and symmetry breaking in a model for short flagella and nodal cilia. *Cell Motil Cytoskeleton* 60, 35–47.

Burgess, S. A. (1995). Rigor and relaxed outer dynein arms in replicas of cryofixed motile flagella. *J Mol Biol* 250, 52–63.

Burgess, S. A. and Knight, P. J. (2004). Is the dynein motor a winch?. *Curr Opin Struct Biol* 14, 138–146.

Burgess, S. A., Walker, M. L., Sakakibara, H., Knight, P. J., and Oiwa, K. (2003). Dynein structure and power stroke. *Nature* 421, 715–718.

Burgess, S. A., Walker, M. L., Sakakibara, H., Oiwa, K., and Knight, P. J. (2004). The structure of dynein-c by negative stain electron microscopy. *J Struct Biol* 146, 205–216.

Burkhard, P., Stetefeld, J., and Strelkov, S. V. (2001). Coiled coils: a highly versatile protein folding motif. *Trends Cell Biol* 11, 82–88.

Cheezum, M. K., Walker, W. F., and Guilford, W. H. (2001). Quantitative comparison of algorithms for tracking single fluorescent particles. *Biophys J* 81, 2378–2388.

Dagenbach, E. M. and Endow, S. A. (2004). A new kinesin tree. *J Cell Sci* 117, 3–7.

Di Bella, L. M. and King, S. M. (2001). Dynein motors of the *Chlamydomonas* flagellum. *Int Rev Cytol* 210, 227–268.

Foth, B. J., Goedecke, M. C., and Soldati, D. (2006). New insights into myosin evolution and classification. *Proc Natl Acad Sci U S A* 103, 3681–3686.

Gee, M. A., Heuser, J. E., and Vallee, R. B. (1997). An extended microtubule-binding structure within the dynein motor domain. *Nature* 390, 636–639.

Gennerich, A., Carter, A. P., Reck-Peterson, S. L., and Vale, R. D. (2007). Force-induced bidirectional stepping of cytoplasmic dynein. *Cell* 131, 952–965.

Gibbons, I. R. (1963). Studies on the protein components of cilia from *Tetrahymena pyriformis*. *Proc Natl Acad Sci U S A* 50, 1002–1010.

Gibbons, I. R., Garbarino, J. E., Tan, C. E., Reck-Peterson, S. L., Vale, R. D., and Carter, A. P. (2005). The affinity of the dynein microtubule-binding domain is modulated by the conformation of its coiled coil stalk. *J Biol Chem* 280, 23960–23965.

Gibbons, I. R., Gibbons, B. H., Mocz, G., and Asai, D. J. (1991). Multiple nucleotide-binding sites in the sequence of dynein β heavy chain. *Nature* 352, 640–643.

Gibbons, I. R., Lee-Eiford, A., Mocz, G., Phillipson, C. A., Tang, W. J., and Gibbons, B. H. (1987). Photosensitized cleavage of dynein heavy chains. Cleavage at the "V1 site" by irradiation at 365 nm in the presence of ATP and vanadate. *J Biol Chem* 262, 2780–2786.

Gibbons, I. R. and Rowe, A. J. (1965). Dynein: a protein with adenosine triphosphatase activity from cilia. *Science* 149, 424–426.

Goodenough, U. and Heuser, J. (1984). Structural comparison of purified dynein proteins with in situ dynein arms. *J Mol Biol* 180, 1083–1118.

Goodenough, U. W., Gebhart, B., Mermall, V., Mitchell, D. R., and Heuser, J. E. (1987). High-pressure liquid chromatography fractionation of *Chlamydomonas* dynein extracts and characterization of inner-arm dynein subunits. *J Mol Biol* 194, 481–494.

Goodenough, U. W. and Heuser, J. E. (1982). Substructure of the outer dynein arm. *J Cell Biol* 95, 798–815.

Goodenough, U. W. and Heuser, J. E. (1985). Outer and inner dynein arms of cilia and flagella. *Cell* 41, 341–342.

Haimo, L. T., Telzer, B. R., and Rosenbaum, J. L. (1979). Dynein binds to and crossbridges cytoplasmic microtubules. *Proc Natl Acad Sci U S A* 76, 5759–5763.

Hastie, A. T., Dicker, D. T., Hingley, S. T., Kueppers, F., Higgins, M. L., and Weinbaum, G. (1986). Isolation of cilia from porcine tracheal epithelium and extraction of dynein arms. *Cell Motil Cytoskeleton* 6, 25–34.

Hirakawa, E., Higuchi, H., and Toyoshima, Y. Y. (2000). Processive movement of single 22S dynein molecules occurs only at low ATP concentrations. *Proc Natl Acad Sci U S A* 97, 2533–2537.

Hirokawa, N. (1998). Kinesin and dynein superfamily proteins and the mechanism of organelle transport. *Science* 279, 519–526.

Hirokawa, N. and Takemura, R. (2004). Molecular motors in neuronal development, intracellular transport and diseases. *Curr Opin Neurobiol* 14, 564–573.

Hisanaga, S. and Pratt, M. M. (1984). Calmodulin interaction with cytoplasmic and flagellar dynein: calcium-dependent binding and stimulation of adenosinetriphosphatase activity. *Biochemistry* 23, 3032–3037.

Höök, P., Mikami, A., Shafer, B., Chait, B. T., Rosenfeld, S. S., and Vallee, R. B. (2005). Long range allosteric control of cytoplasmic dynein ATPase activity by the stalk and C-terminal domains. *J Biol Chem* 280, 33045–33054.

Höök, P. and Vallee, R. B. (2006). The dynein family at a glance. *J Cell Sci* 119, 4369–4371.

Howard, J. (2001). *Mechanics of Motor Proteins and the Cytoskeleton*. Sinauer Associates, Sunderland, MA.

Howard, J. and Spudich, J. A. (1996). Is the lever arm of myosin a molecular elastic element?. *Proc Natl Acad Sci U S A* 93, 4462–4464.

Hua, W., Chung, J., and Gelles, J. (2002). Distinguishing inchworm and hand-over-hand processive kinesin movement by neck rotation measurements. *Science* 295, 844–848.

Inoue, Y. and Shingyoji, C. (2007). The roles of noncatalytic ATP binding and ADP binding in the regulation of dynein motile activity in flagella. *Cell Motil Cytoskeleton* 64, 690–704.

Ishijima, A., Kojima, H., Funatsu, T., Tokunaga, M., Higuchi, H., Tanaka, H., and Yanagida, T. (1998). Simultaneous observation of individual ATPase and mechanical events by a single myosin molecule during interaction with actin. *Cell* 92, 161–171.

Ishikawa, T., Sakakibara, H., and Oiwa, K. (2007). The architecture of outer dynein arms in situ. *J Mol Biol* 368, 1249–1258.

Iyer, L. M., Leipe, D. D., Koonin, E. V., and Aravind, L. (2004). Evolutionary history and higher order classification of AAA+ ATPases. *J Struct Biol* 146, 11–31.

Johnson, K. A. (1985). Pathway of the microtubule-dynein ATPase and the structure of dynein: a comparison with actomyosin. *Annu Rev Biophys Biophys Chem* 14, 161–188.

Johnson, K. A. and Wall, J. S. (1983). Structure and molecular weight of the dynein ATPase. *J Cell Biol* 96, 669–678.

Kagami, O. and Kamiya, R. (1992). Translocation and rotation of microtubules caused by multiple species of *Chlamydomonas* inner-arm dynein. *J Cell Sci* 103, 653–664.

Kamimura, S. and Kamiya, R. (1989). High-frequency nanometre-scale vibration in 'quiescent' flagellar axonemes. *Nature* 340, 476–478.

Kamimura, S. and Takahashi, K. (1981). Direct measurement of the force of microtubule sliding in flagella. *Nature* 293, 566–568.

Kamiya, R. (2002). Functional diversity of axonemal dyneins as studied in *Chlamydomonas* mutants. *Int Rev Cytol* 219, 115–155.

Kikushima, K., Yagi, T., and Kamiya, R. (2004). Slow ADP-dependent acceleration of microtubule translocation produced by an axonemal dynein. *FEBS Lett* 563, 119–122.

King, S. J. and Schroer, T. A. (2000). Dynactin increases the processivity of the cytoplasmic dynein motor. *Nat Cell Biol* 2, 20–24.

King, S. M. (2000a). AAA domains and organization of the dynein motor unit. *J Cell Sci* 113, 2521–2526.

King, S. M. (2000b). The dynein microtubule motor. *Biochim Biophys Acta* 1496, 60–75.

King, S. M., Gatti, J. L., Moss, A. G., and Witman, G. B. (1990). Outer-arm dynein from trout spermatozoa: substructural organization. *Cell Motil Cytoskeleton* 16, 266–278.

Kishino, A. and Yanagida, T. (1988). Force measurements by micromanipulation of a single actin filament by glass needles. *Nature* 334, 74–76.

Kitamura, K., Tokunaga, M., Iwane, A. H., and Yanagida, T. (1999). A single myosin head moves along an actin filament with regular steps of 5.3 nanometres. *Nature* 397, 129–134.

Kojima, H., Ishijima, A., and Yanagida, T. (1994). Direct measurement of stiffness of single actin filaments with and without tropomyosin by in vitro nanomanipulation. *Proc Natl Acad Sci U S A* 91, 12962–12966.

Kon, T., Mogami, T., Ohkura, R., Nishiura, M., and Sutoh, K. (2005). ATP hydrolysis cycle-dependent tail motions in cytoplasmic dynein. *Nat Struct Mol Biol* 12, 513–519.

Kon, T., Nishiura, M., Ohkura, R., Toyoshima, Y. Y., and Sutoh, K. (2004). Distinct functions of nucleotide-binding/hydrolysis sites in the four AAA modules of cytoplasmic dynein. *Biochemistry* 43, 11266–11274.

Koonce, M. P. (2006). Dynein shifts into second gear. *Proc Natl Acad Sci U S A* 103, 17587–17588.

Koonce, M. P., Grissom, P. M., and McIntosh, J. R. (1992). Dynein from *Dictyostelium*: primary structure comparisons between a cytoplasmic motor enzyme and flagellar dynein. *J Cell Biol* 119, 1597–1604.

Koonce, M. P. and Samsó, M. (1996). Overexpression of cytoplasmic dynein's globular head causes a collapse of the interphase microtubule network in *Dictyostelium*. *Mol Biol Cell* 7, 935–948.

Koonce, M. P. and Tikhonenko, I. (2000). Functional elements within the dynein microtubule-binding domain. *Mol Biol Cell* 11, 523–529.

Kotani, N., Sakakibara, H., Burgess, S. A., Kojima, H., and Oiwa, K. (2007). Mechanical properties of inner-arm dynein-f (dynein I1) studied with in vitro motility assays. *Biophys J* 93, 886–894.

Lawrence, C. J., Dawe, R. K., Christie, K. R., Cleveland, D. W., Dawson, S. C., Endow, S. A., Goldstein, L. S., Goodson, H. V., Hirokawa, N., Howard, J., Malmberg, R. L., McIntosh, J. R., Miki, H., Mitchison, T. J., Okada, Y., Reddy, A. S., Saxton, W. M., Schliwa, M., Scholey, J. M., Vale, R. D., Walczak, C. E., and Wordeman, L. (2004). A standardized kinesin nomenclature. *J Cell Biol* 167, 19–22.

Lindemann, C. B. and Hunt, A. J. (2003). Does axonemal dynein push, pull, or oscillate?. *Cell Motil Cytoskeleton* 56, 237–244.

Lindemann, C. B. and Kanous, K. S. (1995). "Geometric clutch" hypothesis of axonemal function: key issues and testable predictions. *Cell Motil Cytoskeleton* 31, 1–8.

Lindemann, C. B. and Mitchell, D. R. (2007). Evidence for axonemal distortion during the flagellar beat of *Chlamydomonas*. *Cell Motil Cytoskeleton* 64, 580–589.

Lupas, A. N. and Martin, J. (2002). AAA proteins. *Curr Opin Struct Biol* 12, 746–753.

Lupetti, P., Lanzavecchia, S., Mercati, D., Cantele, F., Dallai, R., and Mencarelli, C. (2005). Three-dimensional reconstruction of axonemal outer dynein arms in situ by electron tomography. *Cell Motil Cytoskeleton* 62, 69–83.

Mallik, R., Carter, B. C., Lex, S. A., King, S. J., and Gross, S. P. (2004). Cytoplasmic dynein functions as a gear in response to load. *Nature* 427, 649–652.

Mallik, R., Petrov, D., Lex, S. A., King, S. J., and Gross, S. P. (2005). Building complexity: an in vitro study of cytoplasmic dynein with in vivo implications. *Curr Biol* 15, 2075–2085.

Mazumdar, M., Mikami, A., Gee, M. A., and Vallee, R. B. (1996). In vitro motility from recombinant dynein heavy chain. *Proc Natl Acad Sci U S A* 93, 6552–6556.

Mikami, A., Paschal, B. M., Mazumdar, M., and Vallee, R. B. (1993). Molecular cloning of the retrograde transport motor cytoplasmic dynein (MAP 1C). *Neuron* 10, 787–796.

Mizuno, N., Toba, S., Edamatsu, M., Watai-Nishii, J., Hirokawa, N., Toyoshima, Y. Y., and Kikkawa, M. (2004). Dynein and kinesin share an overlapping microtubule-binding site. *EMBO J* 23, 2459–2467.

Mocz, G. and Gibbons, I. R. (2001). Model for the motor component of dynein heavy chain based on homology to the AAA family of oligomeric ATPases. *Structure* 9, 93–103.

Mogami, T., Kon, T., Ito, K., and Sutoh, K. (2007). Kinetic characterization of tail swing steps in the ATPase cycle of *Dictyostelium* cytoplasmic dynein. *J Biol Chem* 282, 21639–21644.

Morris, R. L., Hoffman, M. P., Obar, R. A., McCafferty, S. S., Gibbons, I. R., Leone, A. D., Cool, J., Allgood, E. L., Musante, A. M., Judkins, K. M., Rossetti, B. J., Rawson, A. P., and Burgess, D. R. (2006). Analysis of cytoskeletal and motility proteins in the sea urchin genome assembly. *Dev Biol* 300, 219–237.

Neuwald, A. F., Aravind, L., Spouge, J. L., and Koonin, E. V. (1999). AAA+: a class of chaperone-like ATPases associated with the assembly, operation, and disassembly of protein complexes. *Genome Res* 9, 27–43.

Nicastro, D., McIntosh, J. R., and Baumeister, W. (2005). 3D structure of eukaryotic flagella in a quiescent state revealed by cryo-electron tomography. *Proc Natl Acad Sci U S A* 102, 15889–15894.

Nicastro, D., Schwartz, C., Pierson, J., Gaudette, R., Porter, M. E., and McIntosh, J. R. (2006). The molecular architecture of axonemes revealed by cryoelectron tomography. *Science* 313, 944–948.

Nishiura, M., Kon, T., Shiroguchi, K., Ohkura, R., Shima, T., Toyoshima, Y. Y., and Sutoh, K. (2004). A single-headed recombinant fragment of *Dictyostelium* cytoplasmic dynein can drive the robust sliding of microtubules. *J Biol Chem* 279, 22799–22802.

Nishizaka, T., Oiwa, K., Noji, H., Kimura, S., Muneyuki, E., Yoshida, M., and Kinosita Jr., K. (2004). Chemomechanical coupling in F_1-ATPase revealed by simultaneous observation of nucleotide kinetics and rotation. *Nat Struct Mol Biol* 11, 142–148.

Oda, T., Hirokawa, N., and Kikkawa, M. (2007). Three-dimensional structures of the flagellar dynein-microtubule complex by cryoelectron microscopy. *J Cell Biol* 177, 243–252.

Ogawa, K. (1991). Four ATP-binding sites in the midregion of the β heavy chain of dynein. *Nature* 352, 643–645.

Ogura, T. and Wilkinson, A. J. (2001). AAA+ superfamily ATPases: common structure-diverse function. *Genes Cells* 6, 575–597.

Oiwa, K., Eccleston, J. F., Anson, M., Kikumoto, M., Davis, C. T., Reid, G. P., Ferenczi, M. A., Corrie, J. E., Yamada, A., Nakayama, H., and Trentham, D. R. (2000). Comparative single molecule and ensemble myosin enzymology: sulfoindocyanine ATP and ADP derivatives. *Biophys J* 78, 3048–3071.

Oiwa, K. and Sakakibara, H. (2005). Recent progress in dynein structure and mechanism. *Curr Opin Cell Biol* 17, 98–103.

Oiwa, K. and Takahashi, K. (1988). The force–velocity relationship for microtubule sliding in demembranated sperm flagella of the sea urchin. *Cell Struct Funct* 13, 193–205.

Okada, Y., Higuchi, H., and Hirokawa, N. (2003). Processivity of the single-headed kinesin KIF1A through biased binding to tubulin. *Nature* 424, 574–577.

Okada, Y. and Hirokawa, N. (1999). A processive single-headed motor: kinesin superfamily protein KIF1A. *Science* 283, 1152–1157.

Okten, Z., Churchman, L. S., Rock, R. S., and Spudich, J. A. (2004). Myosin VI walks hand-over-hand along actin. *Nat Struct Mol Biol* 11, 884–887.

Okuno, M. and Hiramoto, Y. (1979). Direct measurements of the stiffness of echinoderm sperm flagella. *J Exp Biol* 79, 235–243.

Omoto, C. K., Yagi, T., Kurimoto, E., and Kamiya, R. (1996). Ability of paralyzed flagella mutants of *Chlamydomonas* to move. *Cell Motil Cytoskeleton* 33, 88–94.

Pallini, V., Bugnoli, M., Mencarelli, C., and Scapigliati, G. (1982). Biochemical properties of ciliary, flagellar and cytoplasmic dyneins. *Symp Soc Exp Biol* 35, 339–352.

Paschal, B. M., Shpetner, H. S., and Vallee, R. B. (1987). MAP 1C is a microtubule-activated ATPase which translocates microtubules in vitro and has dynein-like properties. *J Cell Biol* 105, 1273–1282.

Pazour, G. J., Agrin, N., Walker, B. L., and Witman, G. B. (2006). Identification of predicted human outer dynein arm genes: candidates for primary ciliary dyskinesia genes. *J Med Genet* 43, 62–73.

Pfister, K. K., Fay, R. B., and Witman, G. B. (1982). Purification and polypeptide composition of dynein ATPases from *Chlamydomonas* flagella. *Cell Motil* 2, 525–547.

Piperno, G. and Luck, D. J. (1979). Axonemal adenosine triphosphatases from flagella of *Chlamydomonas reinhardtii*. Purification of two dyneins. *J Biol Chem* 254, 3084–3090.

Piperno, G., Ramanis, Z., Smith, E. F., and Sale, W. S. (1990). Three distinct inner dynein arms in *Chlamydomonas* flagella: molecular composition and location in the axoneme. *J Cell Biol* 110, 379–389.

Purcell, T. J., Morris, C., Spudich, J. A., and Sweeney, H. L. (2002). Role of the lever arm in the processive stepping of myosin V. *Proc Natl Acad Sci U S A* 99, 14159–14164.

Reck-Peterson, S. L. and Vale, R. D. (2004). Molecular dissection of the roles of nucleotide binding and hydrolysis in dynein's AAA domains in *Saccharomyces cerevisiae*. *Proc Natl Acad Sci U S A* 101, 1491–1495.

Reck-Peterson, S. L., Yildiz, A., Carter, A. P., Gennerich, A., Zhang, N., and Vale, R. D. (2006). Single molecule analysis of dynein processivity and stepping behavior. *Cell* 126, 335–348.

Rock, R. S., Ramamurthy, B., Dunn, A. R., Beccafico, S., Rami, B. R., Morris, C., Spink, B. J., Franzini-Armstrong, C., Spudich, J. A., and Sweeney, H. L. (2005). A flexible domain is essential for the large step size and processivity of myosin VI. *Mol Cell* 17, 603–609.

Ross, J. L., Wallace, K., Shuman, H., Goldman, Y. E., and Holzbaur, E. L. F. (2006). Processive bidirectional motion of dynein–dynactin complexes in vitro. *Nat Cell Biol* 8, 562–570.

Sakakibara, H., Kojima, H., Sakai, Y., Katayama, E., and Oiwa, K. (1999). Inner-arm dynein c of *Chlamydomonas* flagella is a single-headed processive motor. *Nature* 400, 586–590.

Sakakibara, H. M., Kunioka, Y., Yamada, T., and Kamimura, S. (2004). Diameter oscillation of axonemes in sea-urchin sperm flagella. *Biophys J* 86, 346–352.

Sakamoto, T., Wang, F., Schmitz, S., Xu, Y., Xu, Q., Molloy, J. E., Veigel, C., and Sellers, J. R. (2003). Neck length and processivity of myosin V. *J Biol Chem* 278, 29201–29207.

Sale, W. S., Goodenough, U. W., and Heuser, J. E. (1985). The substructure of isolated and in situ outer dynein arms of sea urchin sperm flagella. *J Cell Biol* 101, 1400–1412.

Sale, W. S. and Satir, P. (1977). Direction of active sliding of microtubules in *Tetrahymena* cilia. *Proc Natl Acad Sci U S A* 74, 2045–2049.

Samsó, M. and Koonce, M. P. (2004). 25 Å resolution structure of a cytoplasmic dynein motor reveals a seven-member planar ring. *J Mol Biol* 340, 1059–1072.

Samsó, M., Radermacher, M., Frank, J., and Koonce, M. P. (1998). Structural characterization of a dynein motor domain. *J Mol Biol* 276, 927–937.

Schief, W. R. and Howard, J. (2001). Conformational changes during kinesin motility. *Curr Opin Cell Biol* 13, 19–28.

Schnitzer, M. J., Visscher, K., and Block, S. M. (2000). Force production by single kinesin motors. *Nat Cell Biol* 2, 718–723.

Shima, T., Kon, T., Imamula, K., Ohkura, R., and Sutoh, K. (2006). Two modes of microtubule sliding driven by cytoplasmic dynein. *Proc Natl Acad Sci U S A* 103, 17736–17740.

Shingyoji, C., Higuchi, H., Yoshimura, M., Katayama, E., and Yanagida, T. (1998). Dynein arms are oscillating force generators. *Nature* 393, 711–714.

Shiroguchi, K. and Toyoshima, Y. Y. (2001). Regulation of monomeric dynein activity by ATP and ADP concentrations. *Cell Motil Cytoskeleton* 49, 189–199.

Silvanovich, A., Li, M. G., Serr, M., Mische, S., and Hays, T. S. (2003). The third P-loop domain in cytoplasmic dynein heavy chain is essential for dynein motor function and ATP-sensitive microtubule binding. *Mol Biol Cell* 14, 1355–1365.

Spudich, J. A. (2006). Molecular motors take tension in stride. *Cell* 126, 242–244.

Summers, K. E. and Gibbons, I. R. (1971). Adenosine triphosphate-induced sliding of tubules in trypsin-treated flagella of sea-urchin sperm. *Proc Natl Acad Sci U S A* 68, 3092–3096.

Svoboda, K. and Block, S. M. (1994). Biological applications of optical forces. *Annu Rev Biophys Biomol Struct* 23, 247–285.

Takada, S., Wilkerson, C. G., Wakabayashi, K., Kamiya, R., and Witman, G. B. (2002). The outer dynein arm-docking complex: composition and characterization of a subunit (oda1) necessary for outer arm assembly. *Mol Biol Cell* 13, 1015–1029.

Takahashi, K. (1995). Mechanisms of flagellar motility probed with microtechniques. *Cell Motil Cytoskeleton* 32, 110–113.

Takahashi, Y., Edamatsu, M., and Toyoshima, Y. Y. (2004). Multiple ATP-hydrolyzing sites that potentially function in cytoplasmic dynein. *Proc Natl Acad Sci U S A* 101, 12865–12869.

Toba, S., Watanabe, T. M., Yamaguchi-Okimoto, L., Toyoshima, Y. Y., and Higuchi, H. (2006). Overlapping hand-over-hand mechanism of single molecular motility of cytoplasmic dynein. *Proc Natl Acad Sci U S A* 103, 5741–5745.

Toprak, E. and Selvin, P. R. (2007). New fluorescent tools for watching nanometer-scale conformational changes of single molecules. *Annu Rev Biophys Biomol Struct* 36, 349–369.

Vale, R. D. (2000). AAA proteins: lords of the ring. *J Cell Biol* 150, F13–F19.

Vale, R. D. (2003). The molecular motor toolbox for intracellular transport. *Cell* 112, 467–480.

Vale, R. D. and Milligan, R. A. (2000). The way things move: looking under the hood of molecular motor proteins. *Science* 288, 88–95.

Vale, R. D. and Toyoshima, Y. Y. (1988). Rotation and translocation of microtubules in vitro induced by dynein from *Tetrahymena* cilia. *Cell* 52, 459–469.

Vallee, R. B. and Höök, P. (2006). Autoinhibitory and other autoregulatory elements within the dynein motor domain. *J Struct Biol* 156, 175–181.

Vallee, R. B. and Stehman, S. A. (2005). How dynein helps the cell find its center: a servomechanical model. *Trends Cell Biol* 15, 288–294.

Vallee, R. B., Wall, J. S., Paschal, B. M., and Shpetner, H. S. (1988). Microtubule-associated protein 1C from brain is a two-headed cytosolic dynein. *Nature* 332, 561–563.

Vallee, R. B., Williams, J. C., Varma, D., and Barnhart, L. E. (2004). Dynein: an ancient motor protein involved in multiple modes of transport. *J Neurobiol* 58, 189–200.

Wakabayashi, K., Yagi, T., and Kamiya, R. (1997). Ca^{2+}-dependent waveform conversion in the flagellar axoneme of *Chlamydomonas* mutants lacking the central-pair/radial spoke system. *Cell Motil Cytoskeleton* 38, 22–28.

Wang, Z., Khan, S., and Sheetz, M. P. (1995). Single cytoplasmic dynein molecule movements: characterization and comparison with kinesin. *Biophys J* 69, 2011–2023.

Wickstead, B. and Gull, K. (2007). Dyneins across eukaryotes: a comparative genomic analysis. *Traffic* 8, 1708–1721.

Yagi, T. (2000). ADP-dependent microtubule translocation by flagellar inner-arm dyneins. *Cell Struct Funct* 25, 263–267.

Yagi, T. and Kamiya, R. (2000). Vigorous beating of *Chlamydomonas* axonemes lacking central pair/radial spoke structures in the presence of salts and organic compounds. *Cell Motil Cytoskeleton* 46, 190–199.

Yagi, T., Minoura, I., Fujiwara, A., Saito, R., Yasunaga, T., Hirono, M., and Kamiya, R. (2005). An axonemal dynein particularly important for flagellar movement at high viscosity. Implications from a new *Chlamydomonas* mutant deficient in the dynein heavy chain gene DHC9. *J Biol Chem* 280, 41412–41420.

Yamada, A., Yamaga, T., Sakakibara, H., Nakayama, H., and Oiwa, K. (1998). Unidirectional movement of fluorescent microtubules on rows of dynein arms of disintegrated axonemes. *J Cell Sci* 111, 93–98.

Yanagisawa, H. A. and Kamiya, R. (2001). Association between actin and light chains in *Chlamydomonas* flagellar inner-arm dyneins. *Biochem Biophys Res Commun* 288, 443–447.

Yildiz, A., Forkey, J. N., McKinney, S. A., Ha, T., Goldman, Y. E., and Selvin, P. R. (2003). Myosin V walks hand-over-hand: single fluorophore imaging with 1.5-nm localization. *Science* 300, 2061–2065.

Yildiz, A. and Selvin, P. R. (2005). Fluorescence imaging with one nanometer accuracy: application to molecular motors. *Acc Chem Res* 38, 574–582.

Yoneda, M. (1960). Force exerted by a single cilium of *Mytilus edulis*. I. *J Exp Biol* 37, 461–468.

CHAPTER 4

The Bacterial Flagellar Motor

Yoshiyuki Sowa

Clarendon Laboratory, Department of Physics, University of Oxford, Parks Road,
Oxford OX1 3PU, United Kingdom

Richard M. Berry

Clarendon Laboratory, Department of Physics, University of Oxford, Parks Road,
Oxford OX1 3PU, United Kingdom

Summary

The bacterial flagellar motor is a reversible rotary nanomachine, about 45 nm in diameter, embedded in the bacterial cell envelope. It is powered by the flux of H^+ or Na^+ ions across the cytoplasmic membrane driven by an electrochemical gradient, the protonmotive force (PMF), or sodium-motive force (SMF). Each motor rotates a helical filament at several hundreds of revolutions per second (Hz). In many species, the motor switches direction stochastically, with the switching rates controlled by a network of sensory and signaling proteins. The bacterial flagellar motor was confirmed as a rotary motor in 1974, the first direct observation of the function of a single molecular motor. However, due to the large size and complexity of the motor, much remains to be discovered, particularly the structural details of the torque-generating mechanism. This chapter outlines what has been learned about the structure and function of the motor using a combination of genetics, single molecule, and biophysical techniques, with a focus on recent results and single molecule techniques.

Key Words

bacteria; flagellar motor; rotary motor; nanomachine; ion-driven motor

Introduction

The flagellar motor stretches the definition of the term "single molecule." The rotor contains several hundred polypeptide chains, the stator about 50 more. But most molecular machines that are the subject of "single molecule" experiments also contain multiple polypeptide chains. The term is justified because each atom has its place, gives or takes a little variability, in contrast to a macroscopic machine whose components are cut from blocks of bulk material. Also, the motor is a minimal molecular complex with all of its parts working together, and with input (ion transit) and output (rotation) processes on a molecular scale that requires "single molecule" biophysical techniques to observe. Experiments on tethered cells in 1974 were the first observations of the function of single molecular motors. Since then our knowledge of the detailed mechanism of smaller and simpler motors has outstripped the flagellar motor. Because of its large size and location in the membrane, detailed atomic structures of the flagellar motor have been difficult to obtain. Recently, partial X-ray crystal structures of several motor proteins have been combined with site-directed mutagenesis and electron microscopy (EM) to produce credible models of the rotor, but atomic-level structural information on the membrane-bound stators remains elusive. The complex assembly pathway and requirement to anchor stators to the cell wall and locate them in an energized membrane have so far precluded the powerful *in vitro* reconstitution assays that have revealed so much about the other ATP-driven molecular motors in the last one decade or two. Nonetheless, a great deal has been learned about the flagellar motor, including considerable recent progress in the application of single molecule techniques. This chapter summarizes the historical background and recent advances in the field. More comprehensive accounts of the earlier work can be found in several recent reviews (Berg, 2003; Kojima and Blair, 2004a).

Many species of bacteria sense their environment and respond by swimming toward favorable conditions, propelled by rotating flagella that extend from the cell body (Armitage, 1999; Blair, 1995). Each flagellum consists of a long ($\sim 10 \mu m$), thin ($\sim 20 nm$) helical filament turned like a screw by a rotary motor at its base (Berg, 2003; Berry and Armitage, 1999; Namba and Vonderviszt, 1997). The flagellar motor is one of the largest molecular machines in bacteria, with a molecular mass of $\sim 11 MDa$, 13 different component proteins (including the rod and the LP ring but not the export apparatus), and a further approximately 25 proteins required for its expression and assembly (Macnab, 1996). The best studied motors are those of the peritrichously flagellated[1] enteric bacteria *Escherichia coli* and *Salmonella typhimurium*. Unless explicitly stated otherwise, the experiments described in this chapter were performed using motors from one or the other of these species. These motors switch between counterclockwise (CCW, viewed from filament to motor) rotation that allows filaments

---

[1]   That is, with flagella projecting from the cell in all directions.

to form a bundle and propel the cell smoothly and clockwise (CW) rotation that forces a filament out of a bundle and leads to a change in swimming direction, called a tumble. Other flagellated bacteria swim differently, for example, unidirectional motor rotation with cell reorientation when motors stop (*Rhodobacter sphaeroides*) or change speed (*Sinorhizobium meliloti*), polar flagella that push or pull the cell depending on rotation direction (*Vibrio alginolyticus*), or internal periplasmic flagella that drive a helical wave of the whole cell body (spirochetes) (Armitage and Schmitt, 1997; Berry and Armitage, 1999). Much has been written elsewhere about the process of bacterial chemotaxis in *E. coli*, where cells make temporal comparisons of local attractant concentrations and suppress tumbles if conditions are improving, leading to a biased random walk up concentration gradients of attractant (Baker et al., 2006; Blair, 1995; Falke et al., 1997).

True rotation of bacterial flagella, as opposed to propagation of helical waves, was demonstrated in the 1970s (Berg and Anderson, 1973; Silverman and Simon, 1974). Cells were tethered to a surface by filaments containing mutations that prevented them from swimming, and rotation of the cell body, driven by the motor, was observed in a light microscope. This tethered cell assay remained the state-of-the-art in single molecule experiments on the flagellar motor for 25 years until it was surpassed by the attachment of polystyrene spheres to truncated flagella (Ryu et al., 2000), and it revealed a great deal about the mechanism of the motor and the chemotaxis system. Each motor generates a maximum power output on the order of 10^{-15} W, enough to propel cells at speeds up to $\sim 100 \mu m\, s^{-1}$. This high power arises partly from the large size of the motor and the multiple parallel torque-generating stators that it contains, but also because the motor is powered by the flux of ions across the cytoplasmic membrane, not by ATP hydrolysis. Ion flux is driven by an electrochemical gradient, either the protonmotive force (PMF) or sodium-motive force (SMF) in motors driven by H^+ or Na^+ (Hirota and Imae, 1983; Manson et al., 1977; Matsuura et al., 1977), and it is likely that the relatively high speed with which ions can move through the motor, compared to ATP hydrolysis, is essential for the high turnover rates that allow rapid motor rotation at high torque.

In this chapter, we summarize the structure of the flagellar motor and recent progress toward understanding its mechanism. In particular, we focus on (1) new structural information from X-ray crystallography and electron microscopy; (2) measurements of torque and speed; (3) numbers and dynamics of stators, from single molecule fluorescence microscopy of green fluorescent protein (GFP)-labeled motor proteins and careful analysis of rotation speeds; (4) the dependence of motor function on ion-motive force (IMF), from manipulation and measurement of IMF in single cells; and (5) stepping motion in flagellar rotation revealed by particle tracking with nanometer and submillisecond resolution.

Structure

Overview

Like any rotary motor, the bacterial flagellar motor consists of a rotor and a stator. The rotor spins relative to the cell and is attached to the helical filament by a universal joint called the hook; the stator is anchored to the cell wall. Figure 4.1 shows a schematic diagram of the bacterial flagellum of gram-negative bacteria, based on an EM reconstruction of the rotor from *S. typhimurium* (Figure 4.2A) (Francis et al., 1994;

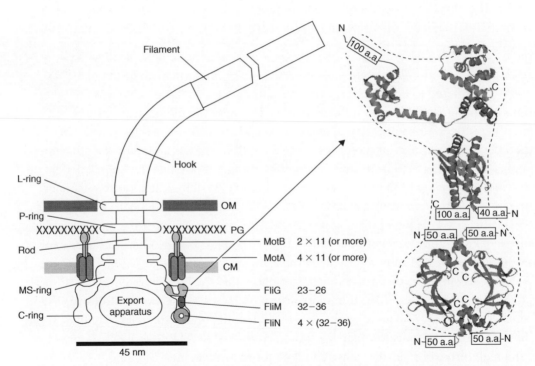

Figure 4.1: Left: Schematic side view of a H$^+$-driven flagellar motor, with the proposed location and copy number of proteins involved in torque generation. MotA and MotB are thought to form stator complexes with stoichiometry A$_4$B$_2$, and FliN a tetramer that has 1:1 stoichiometry with FliM. The motor spans the three layers of the cell envelope: outer membrane (OM), peptidoglycan cell wall (PG), and cytoplasmic membrane (CM). Right: detail of proposed location and orientation of rotor proteins. X-ray crystal structures of truncated rotor proteins, FliG (PDB ID = 1lkv), FliM (PDB ID = 2hp7) and FliN (PDB ID = 1yab), are shown docked into the rotor structure. N- and C-termini and missing amino acids are indicated. Molecular graphics generated using PyMol (http://www.pymol.org)

Thomas et al., 1999, 2006). The core of the motor is called the "basal body" and consists of a set of rings up to ~45 nm in diameter that spans three layers of the cell envelope (DePamphilis and Adler, 1971a,b,c). The L- and P-rings are embedded in the outer lipopolysaccharide membrane and peptidoglycan cell wall, respectively, and are thought to work as a bushing between the rotor and the outer parts of the cell envelope. Whether they rotate relative to the cell envelope, the rotor, or both is not known. The rod connects the hook to the MS-ring located at the cytoplasmic membrane. The MS-ring was once thought to be 2 rings (membrane and supramembrane), but was subsequently shown to consist of 26 copies of single protein, FliF (Suzuki et al., 2004; Ueno et al., 1992, 1994). The MS-ring is the first part of the motor to assemble; thus, it can be thought of as the platform on which the rest of the motor is built (Aizawa, 1996; Macnab, 2003). The cytoplasmic face of the MS-ring is attached to the C-ring, which contains the proteins FliG, FliM, and FliN, and is thought to be the site of torque generation (Francis et al., 1994; Katayama et al., 1996; Khan et al., 1991). Inside the C-ring is the export apparatus that pumps proteins needed to make the hook and filament outside the cell, but is thought to have no role in torque generation.

Propeller and Universal Joint

The hook and filament are thin tubular polymers, each of a single protein. They grow at the distal end by incorporating monomers pumped by the export apparatus through a central channel that spans the entire flagellum. Monomer incorporation is regulated by pentameric cap complexes (Yonekura et al., 2000). Both hook and filament consist of 11 helical protofilaments, each of which has alternative long and short forms that mix to create the helical structures of the hook and filament (Asakura, 1970; Calladine, 1975; Hasegawa et al., 1998). Under steady rotation of the motor the filament is a rigid propeller. Motor switching causes torsionally induced transformations between alternative filament forms with different numbers of long and short protofilaments that lead to cell reorientation in *E. coli* and *R. sphaeroides* (Armitage and Macnab, 1987; Turner et al., 2000). The hook is much more flexible than the filament and works as a universal joint to allow several filaments from motors all over the cell to rotate together in a bundle in peritrichously flagellated species. Atomic structures of straight mutants of hook and filament have been obtained by EM image reconstruction and X-ray crystallography, revealing connections within and between protofilaments that are consistent with the model of helical filament structure (Mimori et al., 1995; Samatey et al., 2001, 2004; Shaikh et al., 2005; Yonekura et al., 2003). Molecular dynamics simulations based on these structures further demonstrated the probable mechanism for switching between long and short protofilament forms in response to force (Furuta et al., 2007; Kitao et al., 2006).

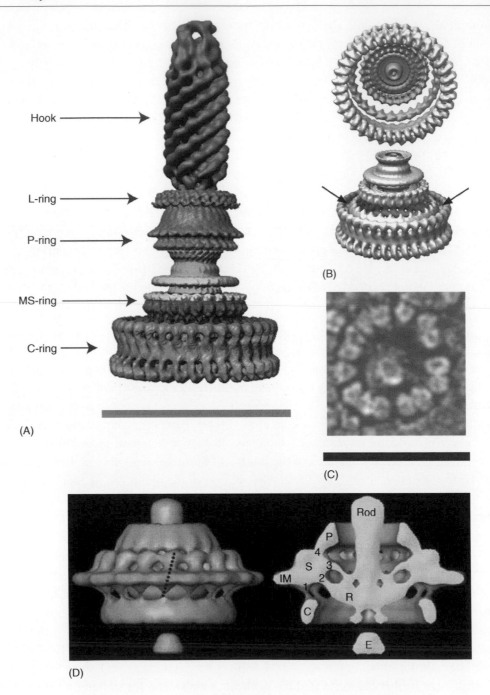

Hook

L-ring

P-ring

MS-ring

C-ring

(A)

(B)

(C)

(D)

Rod

P

4

S   3

IM   2

1

C

R

E

Rotor–Switch Complex

The C-ring (containing FliG, FliM, FliN) is also known as the switch complex, and is the key component of the rotor for torque generation and switching (Yamaguchi et al., 1986a,b). FliG interacts with MotA to generate torque (Garza et al., 1995; Lloyd et al., 1996); FliM binds the active form of the response-regulator CheY, altering the probability of CCW rotation; and FliN may be responsible for the intrinsic bistability of the rotor that gives it relatively stable CW and CCW states. Atomic structures of the middle and C-terminal domains of FliG, middle part of FliM, and C-terminal part of FliN have been resolved by X-ray crystallography (Brown et al., 2002, 2005; Lloyd et al., 1999; Park et al., 2006). Figure 4.1 (inset) shows these crystal structures and a model for where they fit into the C-ring based on a range of biochemical studies (Brown et al., 2007; Lowder et al., 2005; Paul and Blair, 2006; Paul et al., 2006). The overall structure of the C-ring is determined by single-particle reconstruction and cryo-EM of the flagellar basal body. This work has revealed an interesting symmetry mismatch within the rotor. The MS-ring symmetry was first reported to be 26 (Suzuki et al., 2004). A partly functional fusion protein between FliF and FliG is strong evidence that FliG has the same copy number as FliF, the main MS-ring protein (Francis et al., 1992), presumably 26. Early work on the C-ring indicated 33- and 34-fold symmetries (Thomas et al., 1999), with subsequent work extending the range to 31–38 (Young et al., 2003). But most of FliG must be in the C-ring (Francis et al., 1994; Oosawa et al., 1994; Suzuki et al., 2004; Tang et al., 1996; Thomas et al., 1999), raising the question of how 26 copies of FliG can form part of a ~35-fold symmetric C-ring. The most recent reconstructions of the rotor structure (Thomas et al.,

Figure 4.2: EM images of flagellar motors. (A) 3-D reconstruction of the isolated rotor from *S. typhimurium*. Reprinted from DeRosier (2006), with permission from Elsevier. (B) The bottom of the rotor as in (A), but viewed from different angles to show the symmetry of the C-ring. The inner lobe of the C-ring has ~25-fold symmetry (arrow), the remainder of the C-ring ~34-fold. Reprinted from Thomas et al. (2006), with permission from American Society for Microbiology. (C) Freeze-fracture EM image of stators of *Streptococcus*. Fifteen stators can be counted in this motor. Depending on species, the number of stators found by this method varies between 10 and 16. Reprinted from Khan et al. (1988), with permission from Elsevier. (D) 3-D reconstruction of the entire flagellar motor from spirochete *Treponema primitia in vivo*. The dashed line in the left figure shows the axis of a stator. The motor contains 16 stators, each of which makes one link to the P-ring and three to the rotor (right). The rotor is larger than that of *S. typhimurium*. Scale bars (gray: A, black: B–D) are 50 nm. Reprinted from Murphy et al. (2006), with permission from Macmillan Publishers Ltd

2006) offer a resolution of the problem (Figure 4.2B). The inner lobe of the C-ring shares the symmetry of the MS-ring (23–26), presumably identifying it with FliG, whereas the rest of FliG appears to be distributed with the symmetry of the C-ring. Thus there must be N defects in the C-ring, each a missing copy of FliG, where N is the difference between C- and MS-ring symmetries (Brown et al., 2007). (Note that such defects would not be preserved by the image reconstruction procedure used for Figure 4.2A and B). An earlier model predicted this symmetry mismatch and identified $N = 8$ with the number of stator units, postulating that the torque-generating mechanism consisted of each stator unit catalyzing the propagation of a defect around the ring (Thomas et al., 1999). This model predicts that the MS- and C-ring rotate at different speeds, a counter-intuitive prediction that might in future be testable by direct observation of fluorescently labeled rotor proteins. The latest structural evidence shows no correlation between variations in the C- and MS-ring symmetries (Thomas et al., 2006), with the symmetry mismatch N varying between 6 and 13. However, the number of stators has also recently been shown to vary up to at least 11 (Reid et al., 2006), leaving open the intriguing possibility that the symmetry mismatch may be an essential element in the torque-generating mechanism.

Stator

The stators are a complex of two proteins: MotA and MotB in H^+-driven flagellar motors such as those of *E. coli* and *S. typhimurium*, or PomA and PomB in Na^+-driven motors such as the polar motors of *V. alginolyticus* and *Vibrio cholerae*. MotA/MotB and PomA/PomB appear to be very similar in sequence, topology, and function (Asai et al., 1997; Yorimitsu and Homma, 2001). Stator complexes anchor to the peptidoglycan cell wall and span the cytoplasmic membrane, forming ion channels (Blair and Berg, 1990; De Mot and Vanderleyden, 1994; Sato and Homma, 2000). Freeze-fracture cryo-EM shows them located in rings at the periphery of the rotor in *Aquaspirillum serpens*, *Bacillus* species, *E. coli*, and *Streptococcus* (Figure 4.2C), containing between 10 and 16 particles depending on the species and the individual motor (Coulton and Murray, 1978; Khan et al., 1988, 1992). A recent complete structure *in situ* of the flagellar motor of the spirochete *Treponema primitia* at 7 nm resolution shows 16 stators in an interconnected ring around the rotor, with most of the mass in the cytoplasmic membrane and periplasm (Figure 4.2D; Murphy et al., 2006).

Currently, there are no atomic-level structures of any part of the stator complexes. However, extensive analysis using biochemical cross-linking and site-directed mutagenesis has identified their topology, stoichiometry, and likely active regions. MotA/PomA and MotB/PomB have four and one membrane-spanning alpha helices and large cytoplasmic and periplasmic domains, respectively (Chun and Parkinson, 1988; Dean et al., 1984; De Mot and Vanderleyden, 1994). The C-terminal periplasmic domain of MotB contains an essential peptidoglycan-binding motif. The cytoplasmic domain of MotA contains two charged residues that interact with five charged residues in the C-terminal domain of FliG to generate torque (Lloyd and Blair, 1997; Zhou and Blair, 1997; Zhou et al., 1998). No single mutation in these residues completely abolishes torque generation, and charge-reversing mutations in both proteins can compensate each other, indicating an electrostatic interaction at an interface between the two proteins. PomA and FliG of the Na^+-driven flagellar motor of *V. alginolyticus* show a similar pattern (Yakushi et al., 2006), but with differences in which conserved charged residues are most important for function (Yorimitsu et al., 2002, 2003). The stoichiometry of stator complexes deduced from targeted disulfide cross-linking studies and chromatography appears to be A_4B_2 (Braun et al., 2004; Kojima and Blair, 2004b; Sato and Homma, 2000) (Figure 4.1), with the membrane-spanning helices of MotA subunits surrounding a suspected proton-binding site at residue Asp32 of MotB (Sharp et al., 1995a,b). This is the only conserved charged residue in MotA or MotB that is absolutely essential for function, and it is postulated that each stator contains two ion channels, each containing one MotB Asp32 residue (Braun and Blair, 2001). Different patterns of protein digestion indicate a conformational change in MotA between wild-type stators and stators containing the mutation MotB Asp32 to Asn32, which mimics the protonation of Asp32 (Braun et al., 1999; Kojima and Blair, 2001). Thus, a putative mechanism for the motor is that proton flux coordinates conformational changes in MotA via MotB Asp32, and these conformational changes lead to a cyclic interaction with FliG that generates torque. Asp24 of PomB is equivalent to Asp32 of MotB, but correlated conformational change of PomA has not been tested. The Na^+ motor of *V. alginolyticus* requires the additional proteins MotX and MotY for rotation (McCarter, 1994a,b). These form the T-ring in the periplasmic space, which is not found in *E. coli* and *S. typhimurium* (Terashima et al., 2006).

Despite the differences listed above, there is strong evidence, in the form of numerous functional chimeric motors that mix components from motors with different driving ions, that the mechanisms of Na^+ and H^+ motors are very similar. The first such chimera reported was a Na^+-driven chimera made by replacing PomA in *V. alginolyticus* with the highly homologous MotA from *R. sphaeroides* (Asai et al., 1999). Subsequent functional chimeras have swapped both A and B stator proteins, the C-terminal domain of FliG, and the peptidoglycan and membrane-spanning domains of MotB or PomB into species with a motor that runs on a different type of ion (Asai et al., 2000, 2003; Gosink and Hase, 2000; Yorimitsu et al., 2003). These results have demonstrated that there is no single determining component for ion selectivity. MotX and MotY do not always specify Na^+ selectivity, but they are required for function if the periplasmic domain of the stator B protein is from PomB in sodium host, suggesting a role in stabilizing stators.

A particularly useful chimeric motor for single molecule experiments uses stators PomAPotB (PotB is a fusion protein between the periplasmic C-terminal domain of *E. coli* MotB and the membrane-spanning N-terminal domain of PomB from *V. alginolyticus*) to form a Na^+-driven motor in *E. coli* (Asai et al., 2003). Because Na^+ concentration is less important than pH for maintaining the functionality of proteins, and the SMF is not central to the metabolism of *E. coli*, the SMF can be controlled over a wide range without damage to the cells or motors. This has allowed observation of the fundamental stepping motion of the motor at low SMF and measurement of the dependence of motor rotation upon each component of the SMF, as detailed in the following section.

Function

Overview

Whereas the tethered-cell experiments in 1974 were the first observations of the function of a single molecular motor, recent advances in *in vitro* single molecule techniques have revealed much more about the detailed mechanism of other ATP-driven molecular motors than about the flagellar motor. It has not been possible to reconstitute the flagellar motor *in vitro*; this section outlines recent progress in experiments on single flagellar motors in live cells.

Energetics

A molecular motor is a machine that converts chemical or electrical energy to mechanical work. It works at the level of thermal energy, $k_B T$ ($\sim 4 \times 10^{-21}$ J), where k_B is Boltzmann's constant and T the absolute temperature (see Appendix). In the bacterial flagellar motor, the elementary free energy input from a single ion passing through the cytoplasmic membrane is defined as an elementary electric charge times IMF (either PMF or SMF, depending on the driving ion). The IMF consists of an electrical voltage and a chemical component of concentration difference across the membrane, and is defined as

$$IMF = V_m + \frac{k_B T}{e} \ln\left(\frac{C_i}{C_o}\right) \tag{4.1}$$

where V_m is the transmembrane voltage (inside minus to the outside), e the elementary charge, and C_i and C_o the concentration of the coupling ions inside and outside the cell, respectively. With a typical IMF of around $-150\,mV$, the free energy of a single ion transit is $\sim 6k_BT$. Torque is defined as the product of a force and the perpendicular distance to an axis of rotation, and therefore also has dimensions of N m or energy. Because the Reynolds number (see Appendix) for a spinning flagellar motor is $\ll 1$, inertia is negligible and torque can be calculated as

$$M = f\omega \tag{4.2}$$

where f is the rotational drag coefficient and ω angular velocity ($2\pi \times$ rotational speed). In terms of energy, torque is best thought of as the work done per radian, where one radian (equal to about $57°$) is the angle for which the parallel distance moved by a force acting tangentially to a circle is the same as the perpendicular, radial distance to the center.

Single Molecule Methods

In the tethered cell assay, a live cell is tethered to the cover slip by a single flagellar filament, and rotation of the cell body at speeds up to $\sim 20\,Hz$ (revolutions per second) can be observed with video light microscopy. To observe the faster rotation of the motor when driving smaller loads, a variety of techniques have been used. The rotating filaments of stuck or swimming cells have been visualized with conventional dark field (DF), laser DF, differential interference contrast (DIC), and fluorescence microscopy. Conventional DF and DIC studies have been limited to video rates (Block et al., 1991; Hotani, 1976; Macnab and Ornston, 1977). Laser DF has achieved higher time resolution by recording the oscillating light intensity passing through a slit perpendicular to the image of a filament, which appears as a series of bright spots in this method – one spot for each turn of the filament helix (Kudo et al., 1990; Muramoto et al., 1995). The maximum recorded speed of any molecular motor, $1700\,Hz$ in the Na^+-driven motor in *V. alginolyticus* at $37°C$, was measured using this technique (Magariyama et al., 1994). Fluorescent labeling of flagellar filaments combined with stroboscopic laser illumination or high-speed video microscopy has revealed the polymorphic transitions that cause a chemotactic tumble in *E. coli* (Turner et al., 2000), both in swimming cells (Darnton et al., 2007) and in response to external forces applied with optical tweezers (Darnton and Berg, 2007).

More recently, the preferred method of measuring flagellar rotation has been to attach submicrometer polystyrene beads to truncated flagellar filaments of immobilized cells and to record their rotation with either back-focal-plane interferometry (Chen and Berg,

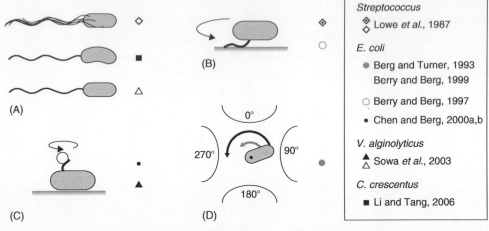

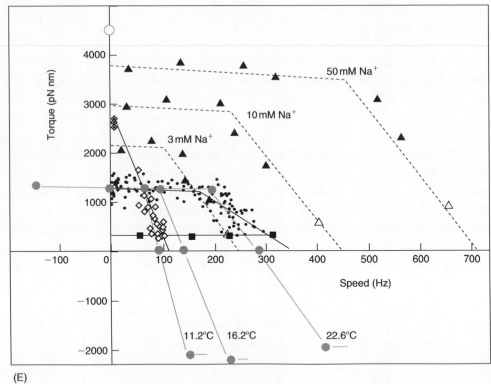

(E)

2000a,b; Gabel and Berg, 2003; Lo et al., 2006, 2007; Reid et al., 2006; Ryu et al., 2000; Sowa et al., 2003, 2005) or high-speed fluorescence microscopy (Sowa et al., 2005). The viscous drag coefficient of a half-micrometer bead is approximately the same as that of a flagellar filament – smaller beads allow measurement of motor rotation at lower loads and higher speeds than those in a swimming cell.

Torque Versus Speed

Fitting the torque–speed relationship under a range of conditions is an important test of models of the mechanochemical cycle of the flagellar motor (Berry, 1993; Elston and Oster, 1997; Läuger, 1988; Oosawa and Hayashi, 1986; Walz and Caplan, 2000; Xing et al., 2006). The speed can be varied either by changing the viscous load or by applying external torque. The latter is technically challenging, but offers the ability to measure the torque generated by the motor when forced to rotate backward or forward faster than its zero-load speed. Figure 4.3 summarizes measurements of the torque–speed relationship of flagellar motors from various species using different methods. Early measurements that varied the viscosity with tethered *Streptococcus* cells showed that the torque was approximately constant at speeds up to about 10 Hz (Manson et al., 1980) and estimated a value of ~2700 pN nm (Lowe et al., 1987). The latter authors also measured the average rotation speed of flagellar bundles, detectable as broad peaks in the frequency spectrum of fluctuations in light intensity scattered by populations of swimming *Streptococcus* cells (Lowe et al., 1987). Comparing these results to the torque in tethered cells gave a linear torque–speed relationship for the *Streptococcus* motor (Figure 4.3, diamonds). More recent experiments using polystyrene beads show a different torque–speed curve for the motors of *E. coli* (Chen and Berg, 2000b) and *V. alginolyticus* (Sowa et al., 2003). There is a plateau of nearly constant torque up to a "knee" speed of several hundred hertz, then

---

Figure 4.3: Torque versus speed. (A–D) Methods of measuring torque–speed relationships. (A) Microscopy of swimming cells. (B) Tethered cells. (C) Beads attached to flagella. (D) Electrorotation of tethered cells – microelectrodes generate a megahertz rotating electric field at the cell that applies an external torque (black arrow) that adds to the motor torque (gray arrow). (E) Torque–speed relationships for flagellar motors of various species measured using various methods. Except where indicated, all measurements were made at room temperature. Symbols shown in (A—D) indicate the methods used to obtain the data in (E). For more details see references indicated in the legend. The *E. coli* experiments using electrorotation (gray circles) and beads (black circles) did not report absolute torques – these curves have been scaled to a stall torque of 1260 pN nm (Reid et al., 2006)

a sharp transition to a regime where torque falls linearly toward the zero-torque speed (Figure 4.3, circles, triangles). The *V. alginolyticus* experiment varied speed by using different sizes of bead and undecorated filaments for the fastest data point, creating some uncertainty in the estimates of the relative viscous drag coefficients – in particular due to unknown filament lengths. The measurements in *E. coli* avoid this uncertainty by varying the viscosity between successive speed measurements with the same bead and motor, using the smallest beads for which a reliable speed signal can be obtained.

These results confirmed the torque–speed relationship measured previously by using electrorotation to apply external torque to tethered *E. coli* cells (Berg and Turner, 1993; Berry and Berg, 1996, 1999; Berry et al., 1995; Washizu et al., 1993). In this method, a rotating electric field (at approximately a megahertz) polarizes the cell body and torque is exerted on the cell due to a phase lag between the field and induced dipole moment (Figure 4.3D). Using microelectrodes and substantial voltages, it was possible to spin the cell body in both directions at speeds of up to ~1 kHz. The electrorotation measurements extend the plateau region to backward speeds of about −100 Hz and the high-speed linear region forward to ~400 Hz, at which point the motor resists rotation with a torque of similar magnitude to the plateau torque (Figure 4.3, gray circles). The absolute magnitude of torque generated by the *E. coli* motor was not estimated in the above experiments, but has subsequently been determined to be 1260±190 pN nm using measurements with polystyrene beads of diameter 1 μm, for which the uncertainty introduced by unknown filament lengths is negligible (Reid et al., 2006). Early electrorotation experiments indicated a ratchet-like mechanism in which considerably more torque is needed to force the motor backward than to stop it rotating forward (Berg and Turner, 1993), but later work showed this to be an artifact of the method (Berry and Berg, 1996, 1999; Berry et al., 1995). This result was confirmed using an optical trap (Ashkin et al., 1987) to demonstrate that the motor torque is similar when rotating very slowly in either direction (Berry and Berg, 1997). The high estimate of torque in this experiment (Figure 4.3, open circles) contradicts the estimate using 1 μm beads, and is probably unreliable due to systematic errors in the estimation of the force exerted by the trap, which was calibrated without the nearby cell body that is present in the torque measurement.

In the torque plateau, transitions linked to ion flux are not rate-limiting; speed is limited mechanically by the load on the motor. The continuity of torque either side of stall indicates that there is no irreversible step in the mechanochemical cycle. Faster than the "knee," transitions in the motor are rate-limiting, and this is the regime where more data are needed to understand the nature of these transitions. This interpretation is supported by the observations that motor speed in the high-speed regime depends on factors that affect absolute transition rates: temperature (Berg and Turner, 1993), hydrogen isotope (Chen and Berg, 2000a), and which component of the SMF is dominant (Lo et al., 2007; Sowa et al., 2003), whereas motor speed in the plateau depends only on the IMF, a thermodynamic quantity (Lo et al., 2007).

It remains to be seen whether the torque–speed relationships of the *E. coli* and *V. alginolyticus* motors are common to all flagellar motors. The *Streptococcus* result would be equivalent if the absolute torque in tethered cells were an overestimate of the true torque. The only other species to date in which the torque–speed relationship has been measured is *Caulobacter crescentus*, where motor torque and speed were inferred from high-speed videos of crescent-shaped free-swimming cells in media of different viscosity (Figure 4.3, squares) (Li and Tang, 2006). These results showed a plateau of low torque extending up to the highest speed measured, $\sim$300 Hz, for motors in cells swimming in aqueous buffer. Torque–speed measurements of the flagellar motors of various species, of chimeras and mutants, and with a range of IMFs are expected to offer further insight in the near future.

Independent Torque-Generating Units

When functional Mot proteins are expressed in a *mot* mutant strain, motor speed increases in discrete increments, a process known as "resurrection." This demonstrates that torque is generated by independent stator units. Early work using tethered *E. coli* cells saw a maximum of eight speed increments (Figure 4.4A (Blair and Berg, 1988; Block and Berg, 1984), at odds with the number of stator particles seen in EM images (10–12) (Khan et al., 1988). This discrepancy has been removed by a recent resurrection of the resurrection experiment, but using 1 μm beads instead of tethered cells (Figure 4.4B) (Reid et al., 2006). Up to 11 or 12 speed increments were seen, consistent with the EM images. Similar results were obtained for the Na^+-driven chimera in *E. coli* (Reid et al., 2006). Up to nine stepwise decreases in the speed of the Na^+-driven motor of an alkalophilic *Bacillus*, upon activation of an irreversible Na^+-channel inhibitor by ultraviolet light, showed that these motors also contain independent torque generators (Muramoto et al., 1994).

Reid et al. (2006) also reported transient speed changes in normally expressed motors, indicating the possibility that stators are not permanent but rather in a process of constant turnover. These results have been confirmed by total internal reflection fluorescence microscopy of GFP-labeled MotB (GFP-MotB) at the single molecule level in spinning motors in live cells (Figure 4.5A and B) (Leake et al., 2006). By comparing the size of stepwise decreases in the fluorescence intensity of single motors (Figure 4.5C, attributable to the photobleaching of single GFP molecules) to the initial intensity of the same motor, Leake et al. (2006) counted an average of 22 GFP-MotB molecules per cell. This is consistent

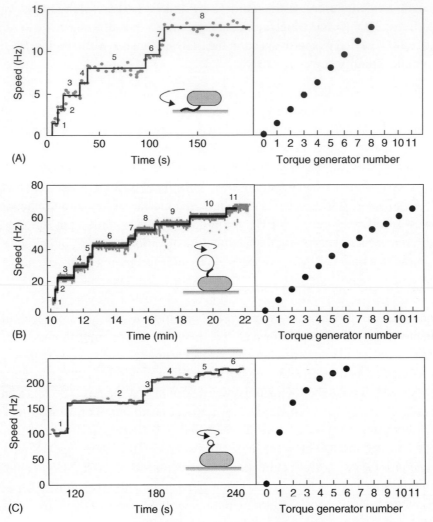

Figure 4.4: Independent torque-generating units in *E. coli*. Left: Speed versus time following "resurrection" of motors with defective stator proteins by induced expression of functional stator proteins. Black lines indicate speed levels. Right: Speed versus number of stators. **(A)** Tethered cells show up to eight equally spaced speeds. Data adapted from Blair and Berg (1988). **(B)** Motors labeled with 1 μm beads show up to 11 or 12 stators. The speed per stator decreases slightly at high stator number. Data adapted from Reid et al. (2006). **(C)** Motors labeled with 0.3 μm beads show a large decrease in the speed per stator as stator number increases. Data adapted from Ryu et al. (2000)

with 11 stator complexes, each with A_4B_2 stoichiometry, as predicted biochemically. Furthermore, a mobile pool of $\sim$200 GFP-MotB molecules was seen in the cell membrane, and fluorescence recovery after photobleaching showed that these exchange with GFP-MotB in the motor on a timescale of minutes, confirming that stator units are dynamic rather than static. This may allow the replacement of damaged stators, or it may be a mechanism for regulating motility in response to environmental conditions. Ion-channel activity of the free stators is very low (Blair and Berg, 1990), as expected if flux is tightly coupled to motor rotation. One model for how this is achieved postulates that the peptidoglycan-binding domain of MotB plugs the channels in stators until they assemble at the motor (Hosking et al., 2006).

Stator turnover explains the observation that motors resurrect after the IMF is transiently removed, for example by removing either the membrane voltage in wild-type *E. coli* (Fung and Berg, 1995) or external Na^+ with the chimeric *E. coli* motor (Sowa et al., 2005). Presumably, the balance between on- and off-rates for stators binding the motor is dependent upon the IMF. It is not known what else affects these rates, but one possibility is that motor rotation itself is the determining factor. This might explain the failure to observe more than eight resurrection steps in tethered cells (reproduced by Reid et al. (2006)), and stepwise decreases in the torque generated by cells stalled for extended periods by electrorotation (R. Berry, personal communication).

Motor speed is proportional to stator number in resurrection experiments with tethered cells, and with 1 μm beads and chimeric motors or wild-type motors at low stator number (Figure 4.4A and B) (Reid et al., 2006). The slight reduction in speed per stator with the wild type at high stator number (Figure 4.4B) may be attributable to steric interference between stators or possibly to local depletion of H^+. Resurrection experiments with smaller beads show a marked nonlinearity even at low stator number (Figure 4.4C) (Ryu et al., 2000). These results were used to show that torque–speed relationships for motors containing between 1 and 5 stators have a plateau and apparently the same knee speed as the wild-type motor, although this conclusion is tentative due to the difficulty of obtaining data at high speed and low load. This was interpreted using a model in which each stator has a high duty ratio, and there is a rate-limiting step in the mechanochemical cycle that cannot be speeded up by the torque exerted by other stators through the common rotor.

Ion Flux and Ion-Motive Force

The IMFs that drive bacterial flagellar motors are significantly different from ATP hydrolysis in a number of ways. They require a membrane and are inherently vectorial in nature – the rotor is oriented in the membrane and ions travel through the motor

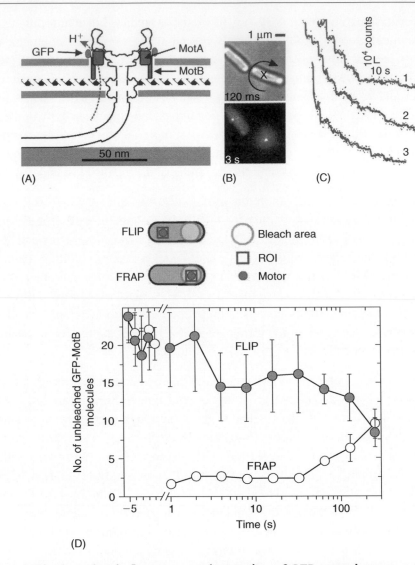

Figure 4.5: Single molecule fluorescence observation of GFP-tagged motor proteins in live cells. (A) GFP-labeled MotB in tethered cells was observed using total internal reflection fluorescence (TIRF) microscopy. TIRF illuminates only a thin layer close to the cover slip, reducing the level of background fluorescence. (B) Cells tethered by a single motor (cell on right) rotated about that motor (lower image: fluorescence, upper image: bright field). The cell on the left is tethered by two motors, and does not rotate. (C) Fluorescence intensity versus time for regions of interest (ROI) centered on three separate motors, showing photobleaching. Stepwise photobleaching toward the end of the traces corresponds to

in a particular direction. This has led to the proposal of several models of the motor mechanism that are based on geometric constraints with no real equivalent in ATP-driven motors (Berry, 1993, 2000; Khan and Berg, 1983; Läuger, 1988; Meister et al., 1989). Ions are smaller and more symmetric than ATP, which may explain the very high stator turnover speeds and correspondingly high power output of the flagellar motor (Ryu et al., 2000). Also the quantum of free energy, corresponding to one ion crossing the membrane, is both smaller than the free energy of hydrolysis of ATP ($\sim 6k_BT$ vs. $\sim 20k_BT$) and more variable by virtue of the fact that the enthalpic contribution, proportional to membrane voltage, is continuously variable and even reversible. Thus, experiments to understand the effects of IMF on flagellar rotation are more difficult than their equivalents in ATP motors, and are likely to lead to different types of conclusion.

While appealing in principle, patch-clamp techniques developed for single ion-channel recordings (Sakmann and Neher, 1995) (see also Chapter 8, this volume) are not practical for the flagellar motor. The estimated current through the flagellar motor is on the order of ~ 0.01 pA (see below), two orders of magnitude smaller than a typical single-channel current. Combined with difficulties in obtaining a tight electrical connection between an external electrode and the cell interior that result from the small size of bacteria, their cell wall, and the outer membrane in gram-positive species such as *E. coli*, this has so far ruled out the direct measurement of ion fluxes through single flagellar motors. The only measurement of ion flux was based on changes in the rate of pH change of a weakly buffered dense suspension of swimming *Streptococcus* when motors were stopped by cross-linking their filaments with antifilament antibody (Meister et al., 1987). The estimated flux was around 1200 H^+ ions per revolution per motor over a speed range of ~ 20–60 Hz.

Direct control of the membrane voltage at the flagellar motor by voltage clamp was achieved in 1995 by pulling filamentous *E. coli* cells (grown with the antibiotic cephalexin to prevent cell division) into custom-made micropipettes containing the proton ionophore gramicidin

single GFP molecules. Comparing the initial fluorescence intensity to the single molecule photobleaching steps allows the number of MotB molecules per motor to be estimated as ~ 22. (D) Fluorescence loss in photobleaching (FLIP) and recovery after photobleaching (FRAP) show that MotB exchanges between the motor and a mobile pool in the membrane on a timescale of minutes. The graph shows the average intensity in an ROI surrounding a motor versus time after localized photobleaching of either a remote part of the cell (FLIP) or the motor itself (FRAP). Data adapted from Leake et al. (2006)

S to establish electrical contact between the pipette and the cell interior (Figure 4.6A, left) (Fung and Berg, 1995). Motor rotation was monitored by video microscopy of dead cells attached to motors, a viscous load equivalent to tethered cells. Speed was proportional to the applied voltage up to $-150\,\mathrm{mV}$ (Figure 4.6A, right), consistent with earlier measurements of the speed of tethered gram-negative bacteria, *Streptococcus* and *Bacillus*, energized by a K^+ diffusion potential (Khan et al., 1985; Manson et al., 1980; Meister and Berg, 1987). Tethered cell experiments using diffusion potentials and variations in the concentration of driving ions also demonstrated that the electrical and chemical components of the PMF (see Eq. (4.1)) are equivalent at high load in H^+-driven motors (Manson et al., 1980).

More recent measurements of the dependence of motor rotation on IMF have relied upon measuring the IMF in individual bacteria in response to different perturbations, rather than attempting to achieve a specific IMF using diffusion potentials or voltage clamp. Gabel and Berg (2003) exploited the proportionality between PMF and tethered cell rotation rate, using the speed of a tethered *E. coli* cell to indicate the PMF in response to perturbation by sodium azide or carbonyl cyanide *m*-chlorophenylhydrazone (CCCP) while simultaneously recording the speed of a $0.4\,\mu\mathrm{m}$ bead attached to another motor of the same cell. The speeds of the two motors were proportional (Figure 4.6B), thus the speed of the fast motor at low load is also proportional to IMF. Lo et al. (2006, 2007) developed fluorescence methods to measure both components of the SMF in single *E. coli* cells expressing the chimeric flagellar motor. They found that the membrane voltage (V_m) and Na^+ concentration gradient ($\Delta p\mathrm{Na}$) could be independently controlled over the ranges $V_\mathrm{m} = -140$ to $-85\,\mathrm{mV}$ and $\Delta p\mathrm{Na} = -50$ to $+40\,\mathrm{mV}$ by variation of pH between 5 and 7 and external Na^+ concentration between 1 and 85 mM. Chimeric motor speed at high load (1 μm beads) was proportional to SMF, and V_m and $\Delta p\mathrm{Na}$ were equivalent, as with tethered gram-negative cells. At low load ($0.36\,\mu$m beads) V_m and $\Delta p\mathrm{Na}$ were not equivalent. For a given external sodium concentration, speed was proportional to SMF, but the constant of proportionality was larger with higher Na^+ concentration and correspondingly larger $\Delta p\mathrm{Na}$ (Figure 4.6C). A similar result was obtained for the Na^+-driven motor of *V. alginolyticus*; reduction of sodium concentration from 50 to 3 mM reduced the speed at low load about threefold but the plateau torque, presumably proportional to SMF, only about two fold (Sowa et al., 2003). The SMF variation at a given sodium concentration in the *E. coli* experiment is mostly in V_m, with only a small change in $\Delta p\mathrm{Na}$. If the chimeric and wild-type motors are the same, this would imply that PMF changes in the experiments of Gabel and Berg (2003) were also dominated by changes in V_m. One possible interpretation of these results is that ion binding is rate-limiting at low load. In the near future, a systematic study of the effects of

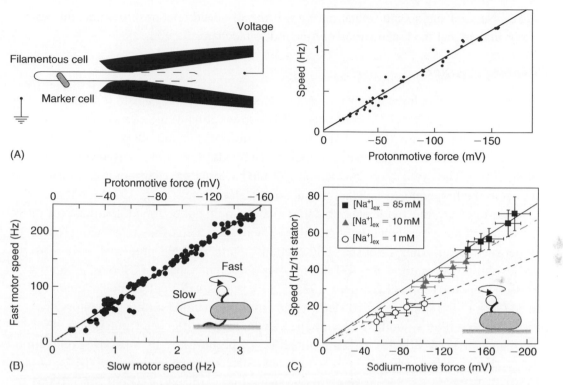

Figure 4.6: Torque versus ion-motive force. (A) Left: Schematic of a voltage-clamp method using filamentous *E. coli* cells held in custom made micropipettes. The part of the membrane inside the pipette (indicated by the dashed line) is made permeable using the ionophore gramicidin S. Motor speed was monitored by video microscopy of a dead cell attached to the motor. Right: Motor speed is proportional to membrane voltage (PMF) between 0 and −150 mV. Data adapted from Fung and Berg (1995). **(B)** Using the result of (A), the speed of a tethered *E. coli* motor (lower axis) was used as a proxy for PMF (upper axis, absolute value shown). The speed of a second motor on the same cell, attached to a submicrometer bead, was found to be proportional to PMF. PMF was varied between −150 mV and 0 by addition of CCCP or sodium azide. Data adapted from Gabel and Berg (2003). **(C)** The speed of single-stator chimeric motors driving small loads is proportional to SMF at a given external Na$^+$ concentration, but motors spin faster in high Na$^+$ even at the same SMF. The membrane voltage was varied via external pH, and the effects of pH and Na$^+$ concentration on both components of the SMF were measured using fluorescence methods. Data adapted from Lo et al. (2007)

V_m, ΔpNa, and site-specific mutations on the torque–speed relationship of the chimeric motor may reveal the kinetic details of the motor mechanism.

Stepping Rotation

An important recent breakthrough toward understanding the mechanism of flagellar rotation at the microscopic level has been direct observation of the elementary process, that is, stepping rotation (Sowa et al., 2005), in a fashion comparable to recent experiments on ATP-driven molecular motors (Svoboda et al., 1993; Yasuda et al., 1998, 2001). The quest to observe stepping of the bacterial flagellar motor dates back almost to the first confirmation that the motor actually rotates (Berg, 1976). At that time, despite careful experiments, stepping motion could not be found. Estimates of the resolution of the experiments predicted that if there were steps, they must be more than 10 per revolution. Analysis of speed fluctuations in tethered cells predicted ~50 steps per revolution per stator and ~400 steps per revolution in a wild-type motor, assuming Poisson stepping (Samuel and Berg, 1995, 1996), setting a difficult technical challenge. In addition to the small expected step size and multiple parallel torque generators, the hook acts as a filter smoothing out any steps in the rotation of a bead attached to the filament. The hook stiffness was determined as $400\,\mathrm{pN\,nm\,rad^{-1}}$ by measuring relaxation times after applying external torque to tethered cells using optical tweezers (Block et al., 1989). This means that even a single stator generating ~$150\,\mathrm{pN\,nm}$ (Reid et al., 2006) twists the hook 0.35 rad, or 20°. If the motor takes a step, the subsequent motion of the bead will be damped with a relaxation time equal to the viscous drag coefficient divided by the spring constant of the hook. The time between steps must be larger than the relaxation time if they are to be detected. Using smaller beads reduces the drag coefficient and thus the relaxation time, but also leads to faster rotation of the motor and thus less time per step.

These problems were overcome by Sowa et al. (2005) by using small beads attached to Na^+-driven chimeric motors in *E. coli* (Asai et al., 2003). Stators were expressed at low levels and the SMF was reduced by lowering external sodium concentration. This achieved very slow rotation (~$10\,\mathrm{Hz}$ or below) combined with fast bead response, although the actual SMF was unknown and the rotation rate was not stable because of the extremely low SMF. Rotation was detected by tracking either $0.5\,\mu\mathrm{m}$ beads using back-focal-plane interferometry (Ryu et al., 2000) or $0.2\,\mu\mathrm{m}$ fluorescent beads using a high-speed electron-multiplying CCD (EMCCD) camera. Steps were resolved in single-stator motors at speeds below $10\,\mathrm{Hz}$ (Figure 4.7A and B). The distribution of step sizes was fitted by a multiple Gaussian, giving an angle of most probable step size of 13.7° (Figure 4.7C), or 26

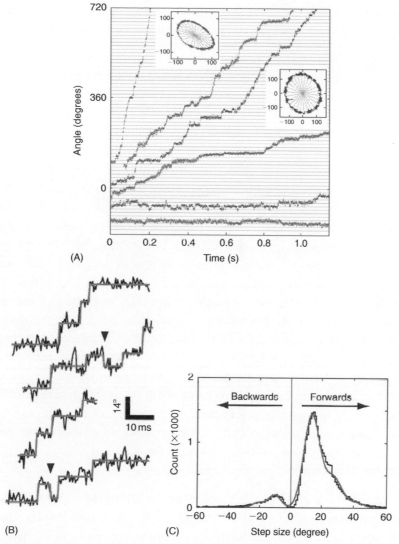

Figure 4.7: Steps in slow flagellar rotation. (A) Stepping rotation of flagellar motors with a range of average speeds depending on different SMFs. Insets show the positions of beads attached to flagellar filaments, scales in nanometers. Beads were tracked by optical interferometry (left inset, gray traces) or high-speed fluorescence microscopy (right inset, black traces). Horizontal and radial lines indicate 1/26th revolutions. (B) Expanded traces of stepping behavior. Backward steps can also be observed clearly (arrowheads). (C) Step-size distribution (black) and multiple Gaussian fit (gray). The peak of forward steps is 13.7°, indicating 26 steps per revolution. Data adapted from Sowa et al. (2005)

steps per revolution ($360°/13.7° = 26.3$). Although average speeds varied from cell to cell because of uncontrolled low SMF, the step size was the same for all speeds.

Stepping motion in ATP-driven molecular motors reflects both the discrete molecular nature of the fuel and the periodicity of the track along which the motor runs (Svoboda et al., 1993; Yasuda et al., 1998). Twenty-six steps per revolution is consistent with the periodicity of the ring of FliG, the track on the rotor where rotational torque is believed to be generated (Suzuki et al., 2004; Thomas et al., 2006). Whether each step corresponds to a single ion transit is not clear. However, previous data indicate that single ions in fully energized wild-type motors cannot drive steps as large as $14°$. If about 10 torque-generating units pass about 1200 ions per revolution independently, then 1 ion in 1 unit should step about $3°$, assuming tight coupling (Meister et al., 1987). Energy conservation sets an upper bound to the angle coupled to one ion transit, equal to (free energy per ion)/(average torque per unit). Taking a PMF of around $-150\,\text{mV}$ and a torque, driving a $1\,\mu\text{m}$ bead, of $\sim150\,\text{pN}\,\text{nm}$ per unit (Gabel and Berg, 2003; Reid et al., 2006) gives an upper bound of $\sim10°$ per ion. Either the coupling ratio is different at low SMF and low load or more than one ion is needed per step. Interestingly, backward steps were common, particularly at the lowest speeds (Figure 4.7A). Recent theoretical work indicates that careful statistical analysis of forward and backward steps can reveal the free energy change driving a step and the details of substeps in the mechanochemical cycle (Linden and Wallin, 2007). What is needed are large quantities of stepping data under well controlled conditions of known SMF (Lo et al., 2006, 2007). But at least now, after 30 years, it is possible to contemplate experiments on the elementary steps of bacterial flagellar motors at a single molecule level, similar to those on ATP-driven motors.

Reversibility and Switching

The flagellar motors of *E. coli* and *V. alginolyticus* are reversible in two senses. Under natural conditions, with an IMF of around $-150\,\text{mV}$, motors switch direction stochastically every second or so, under the control of the chemotactic signaling system. Nonswitching mutants also rotate in the opposite direction when the PMF is reversed. This was achieved using a K^+ diffusion potential in *Streptococcus* (Berg et al., 1982), and a voltage clamp in *E. coli* (Fung and Berg, 1995). In both cases, only a fraction of motors rotated when the PMF was reversed, and in the *E. coli* experiment only for a few revolutions, presumably due to detachment of stators caused by removal of the normal PMF. These results indicate that the mechanochemical cycle of the flagellar motor is essentially reversible, but the robustness of the motor to changes in PMF depends on species.

Chemotactic switching is induced by binding of the active phosphorylated form of the response regulator CheY (CheY-P) to FliM on the rotor (Lee et al., 2001; Toker and Macnab, 1997; Welch et al., 1993). CheY-P concentration in turn is controlled by the chemotactic signaling system, which has been the subject of extensive study (see recent reviews by Baker et al., 2006; Berg, 2003; Parkinson et al., 2005; Sourjik et al., 2007). Using a GFP-labeled CheY in a mutant strain, where all CheY is predicted to be phosphorylated, to quantify the concentration of CheY-P in single cells, Cluzel et al. (2000) discovered a very steep dependence of motor bias (probability of rotation in a particular direction) upon CheY-P, with a Hill coefficient of $\sim$10 (Cluzel et al., 2000). This cannot be explained by cooperative binding, which was shown to be absent using Förster resonance energy transfer (FRET) to quantify the binding between the CheY and FliM labeled with spectral variants of GFP (Sourjik and Berg, 2002). The best candidate to explain the switch mechanism is the conformational spread model (Duke et al., 2001) in which the rotor contains $\sim$34 bistable protomers, each consisting of a tetramer of FliN and 1 copy of FliM (Brown et al., 2007). The steep dependence and lack of cooperative binding are predicted if there is a free energy penalty for adjacent protomers in different states so that the entire rotor is most stable with all protomers in either the CW or CCW states. Recent improvements in the time resolution of single molecule measurements of flagellar rotation will allow testing of the detailed predictions of the conformational spread model in the near future.

Models of the Mechanism

Many models have been proposed for the mechanism of the bacterial flagellar motor (Berg and Turner, 1993; Berry, 2000; Caplan and Kara-Ivanov, 1993; Oosawa and Hayashi, 1986). The better studied models, those which have been formulated within a mathematical framework that provides quantitative predictions for comparison with data, can be divided into three categories: ion turbines, ion turnstiles, and binding with conformational change. In an ion turbine model, the path of ions across the membrane is formed partly by elements in the stator and partly by elements in the rotor, and these elements are arranged in lines that are tilted with respect to each other. The "elements" can be half-binding sites on rotor and stator that need to be aligned to bind a permeant ion (Läuger, 1977) or ion channels in the stator that interact with tilted lines of charge on the rotor by long-range electrostatic interactions (Berry, 1993; Elston and Oster, 1997; Walz and Caplan, 2000). The charged residues on the surface of FliG that are involved in torque generation could in principle be arranged in such a way as to provide the electrostatic interactions that are proposed in the latter models. In a turnstile model, ions are deposited onto the rotor from outside the cell by one type of stator channel and can only complete a transit if the rotor rotates, carrying them to a second type of stator channel that connects to the cell interior (Khan and Berg, 1983; Meister et al., 1989). This type of model is believed to describe well the mechanism of F_O-ATPase, with an essential conserved residue on the c-subunit providing a probable binding site for ions halfway across the membrane (Elston et al., 1998; Vik and Antonio, 1994). However, the lack of an equivalent essential residue in the flagellar motor is evidence against this type of model. In a conformational change model, ion transit through a stator is coupled to a cycle of conformational changes of the stator that exerts torque on the rotor, either by long-range electrostatic or short-range steric interactions (Läuger, 1988). This type of mechanism is believed to describe the ATP-driven molecular motors myosin and F_1-ATPase. Conformational changes in MotA linked to the proposed ion-binding site in MotB provide indirect evidence for this type of model in the flagellar motor (Kojima and Blair, 2001). The most recent detailed study of a flagellar motor model, and the only one to date to reproduce successfully the knee in the torque–speed relationship, falls into the conformational change category (Xing et al., 2006).

Outlook

Experiments in 2005 and 2006 that have applied *in vitro* single molecule techniques to flagellar motors *in vivo* promise substantial advances in our understanding of the flagellar motor in the near future. Early discoveries using these methods are that flagellar motors with a single stator take 26 steps per revolution (Sowa et al., 2005), and that stators are in a constant process of turnover (Leake et al., 2006). The next step is clearly to make a systematic study of flagellar steps: their statistical properties, dependence on driving force, and the possible existence of substeps. These measurements will offer the same considerable insights into the mechanism as do equivalent single molecule experiments on ATP-driven molecular motors in the last 10 years. The flagellar motor is the first ion-driven molecular machine that can be studied at this level of mechanical detail. The chimeric sodium-driven motor in *E. coli* and the newly developed fluorescent methods to measure both components of the SMF offer exciting new possibilities. Because the enthalpic (membrane-voltage) and entropic (ion-concentration gradient) components of the IMF are of similar magnitude, unlike ATP hydrolysis where the dominant free energy change is usually enthalpic, this work will offer insight into the fundamental question of how molecular motors convert the different components of free energy into mechanical work. An important factor will be increased time resolution in the techniques for detecting flagellar rotation. This is currently limited by the need to observe a relatively large polystyrene bead connected to the motor by a relatively flexible hook.

Smaller labels, for example, fluorescent proteins, quantum dots or gold nanoparticles, and possibly stiffer hooks are likely to be the way forward.

Finally, without detailed atomic-level structures of the rotor and stator, it will not be possible to understand the precise structural and mechanical details of the coupling mechanism, even if single molecule mechanical experiments reveal the kinetics and energetics. Recent successes in obtaining partial atomic structures of rotor proteins, and in docking these into ever better EM reconstructions of the rotor, show the way forward. The biggest challenge will be to obtain atomic-level structures of stators, as these are large transmembrane complexes that are now known to interact dynamically with the cell wall as well as the rotor.

References

Aizawa, S. I. (1996). Flagellar assembly in *Salmonella typhimurium*. *Mol Microbiol* 19, 1–5.

Armitage, J. P. (1999). Bacterial tactic responses. *Adv Microb Physiol* 41, 229–289.

Armitage, J. P. and Macnab, R. M. (1987). Unidirectional, intermittent rotation of the flagellum of *Rhodobacter sphaeroides*. *J Bacteriol* 169, 514–518.

Armitage, J. P. and Schmitt, R. (1997). Bacterial chemotaxis: *Rhodobacter sphaeroides* and *Sinorhizobium meliloti* – variations on a theme?. *Microbiology* 143, 3671–3682.

Asai, Y., Kawagishi, I., Sockett, R. E., and Homma, M. (1999). Hybrid motor with H^+- and Na^+-driven components can rotate *Vibrio* polar flagella by using sodium ions. *J Bacteriol* 181, 6332–6338.

Asai, Y., Kawagishi, I., Sockett, R. E., and Homma, M. (2000). Coupling ion specificity of chimeras between H^+- and Na^+-driven motor proteins, MotB and PomB, in *Vibrio* polar flagella. *Embo J* 19, 3639–3648.

Asai, Y., Kojima, S., Kato, H., Nishioka, N., Kawagishi, I., and Homma, M. (1997). Putative channel components for the fast-rotating sodium-driven flagellar motor of a marine bacterium. *J Bacteriol* 179, 5104–5110.

Asai, Y., Yakushi, T., Kawagishi, I., and Homma, M. (2003). Ion-coupling determinants of Na^+-driven and H^+-driven flagellar motors. *J Mol Biol* 327, 453–463.

Asakura, S. (1970). Polymerization of flagellin and polymorphism of flagella. *Adv Biophys* 1, 99–155.

Ashkin, A., Dziedzic, J. M., and Yamane, T. (1987). Optical trapping and manipulation of single cells using infrared laser beams. *Nature* 330, 769–771.

Baker, M. D., Wolanin, P. M., and Stock, J. B. (2006). Signal transduction in bacterial chemotaxis. *Bioessays* 28, 9–22.

Berg, H. C. (1976). Does the flagellar rotary motor step? In: *Cell motility* (R. Goldman, T. Pollad, and J. Rosenbaum, Eds.), pp. 47–56. Cold Spring Harbor Laboratory, New York.

Berg, H. C. (2003). The rotary motor of bacterial flagella. *Annu Rev Biochem* 72, 19–54.

Berg, H. C. and Anderson, R. A. (1973). Bacteria swim by rotating their flagellar filaments. *Nature* 245, 380–382.

Berg, H. C., Manson, M. D., and Conley, M. P. (1982). Dynamics and energetics of flagellar rotation in bacteria. *Symp Soc Exp Biol* 35, 1–31.

Berg, H. C. and Turner, L. (1993). Torque generated by the flagellar motor of *Escherichia coli*. *Biophys J* 65, 2201–2216.

Berry, R. M. (1993). Torque and switching in the bacterial flagellar motor. An electrostatic model. *Biophys J* 64, 961–973.

Berry, R. M. (2000). Theories of rotary motors. *Philos Trans R Soc Lond B Biol Sci* 355, 503–509.

Berry, R. M. and Armitage, J. P. (1999). The bacterial flagella motor. *Adv Microb Physiol* 41, 291–337.

Berry, R. M. and Berg, H. C. (1996). Torque generated by the bacterial flagellar motor close to stall. *Biophys J* 71, 3501–3510.

Berry, R. M. and Berg, H. C. (1997). Absence of a barrier to backwards rotation of the bacterial flagellar motor demonstrated with optical tweezers. *Proc Natl Acad Sci U S A* 94, 14433–14437.

Berry, R. M. and Berg, H. C. (1999). Torque generated by the flagellar motor of *Escherichia coli* while driven backward. *Biophys J* 76, 580–587.

Berry, R. M., Turner, L., and Berg, H. C. (1995). Mechanical limits of bacterial flagellar motors probed by electrorotation. *Biophys J* 69, 280–286.

Blair, D. F. (1995). How bacteria sense and swim. *Annu Rev Microbiol* 49, 489–522.

Blair, D. F. and Berg, H. C. (1988). Restoration of torque in defective flagellar motors. *Science* 242, 1678–1681.

Blair, D. F. and Berg, H. C. (1990). The MotA protein of *E. coli* is a proton-conducting component of the flagellar motor. *Cell* 60, 439–449.

Block, S. M. and Berg, H. C. (1984). Successive incorporation of force-generating units in the bacterial rotary motor. *Nature* 309, 470–472.

Block, S. M., Blair, D. F., and Berg, H. C. (1989). Compliance of bacterial flagella measured with optical tweezers. *Nature* 338, 514–518.

Block, S. M., Fahrner, K. A., and Berg, H. C. (1991). Visualization of bacterial flagella by video-enhanced light microscopy. *J Bacteriol* 173, 933–936.

Braun, T. F., Al-Mawsawi, L. Q., Kojima, S., and Blair, D. F. (2004). Arrangement of core membrane segments in the MotA/MotB proton-channel complex of *Escherichia coli*. *Biochemistry* 43, 35–45.

Braun, T. F. and Blair, D. F. (2001). Targeted disulfide cross-linking of the MotB protein of *Escherichia coli*: evidence for two H$^+$ channels in the stator complex. *Biochemistry* 40, 13051–13059.

Braun, T. F., Poulson, S., Gully, J. B., Empey, J. C., Van Way, S., Putnam, A., and Blair, D. F. (1999). Function of proline residues of MotA in torque generation by the flagellar motor of *Escherichia coli*. *J Bacteriol* 181, 3542–3551.

Brown, P. N., Hill, C. P., and Blair, D. F. (2002). Crystal structure of the middle and C-terminal domains of the flagellar rotor protein FliG. *EMBO J* 21, 3225–3234.

Brown, P. N., Mathews, M. A., Joss, L. A., Hill, C. P., and Blair, D. F. (2005). Crystal structure of the flagellar rotor protein FliN from *Thermotoga maritima*. *J Bacteriol* 187, 2890–2902.

Brown, P. N., Terrazas, M., Paul, K., and Blair, D. F. (2007). Mutational analysis of the flagellar protein FliG: sites of interaction with FliM and implications for organization of the switch complex. *J Bacteriol* 189, 305–312.

Calladine, C. R. (1975). Construction of bacterial flagella. *Nature* 255, 121–124.

Caplan, S. R. and Kara-Ivanov, M. (1993). The bacterial flagellar motor. *Int Rev Cytol* 147, 97–164.

Chen, X. and Berg, H. C. (2000a). Solvent-isotope and pH effects on flagellar rotation in *Escherichia coli*. *Biophys J* 78, 2280–2284.

Chen, X. and Berg, H. C. (2000b). Torque-speed relationship of the flagellar rotary motor of *Escherichia coli*. *Biophys J* 78, 1036–1041.

Chun, S. Y. and Parkinson, J. S. (1988). Bacterial motility: membrane topology of the *Escherichia coli* MotB protein. *Science* 239, 276–278.

Cluzel, P., Surette, M., and Leibler, S. (2000). An ultrasensitive bacterial motor revealed by monitoring signaling proteins in single cells. *Science* 287, 1652–1655.

Coulton, J. W. and Murray, R. G. (1978). Cell envelope associations of *Aquaspirillum serpens* flagella. *J Bacteriol* 136, 1037–1049.

Darnton, N. C. and Berg, H. C. (2007). Force-extension measurements on bacterial flagella: triggering polymorphic transformations. *Biophys J* 92, 2230–2236.

Darnton, N. C., Turner, L., Rojevsky, S., and Berg, H. C. (2007). On torque and tumbling in swimming *Escherichia coli*. *J Bacteriol* 189, 1756–1764.

Dean, G. E., Macnab, R. M., Stader, J., Matsumura, P., and Burks, C. (1984). Gene sequence and predicted amino acid sequence of the motA protein, a membrane-associated protein required for flagellar rotation in *Escherichia coli*. *J Bacteriol* 159, 991–999.

De Mot, R. and Vanderleyden, J. (1994). The C-terminal sequence conservation between OmpA-related outer membrane proteins and MotB suggests a common function in both gram-positive and gram-negative bacteria, possibly in the interaction of these domains with peptidoglycan. *Mol Microbiol* 12, 333–334.

DePamphilis, M. L. and Adler, J. (1971a). Attachment of flagellar basal bodies to the cell envelope: specific attachment to the outer, lipopolysaccharide membrane and the cyoplasmic membrane. *J Bacteriol* 105, 396–407.

DePamphilis, M. L. and Adler, J. (1971b). Fine structure and isolation of the hook-basal body complex of flagella from *Escherichia coli* and *Bacillus subtilis*. *J Bacteriol* 105, 384–395.

DePamphilis, M. L. and Adler, J. (1971c). Purification of intact flagella from *Escherichia coli* and *Bacillus subtilis*. *J Bacteriol* 105, 376–383.

DeRosier, D. (2006). Bacterial flagellum: visualizing the complete machine *in situ*. *Curr Biol* 16, R928–R930.

Duke, T. A., Le Novere, N., and Bray, D. (2001). Conformational spread in a ring of proteins: a stochastic approach to allostery. *J Mol Biol* 308, 541–553.

Elston, T., Wang, H., and Oster, G. (1998). Energy transduction in ATP synthase. *Nature* 391, 510–513.

Elston, T. C. and Oster, G. (1997). Protein turbines. I: The bacterial flagellar motor. *Biophys J* 73, 703–721.

Falke, J. J., Bass, R. B., Butler, S. L., Chervitz, S. A., and Danielson, M. A. (1997). The two-component signaling pathway of bacterial chemotaxis: a molecular view of signal transduction by receptors, kinases, and adaptation enzymes. *Annu Rev Cell Dev Biol* 13, 457–512.

Francis, N. R., Irikura, V. M., Yamaguchi, S., DeRosier, D. J., and Macnab, R. M. (1992). Localization of the *Salmonella typhimurium* flagellar switch protein FliG to the cytoplasmic M-ring face of the basal body. *Proc Natl Acad Sci U S A* 89, 6304–6308.

Francis, N. R., Sosinsky, G. E., Thomas, D., and DeRosier, D. J. (1994). Isolation, characterization and structure of bacterial flagellar motors containing the switch complex. *J Mol Biol* 235, 1261–1270.

Fung, D. C. and Berg, H. C. (1995). Powering the flagellar motor of *Escherichia coli* with an external voltage source. *Nature* 375, 809–812.

Furuta, T., Samatey, F. A., Matsunami, H., Imada, K., Namba, K., and Kitao, A. (2007). Gap compression/extension mechanism of bacterial flagellar hook as the molecular universal joint. *J Struct Biol* 157, 481–490.

Gabel, C. V. and Berg, H. C. (2003). The speed of the flagellar rotary motor of *Escherichia coli* varies linearly with protonmotive force. *Proc Natl Acad Sci U S A* 100, 8748–8751.

Garza, A. G., Harris-Haller, L. W., Stoebner, R. A., and Manson, M. D. (1995). Motility protein interactions in the bacterial flagellar motor. *Proc Natl Acad Sci U S A* 92, 1970–1974.

Gosink, K. K. and Hase, C. C. (2000). Requirements for conversion of the Na^+-driven flagellar motor of *Vibrio cholerae* to the H^+-driven motor of *Escherichia coli*. *J Bacteriol* 182, 1234–1240.

Hasegawa, K., Yamashita, I., and Namba, K. (1998). Quasi- and nonequivalence in the structure of bacterial flagellar filament. *Biophys J* 74, 569–575.

Hirota, N. and Imae, Y. (1983). Na^+-driven flagellar motors of an alkalophilic *Bacillus* strain YN-1. *J Biol Chem* 258, 10577–10581.

Hosking, E. R., Vogt, C., Bakker, E. P., and Manson, M. D. (2006). The *Escherichia coli* MotAB proton channel unplugged. *J Mol Biol* 364, 921–937.

Hotani, H. (1976). Light microscope study of mixed helices in reconstituted *Salmonella* flagella. *J Mol Biol* 106, 151–166.

Katayama, E., Shiraishi, T., Oosawa, K., Baba, N., and Aizawa, S. (1996). Geometry of the flagellar motor in the cytoplasmic membrane of *Salmonella typhimurium* as determined by stereo-photogrammetry of quick-freeze deep-etch replica images. *J Mol Biol* 255, 458–475.

Khan, S. and Berg, H. C. (1983). Isotope and thermal effects in chemiosmotic coupling to the flagellar motor of Streptococcus. *Cell* 32, 913–919.

Khan, S., Dapice, M., and Reese, T. S. (1988). Effects of *mot* gene expression on the structure of the flagellar motor. *J Mol Biol* 202, 575–584.

Khan, S., Ivey, D. M., and Krulwich, T. A. (1992). Membrane ultrastructure of alkaliphilic *Bacillus* species studied by rapid-freeze electron microscopy. *J Bacteriol* 174, 5123–5126.

Khan, S., Khan, I. H., and Reese, T. S. (1991). New structural features of the flagellar base in *Salmonella typhimurium* revealed by rapid-freeze electron microscopy. *J Bacteriol* 173, 2888–2896.

Khan, S., Meister, M., and Berg, H. C. (1985). Constraints on flagellar rotation. *J Mol Biol* 184, 645–656.

Kitao, A., Yonekura, K., Maki-Yonekura, S., Samatey, F. A., Imada, K., Namba, K., and Go, N. (2006). Switch interactions control energy frustration and multiple flagellar filament structures. *Proc Natl Acad Sci U S A* 103, 4894–4899.

Kojima, S. and Blair, D. F. (2001). Conformational change in the stator of the bacterial flagellar motor. *Biochemistry* 40, 13041–13050.

Kojima, S. and Blair, D. F. (2004a). The bacterial flagellar motor: structure and function of a complex molecular machine. *Int Rev Cytol* 233, 93–134.

Kojima, S. and Blair, D. F. (2004b). Solubilization and purification of the MotA/MotB complex of *Escherichia coli*. *Biochemistry* 43, 26–34.

Kudo, S., Magariyama, Y., and Aizawa, S. (1990). Abrupt changes in flagellar rotation observed by laser dark-field microscopy. *Nature* 346, 677–680.

Läuger, P. (1977). Ion transport and rotation of bacterial flagella. *Nature* 268, 360–362.

Läuger, P. (1988). Torque and rotation rate of the bacterial flagellar motor. *Biophys J* 53, 53–65.

Leake, M. C., Chandler, J. H., Wadhams, G. H., Bai, F., Berry, R. M., and Armitage, J. P. (2006). Stoichiometry and turnover in single, functioning membrane protein complexes. *Nature* 443, 355–358.

Lee, S. Y., Cho, H. S., Pelton, J. G., Yan, D., Henderson, R. K., King, D. S., Huang, L., Kustu, S., Berry, E. A., and Wemmer, D. E. (2001). Crystal structure of an activated response regulator bound to its target. *Nat Struct Biol* 8, 52–56.

Li, G. and Tang, J. X. (2006). Low flagellar motor torque and high swimming efficiency of *Caulobacter crescentus* swarmer cells. *Biophys J* 91, 2726–2734.

Linden, M. and Wallin, M. (2007). Dwell time symmetry in random walks and molecular motors. *Biophys J* 92, 3804–3816.

Lloyd, S. A. and Blair, D. F. (1997). Charged residues of the rotor protein FliG essential for torque generation in the flagellar motor of *Escherichia coli*. *J Mol Biol* 266, 733–744.

Lloyd, S. A., Tang, H., Wang, X., Billings, S., and Blair, D. F. (1996). Torque generation in the flagellar motor of *Escherichia coli*: evidence of a direct role for FliG but not for FliM or FliN. *J Bacteriol* 178, 223–231.

Lloyd, S. A., Whitby, F. G., Blair, D. F., and Hill, C. P. (1999). Structure of the C-terminal domain of FliG, a component of the rotor in the bacterial flagellar motor. *Nature* 400, 472–475.

Lo, C. J., Leake, M. C., and Berry, R. M. (2006). Fluorescence measurement of intracellular sodium concentration in single *Escherichia coli* cells. *Biophys J* 90, 357–365.

Lo, C. J., Leake, M. C., Pilizota, T., and Berry, R. M. (2007). Nonequivalence of membrane voltage and ion-gradient as driving forces for the bacterial flagellar motor at low load. *Biophys J* 93, 294–302.

Lowder, B. J., Duyvesteyn, M. D., and Blair, D. F. (2005). FliG subunit arrangement in the flagellar rotor probed by targeted cross-linking. *J Bacteriol* 187, 5640–5647.

Lowe, G., Meister, M., and Berg, H. C. (1987). Rapid rotation of flagellar bundles in swimming bacteria. *Nature* 325, 637–640.

Macnab, R. M. (1996). Flagella and motility. *In Eschericia coli* and *Salmonella*: Cellular and Molecular Biology (Neidhart, F.C., Curtiss, R., Ingraham, J.L., Lin, E.C.C., Low, K.B., Magasanik, B., Reznikoff, W.S., Reily, M., Schaechter, M., and Umbarger, H.E., Eds.). pp. 123–145. American Society for Microbiology, Washington DC.

Macnab, R. M. (2003). How bacteria assemble flagella. *Annu Rev Microbiol* 57, 77–100.

Macnab, R. M. and Ornston, M. K. (1977). Normal-to-curly flagellar transitions and their role in bacterial tumbling. Stabilization of an alternative quaternary structure by mechanical force. *J Mol Biol* 112, 1–30.

Magariyama, Y., Sugiyama, S., Muramoto, K., Maekawa, Y., Kawagishi, I., Imae, Y., and Kudo, S. (1994). Very fast flagellar rotation. *Nature* 371, 752.

Manson, M. D., Tedesco, P., Berg, H. C., Harold, F. M., and Van der Drift, C. (1977). A protonmotive force drives bacterial flagella. *Proc Natl Acad Sci U S A* 74, 3060–3064.

Manson, M. D., Tedesco, P. M., and Berg, H. C. (1980). Energetics of flagellar rotation in bacteria. *J Mol Biol* 138, 541–561.

Matsuura, S., Shioi, J., and Imae, Y. (1977). Motility in *Bacillus subtilis* driven by an artificial protonmotive force. *FEBS Lett* 82, 187–190.

McCarter, L. L. (1994a). MotX, the channel component of the sodium-type flagellar motor. *J Bacteriol* 176, 5988–5998.

McCarter, L. L. (1994b). MotY, a component of the sodium-type flagellar motor. *J Bacteriol* 176, 4219–4225.

Meister, M. and Berg, H. C. (1987). The stall torque of the bacterial flagellar motor. *Biophys J* 52, 413–419.

Meister, M., Caplan, S. R., and Berg, H. C. (1989). Dynamics of a tightly coupled mechanism for flagellar rotation. Bacterial motility, chemiosmotic coupling, protonmotive force. *Biophys J* 55, 905–914.

Meister, M., Lowe, G., and Berg, H. C. (1987). The proton flux through the bacterial flagellar motor. *Cell* 49, 643–650.

Mimori, Y., Yamashita, I., Murata, K., Fujiyoshi, Y., Yonekura, K., Toyoshima, C., and Namba, K. (1995). The structure of the R-type straight flagellar filament of *Salmonella* at 9 Å resolution by electron cryomicroscopy. *J Mol Biol* 249, 69–87.

Muramoto, K., Kawagishi, I., Kudo, S., Magariyama, Y., Imae, Y., and Homma, M. (1995). High-speed rotation and speed stability of the sodium-driven flagellar motor in *Vibrio alginolyticus*. *J Mol Biol* 251, 50–58.

Muramoto, K., Sugiyama, S., Cragoe Jr., E. J., and Imae, Y. (1994). Successive inactivation of the force-generating units of sodium-driven bacterial flagellar motors by a photoreactive amiloride analog. *J Biol Chem* 269, 3374–3380.

Murphy, G. E., Leadbetter, J. R., and Jensen, G. J. (2006). In situ structure of the complete *Treponema primitia* flagellar motor. *Nature* 442, 1062–1064.

Namba, K. and Vonderviszt, F. (1997). Molecular architecture of bacterial flagellum. *Q Rev Biophys* 30, 1–65.

Oosawa, F. and Hayashi, S. (1986). The loose coupling mechanism in molecular machines of living cells. *Adv Biophys* 22, 151–183.

Oosawa, K., Ueno, T., and Aizawa, S. (1994). Overproduction of the bacterial flagellar switch proteins and their interactions with the MS ring complex *in vitro*. *J Bacteriol* 176, 3683–3691.

Park, S. Y., Lowder, B., Bilwes, A. M., Blair, D. F., and Crane, B. R. (2006). Structure of FliM provides insight into assembly of the switch complex in the bacterial flagella motor. *Proc Natl Acad Sci U S A* 103, 11886–11891.

Parkinson, J. S., Ames, P., and Studdert, C. A. (2005). Collaborative signaling by bacterial chemoreceptors. *Curr Opin Microbiol* 8, 116–121.

Paul, K. and Blair, D. F. (2006). Organization of FliN subunits in the flagellar motor of *Escherichia coli*. *J Bacteriol* 188, 2502–2511.

Paul, K., Harmon, J. G., and Blair, D. F. (2006). Mutational analysis of the flagellar rotor protein FliN: identification of surfaces important for flagellar assembly and switching. *J Bacteriol* 188, 5240–5248.

Reid, S. W., Leake, M. C., Chandler, J. H., Lo, C. J., Armitage, J. P., and Berry, R. M. (2006). The maximum number of torque-generating units in the flagellar motor of *Escherichia coli* is at least 11. *Proc Natl Acad Sci U S A* 103, 8066–8071.

Ryu, W. S., Berry, R. M., and Berg, H. C. (2000). Torque-generating units of the flagellar motor of *Escherichia coli* have a high duty ratio. *Nature* 403, 444–447.

Sakmann, B. and Neher, E. (1995). *Single-Channel Recording*. Plenum Press, New York.

Samatey, F. A., Imada, K., Nagashima, S., Vonderviszt, F., Kumasaka, T., Yamamoto, M., and Namba, K. (2001). Structure of the bacterial flagellar protofilament and implications for a switch for supercoiling. *Nature* 410, 331–337.

Samatey, F. A., Matsunami, H., Imada, K., Nagashima, S., Shaikh, T. R., Thomas, D. R., Chen, J. Z., Derosier, D. J., Kitao, A., and Namba, K. (2004). Structure of the bacterial flagellar hook and implication for the molecular universal joint mechanism. *Nature* 431, 1062–1068.

Samuel, A. D. and Berg, H. C. (1995). Fluctuation analysis of rotational speeds of the bacterial flagellar motor. *Proc Natl Acad Sci U S A* 92, 3502–3506.

Samuel, A. D. and Berg, H. C. (1996). Torque-generating units of the bacterial flagellar motor step independently. *Biophys J* 71, 918–923.

Sato, K. and Homma, M. (2000). Functional reconstitution of the Na^+-driven polar flagellar motor component of *Vibrio alginolyticus*. *J Biol Chem* 275, 5718–5722.

Shaikh, T. R., Thomas, D. R., Chen, J. Z., Samatey, F. A., Matsunami, H., Imada, K., Namba, K., and Derosier, D. J. (2005). A partial atomic structure for the flagellar hook of *Salmonella typhimurium*. *Proc Natl Acad Sci U S A* 102, 1023–1028.

Sharp, L. L., Zhou, J., and Blair, D. F. (1995a). Features of MotA proton channel structure revealed by tryptophan-scanning mutagenesis. *Proc Natl Acad Sci U S A* 92, 7946–7950.

Sharp, L. L., Zhou, J., and Blair, D. F. (1995b). Tryptophan-scanning mutagenesis of MotB, an integral membrane protein essential for flagellar rotation in *Escherichia coli*. *Biochemistry* 34, 9166–9171.

Silverman, M. and Simon, M. (1974). Flagellar rotation and the mechanism of bacterial motility. *Nature* 249, 73–74.

Sourjik, V. and Berg, H. C. (2002). Binding of the *Escherichia coli* response regulator CheY to its target measured *in vivo* by fluorescence resonance energy transfer. *Proc Natl Acad Sci U S A* 99, 12669–12674.

Sourjik, V., Vaknin, A., Shimizu, T. S., and Berg, H. C. (2007). *In vivo* measurement by FRET of pathway activity in bacterial chemotaxis. *Methods Enzymol* 423, 363–391.

Sowa, Y., Hotta, H., Homma, M., and Ishijima, A. (2003). Torque-speed relationship of the Na$^+$-driven flagellar motor of *Vibrio alginolyticus*. *J Mol Biol* 327, 1043–1051.

Sowa, Y., Rowe, A. D., Leake, M. C., Yakushi, T., Homma, M., Ishijima, A., and Berry, R. M. (2005). Direct observation of steps in rotation of the bacterial flagellar motor. *Nature* 437, 916–919.

Suzuki, H., Yonekura, K., and Namba, K. (2004). Structure of the rotor of the bacterial flagellar motor revealed by electron cryomicroscopy and single-particle image analysis. *J Mol Biol* 337, 105–113.

Svoboda, K., Schmidt, C. F., Schnapp, B. J., and Block, S. M. (1993). Direct observation of kinesin stepping by optical trapping interferometry. *Nature* 365, 721–727.

Tang, H., Braun, T. F., and Blair, D. F. (1996). Motility protein complexes in the bacterial flagellar motor. *J Mol Biol* 261, 209–221.

Terashima, H., Fukuoka, H., Yakushi, T., Kojima, S., and Homma, M. (2006). The *Vibrio* motor proteins, MotX and MotY, are associated with the basal body of Na$^+$-driven flagella and required for stator formation. *Mol Microbiol* 62, 1170–1180.

Thomas, D. R., Francis, N. R., Xu, C., and DeRosier, D. J. (2006). The three-dimensional structure of the flagellar rotor from a clockwise-locked mutant of *Salmonella enterica* serovar Typhimurium. *J Bacteriol* 188, 7039–7048.

Thomas, D. R., Morgan, D. G., and DeRosier, D. J. (1999). Rotational symmetry of the C ring and a mechanism for the flagellar rotary motor. *Proc Natl Acad Sci U S A* 96, 10134–10139.

Toker, A. S. and Macnab, R. M. (1997). Distinct regions of bacterial flagellar switch protein FliM interact with FliG, FliN and CheY. *J Mol Biol* 273, 623–634.

Turner, L., Ryu, W. S., and Berg, H. C. (2000). Real-time imaging of fluorescent flagellar filaments. *J Bacteriol* 182, 2793–2801.

Ueno, T., Oosawa, K., and Aizawa, S. (1992). M ring, S ring and proximal rod of the flagellar basal body of *Salmonella typhimurium* are composed of subunits of a single protein, FliF. *J Mol Biol* 227, 672–677.

Ueno, T., Oosawa, K., and Aizawa, S. (1994). Domain structures of the MS ring component protein (FliF) of the flagellar basal body of *Salmonella typhimurium*. *J Mol Biol* 236, 546–555.

Vik, S. B. and Antonio, B. J. (1994). A mechanism of proton translocation by F_1F_O ATP synthases suggested by double mutants of the a subunit. *J Biol Chem* 269, 30364–30369.

Walz, D. and Caplan, S. R. (2000). An electrostatic mechanism closely reproducing observed behavior in the bacterial flagellar motor. *Biophys J* 78, 626–651.

Washizu, M., Kurahashi, Y., Iochi, H., Kurosawa, O., Aizawa, S., Kudo, S., Magariyama, Y., and Hotani, H. (1993). Dielectrophoretic measurement of bacterial motor characteristics. *IEEE Trans Ind Appl* 29, 286–294.

Welch, M., Oosawa, K., Aizawa, S., and Eisenbach, M. (1993). Phosphorylation-dependent binding of a signal molecule to the flagellar switch of bacteria. *Proc Natl Acad Sci U S A* 90, 8787–8791.

Xing, J., Bai, F., Berry, R., and Oster, G. (2006). Torque-speed relationship of the bacterial flagellar motor. *Proc Natl Acad Sci U S A* 103, 1260–1265.

Yakushi, T., Yang, J., Fukuoka, H., Homma, M., and Blair, D. F. (2006). Roles of charged residues of rotor and stator in flagellar rotation: comparative study using H^+-driven and Na^+-driven motors in *Escherichia coli*. *J Bacteriol* 188, 1466–1472.

Yamaguchi, S., Aizawa, S., Kihara, M., Isomura, M., Jones, C. J., and Macnab, R. M. (1986a). Genetic evidence for a switching and energy-transducing complex in the flagellar motor of *Salmonella typhimurium*. *J Bacteriol* 168, 1172–1179.

Yamaguchi, S., Fujita, H., Ishihara, A., Aizawa, S., and Macnab, R. M. (1986b). Subdivision of flagellar genes of *Salmonella typhimurium* into regions responsible for assembly, rotation, and switching. *J Bacteriol* 166, 187–193.

Yasuda, R., Noji, H., Kinosita Jr., K., and Yoshida, M. (1998). F_1-ATPase is a highly efficient molecular motor that rotates with discrete 120 degree steps. *Cell* 93, 1117–1124.

Yasuda, R., Noji, H., Yoshida, M., Kinosita Jr., K., and Itoh, H. (2001). Resolution of distinct rotational substeps by submillisecond kinetic analysis of F_1-ATPase. *Nature* 410, 898–904.

Yonekura, K., Maki, S., Morgan, D. G., DeRosier, D. J., Vonderviszt, F., Imada, K., and Namba, K. (2000). The bacterial flagellar cap as the rotary promoter of flagellin self-assembly. *Science* 290, 2148–2152.

Yonekura, K., Maki-Yonekura, S., and Namba, K. (2003). Complete atomic model of the bacterial flagellar filament by electron cryomicroscopy. *Nature* 424, 643–650.

Yorimitsu, T. and Homma, M. (2001). Na^+-driven flagellar motor of *Vibrio*. *Biochim Biophys Acta* 1505, 82–93.

Yorimitsu, T., Mimaki, A., Yakushi, T., and Homma, M. (2003). The conserved charged residues of the C-terminal region of FliG, a rotor component of the Na^+-driven flagellar motor. *J Mol Biol* 334, 567–583.

Yorimitsu, T., Sowa, Y., Ishijima, A., Yakushi, T., and Homma, M. (2002). The systematic substitutions around the conserved charged residues of the cytoplasmic loop of Na^+-driven flagellar motor component PomA. *J Mol Biol* 320, 403–413.

Young, H. S., Dang, H., Lai, Y., DeRosier, D. J., and Khan, S. (2003). Variable symmetry in *Salmonella typhimurium* flagellar motors. *Biophys J* 84, 571–577.

Zhou, J. and Blair, D. F. (1997). Residues of the cytoplasmic domain of MotA essential for torque generation in the bacterial flagellar motor. *J Mol Biol* 273, 428–439.

Zhou, J., Lloyd, S. A., and Blair, D. F. (1998). Electrostatic interactions between rotor and stator in the bacterial flagellar motor. *Proc Natl Acad Sci U S A* 95, 6436–6441.

CHAPTER 5

Single Molecule Studies of Chromatin Structure and Dynamics

Sanford H. Leuba

Department of Cell Biology and Physiology, University of Pittsburgh School of Medicine,
Petersen Institute of NanoScience and Engineering, Department of Bioengineering, Swanson School of Engineering,
2.26g Hillman Cancer Center, University of Pittsburgh Cancer Institute, Pittsburgh, PA 15213-1863, USA

Laurence R. Brewer

School of Chemical Engineering and Bioengineering, Center for Reproductive Biology,
Washington State University, Pullman, WA 99164-2710, USA

Summary

In the eukaryotic cell nucleus, DNA is packaged into chromatin. In most cells, the DNA is wrapped around core histones to form nucleosomes. In the sperm cell, however, protamines replace the histones. Understanding how histones in somatic cells package chromatin and regulate access for processes such as transcription, replication, recombination, and repair to the underlying DNA template is of current interest. Determining how protamine repackages chromatin in the sperm cell is important for understanding how the genome is inactivated and protected prior to fertilization of the egg. Single molecule approaches in which one molecule is studied at a time are providing new information about chromatin. These assays are able to ascertain new data about structure and dynamics that is not possible to discern from the average milieu of bulk measurements. This chapter provides an overview of the field including the authors' own research.

Key Words

chromatin; nucleosome; histone; protamine; toroid; optical tweezers; single-pair fluorescence resonance energy transfer

Introduction

In this chapter, we discuss two distinct subunits of eukaryotic chromatin: the nucleosome in somatic cells and the toroid in the sperm cell. Although they are of similar dimensions, approximately tens of nanometers, they function in dramatically different ways. The nucleosome contains 147 base pairs (bp) of DNA that is wrapped 1.65 turns around a histone octamer. Understanding how these proteins regulate access to the DNA they are in contact with for fundamental cellular processes is just beginning to be understood. Toroids, on the other hand, contain 50 kilobase pairs (kb) of DNA that has been packaged by sperm nuclear proteins into such a dense form (the DNA loops that form the toroid are separated by approximately 2.4 nm, forming a hexagonal lattice) that access to the DNA by an enzyme such as RNA polymerase is impossible. The genome is "asleep" and inactive until it is "reawakened" following fertilization. In the following pages, we describe what has been recently learned about these two remarkable subunits of eukaryotic chromatin using experimental techniques that allow insight into mechanisms that govern their function at the single molecule level.

Sperm Chromatin

Sperm chromatin in mammals is densely packaged, approaching that of DNA in viruses, to protect it from damaging nucleases prior to fertilizing the egg. In 1953 Wilkins and Randall obtained the first X-ray scattering measurements from sperm heads showing that the DNA within was crystalline in form. Although these measurements established the molecular structure of DNA, for which Wilkins, Crick, and Watson won the Nobel Prize for Medicine, we are only just beginning to learn how sperm chromatin is packaged into this remarkably compact form. Using knockout mice and optical microscopy, reproductive biologists have determined that the DNA is packaged by sperm nuclear proteins, as part of a process known as spermiogenesis (Gagnon, 1999) (Figure 5.1), to protect it from damaging biological agents and to cause the cessation of transcription, inactivating the genome. Histones are displaced from nucleosomes, and the DNA is organized by proteins called protamines into small donut-shaped "toroids" (Hud et al., 1993) (Figure 5.2) coincident with a 40-fold reduction in the cell's volume. Toroids are the primary subunit of sperm chromatin (Hud et al., 1993) and contain approximately 50 kb of DNA. In human sperm, approximately 85% of the genome is repackaged as toroids and 15% remains in the form of nucleosomes.

To understand just how tightly compressed the sperm genome is, we can calculate its fractional volume in the sperm nucleus. The sperm genome can be modeled as a cylinder

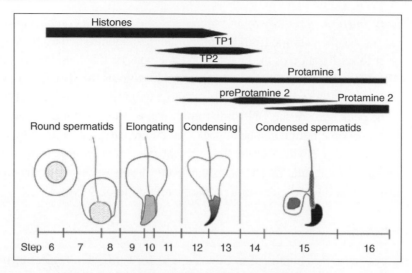

Figure 5.1: Steps of spermiogenesis in the mouse. Both the changes in shape of the cell and chromatin density are shown as well as the order of appearance of the sperm nuclear proteins (Zhao et al., 2004b)

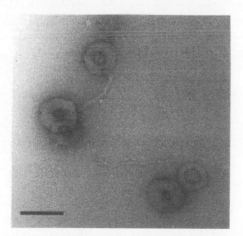

Figure 5.2: DNA–protamine toroids imaged using electron microscopy. The scale bar is 100 nm (Hud et al., 1993)

that is 1 nm in radius, with length equal to 1.5 Gbp times the distance between base pairs (0.34 nm/bp) giving a volume of $1.6 \mu m^3$. The volume of the hydrated mouse sperm nucleus is $12 \mu m^3$ (Allen et al., 1996). The fractional volume taken up by the sperm genome in the nucleus is then 13%, comparable to the fractional volume of DNA in viruses (Purohit et al.,

2005). *What is the purpose of packaging the sperm genome in such a compact state?* The first reason is to protect the genome from harmful biological agents such as nucleases and free radicals prior to fertilization of the egg. Second, the volume taken up by the genome is reduced to allow the sperm head to be compressed. This allows the sperm to assume a hydrodynamic shape and swim as quickly as possible to maximize its chances of fertilizing the egg. Third, the packaging of the DNA into toroids causes the cessation of transcription and inactivates the genome until it can be unpackaged and reactivated following fertilization. In the following sections, we will describe the different steps of spermiogenesis, what has been learned about the molecular mechanisms that occur during each step using single molecule experiments and what interesting problems remain to be solved.

Spermiogenesis

Spermiogenesis (Figure 5.1) occurs after meiosis, and for Eutherian (placental) mammals, it is initiated by the synthesis and binding of the transition proteins (TP1 and TP2) to DNA, coincident with the first phase of sperm chromatin condensation and the displacement of histones from nucleosomes (Oko et al., 1996). The transition proteins are then replaced by the protamines, P1 and P2, which are highly basic arginine-rich proteins containing 21 positive charges that bind to DNA, completely neutralizing the negatively charged phosphodiester backbone and forming DNA–protamine toroids (Figure 5.2). The toroid is a donut-shaped structure containing 50 kb of DNA in an extremely compact form that is 50–100 nm in diameter (Hud et al., 1993). The position of chromosomes and the higher-order structure of toroids formed from these chromosomes within the sperm nucleus are unknown.

Previous Studies of Toroid Structure

Hud and Balhorn (1993) were the first to identify the toroid as the fundamental unit of sperm chromatin in mammals using both atomic force microscopy and electron microscopy. Hud and Downing (2001) were able to observe the structure of DNA within toroids formed using λ-phage DNA and cobalt-hexamine (3+) and imaged using cryoelectron microscopy. This technique, while providing exquisite detail of DNA in the interior of these toroids, is not amenable to understanding the kinetics of toroid formation.

Single Molecule Experiments

We have studied the formation of DNA–protamine toroids in real time using single DNA molecules to understand how the sperm genome is repackaged during spermiogenesis. Our method avoids the problems of aggregation and precipitation of DNA–protein complexes that are inherent to bulk biochemistry approaches. An optical trap (laser-tweezers) was used to move a single DNA molecule from one channel of a microfluidic flow cell into another (Figure 5.3) containing a specific sperm nuclear protein. The DNA molecule

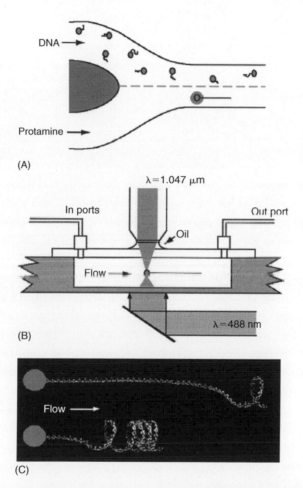

Figure 5.3: Experimental diagram (A) shows how a single DNA molecule is pulled into protamine solution using an optical trap. The DNA is stained with the dye YOYO-1 at low concentrations so that binding of the sperm nuclear proteins is not affected and illuminated with light from an argon-ion laser ($\lambda = 488$ nm) (B). Fluorescence from the YOYO-1 allowed the DNA molecule to be imaged using an image-intensified CCD camera. An image of how DNA toroids are thought to form is shown in (C) (Brewer et al., 1999)

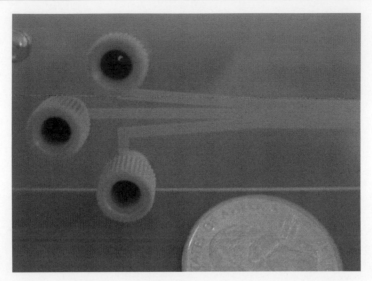

Figure 5.4: A three-channel flow cell was fabricated using a glass microscope cover slide, a glass microscope slide into which channels have been formed, and glass frit (manufactured by MMR Technologies). The three parts form a sandwich and were bonded together in an oven at high temperature (Brewer and Bianco, 2008)

was stained using low concentrations of the intercalating dye, YOYO-1. Fluorescence from the dye was stimulated using an argon-ion laser (wavelength, $\lambda = 488\,\text{nm}$) and the structural changes to the DNA molecule were imaged using an image-intensified CCD camera. Using this technique, we were able to measure the rates of binding and dissociation of sperm nuclear proteins and directly measure the equilibrium constant for a specific protein–DNA complex without needing to fit the data to a theoretical model. Figure 5.4 shows a three-channel flow cell. Glass frit was used to bond a glass cover slip to a microscope slide at high temperature. The microscope slide had channels etched into it using a "dry-etch" technique. Use of these fabrication methods creates a very strong and mechanically stable flow cell that is impervious to solvents such as methanol, used for releasing bubbles from the flow cell, and denaturants such as guanidinium hydrochloride, used for cleaning DNA and proteins from the cell after each use.

The Protamines P1 and P2

The primary function of the protamines is to reduce the volume of the sperm genome (Dooher and Bennett, 1973) by compacting it into DNA–protamine toroids (Figures 5.1 and 5.2) (Gagnon, 1999). P1 is present in all vertebrates and P2 is additionally present

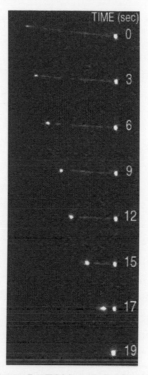

Figure 5.5: Real-time formation of a DNA–protamine toroid starting at the left end of a 50 μm concatemer of λ-phage DNA. The DNA molecule was extended by flowing buffer and attached to a 1 μm bead (right end) held by an optical trap. The protamine concentration is 1.1 μM (Brewer et al., 1999)

in some mammalian species, including human and mouse (Oliva, 2006). Both P1 and P2 are arginine-rich, highly basic proteins (Brewer et al., 2002) that bind to DNA in a nonsequence dependent manner. P2 is synthesized as a precursor (preprotamine 2, Figure 5.1) and processed after it binds to DNA to form P2 (Evenson et al., 2002). The P1/P2 ratio is used extensively as a metric for human male fertility. Increased P1/P2 ratios are strongly correlated with male infertility and low levels of protamine 2 (Cho et al., 2003), abnormal sperm morphology including round headed sperm, and high levels of DNA fragmentation (Oliva, 2006; Torregrosa et al., 2006).

We were the first group to study the condensation of single DNA molecules by salmon protamine (salmine) (Brewer et al., 1999) as well as P1 and P2 in the bull and hamster (Brewer et al., 2002, 2003). We were able to measure that the off-rate of protamine from

DNA was less than one molecule per second, indicating that the genome would take approximately 6 years to completely decondense in the absence of other mechanisms. However, once the sperm cell fertilizes the egg, the sperm genome must be unpackaged from its very dense state within 10 min or so. We concluded that protamine must be actively removed from DNA once sperm chromatin enters the egg's cytoplasm, in agreement with the work of others (Ruiz-Lara et al., 1996).

The Transition Proteins TP1 and TP2

A number of groups have studied the ability of the transition proteins TP1 and TP2 to condense DNA in vitro (Caron et al., 2001; Kundu and Rao, 1995, 1996; Levesque et al., 1998; Meetei et al., 2002; Sato et al., 1999). However, interpretation of their results was hampered by the formation of DNA–protein aggregates. We repeated these experiments on single DNA molecules, additionally studying the binding kinetics of Syrian hamster TP1 and TP2 (Brewer et al., 2002). In contrast to the measurements of Kundu and Rao (1995, 1996) and Meetei et al. (2002), we found that both TP1 and TP2 condensed DNA with rates similar to the protamines, P1 and P2. Measurements of the DNA condensation rate using a 25-residue peptide sequence corresponding to the C-terminal of TP2's binding domain were identical to the DNA condensation rate using the entire protein and allowed us to determine that this portion of the protein was responsible for condensing DNA.

Experiments performed by Zhao et al. (2001, 2004a,b) showed that knockout mice lacking either TP1 or TP2 remained fertile. They hypothesized that each protein could compensate if the other was absent and that there was a redundancy in the function of these proteins. Knockout mice lacking both proteins were sterile, and chromatin condensation was abnormal in all spermatids (Figure 5.1, step 8) and many late spermatids showed DNA strand breaks. The transition proteins are required for normal chromatin condensation in mammals in order to minimize the number of DNA strand breaks and maintain the integrity of the genome. Interestingly, histones were still displaced from nucleosomes during spermiogenesis for mice lacking both TP1 and TP2. It is still not understood how histones are displaced from nucleosomes during spermiogenesis.

Shaping of the Sperm Head and the Role of the Manchette

The mechanism for shaping the sperm head during spermiogenesis – reducing cell volume by a factor of 40 and altering its morphology – has long been a controversial subject. Studies of sperm cell development via microscopy by Kierszenbaum and others

(Cole et al., 1988; Kierszenbaum et al., 2003; Kierszenbaum and Tres, 2004; Meistrich et al., 1990; Russell et al., 1991) have hypothesized that the manchette, a sleeve-like collection of microtubules exterior to the sperm cell, is responsible for changing the cell's shape during spermiogenesis (Figure 5.1). However, Fawcett et al. (1971) argued that the sperm head shape is a direct result of the condensation and compaction of the sperm genome. They noted that if the manchette were responsible for the large variation in sperm head shapes, then there would be equally large variations in manchette microtubule arrangement between species, which had not been observed. They also argued that the circular shape of the manchette could not produce the high degree of flattening observed in most mammalian sperm, parallel to the long axis of the cell. Dooher and Bennett (1973) concluded that the shape of the sperm nucleus is due to a species-specific location, attachment and condensation of chromosomes within the nucleus. Single molecule experiments will be important for understanding if biomechanical forces play a role in both repackaging the sperm genome and shaping the sperm head during spermiogenesis.

Posttranscriptional Modifications of Sperm Nuclear Proteins

Phosphorylation–dephosphorylation cycles are thought to regulate the correct binding of sperm nuclear proteins to DNA as well as to initiate DNA condensation during spermiogenesis (D'Occhio et al., 2007). A number of different labs have reported that P1, P2, and TP2 are all strongly phosphorylated in vivo and in vitro (Green et al., 1994; Levesque et al., 1998; Meetei et al., 2002), however, it is not clear what function this serves. Meetei et al. (2002) have proposed that the phosphorylation of TP2 inhibits its ability to condense DNA, allowing it to bind optimally, while subsequent dephosphorylation initiates DNA condensation. Phosphorylation of the protamines occurs at serine and threonine residues and is thought to ensure the proper formation of DNA–protamine toroids (Lewis et al., 2003).

The internal structure of DNA–protamine toroids, although never directly observed, has been hypothesized to consist of a series of loops of DNA that are stacked on top of each other (Hud et al., 1995). Disulfide bonds form between cysteine residues in adjacent mammalian protamine molecules (Raukas and Mikelsaar, 1999) after toroid formation and may greatly increase the toroid's structural stability. It is unclear whether disulfide bonds form only between protamine molecules within the same loop of DNA, or if they can also form between protamine molecules on adjacent loops of DNA, firmly locking the toroid together.

Zinc is essential for male fertility (Abbasi et al., 1980; Dincer and Oz, 1990). Experiments conducted by our group showed that the binding rate of protamine 2 to single DNA molecules in the presence of zinc increased by threefold (Brewer et al., 2002) compared to the binding rate without zinc. These results were consistent with previous studies showing that when zinc binds to protamine 2, it may cause a structural change in the protein, such as a zinc finger, that helps the protein bind to DNA (Bianchi et al., 1992, 1994; Gatewood et al., 1990). Kundu and Rao (1996) and Meetei et al. (2000) have also shown that TP2 contains zinc finger motifs.

Conclusions: Sperm Chromatin

Understanding how sperm chromatin is repackaged and compacted during spermiogenesis is important for determining the organization of the sperm genome, an architecture that is crucial for protecting its genetic integrity and fertility. Chromatin compaction may also play a role in reducing the volume of the sperm nucleus and shaping the cell. It has been difficult to study how histones are displaced during spermiogenesis using mouse knockout models. Single molecule techniques should be particularly amenable to determining the mechanism of displacement and the role that posttranscriptional modifications of chromatin play in this process. Many fundamental problems remain to be studied to understand how the sperm genome is organized and the sperm cell is reformed during spermiogenesis. Single molecule techniques, which allow kinetic and structural information to be elucidated without the problems of intermolecular aggregation and precipitation, will be instrumental in solving them.

Somatic Chromatin

The nucleosome is the fundamental portion of protein–DNA packaging in the somatic eukaryotic cell nucleus (Tsanev et al., 1992; Turner, 2002; van Holde, 1988; Wolffe, 1998; Zlatanova and Leuba, 2004b). In a nucleosome, there are about two wraps of DNA around a histone octamer of the core histones H3, H4, H2B, and H2A. A fifth histone, linker histone H1, binds to the DNA between adjacent nucleosomes. Although we have had a general understanding of nucleosome structure for over three decades, questions remain how biological processes such as transcription, replication, recombination, and repair access the underlying DNA template in the chromatin milieu. Regulation of access to these biological processes is thought to lie in the complexities of (i) posttranslational modifications to the histones and the DNA, (ii) the histone variants, and (iii) external proteins that remodel chromatin (Zlatanova and Leuba, 2004b).

Recently, a number of single molecule approaches (Leuba and Zlatanova, 2001) have been developed to probe dynamic changes occurring in individual nucleosomes, fibers of nucleosomes, or individual chromosomes in solution (reviewed in Cairns, 2007;

Claudet and Bednar, 2006; Langowski and Schiessel, 2004; Leuba and Zlatanova, 2002; Leuba et al., 2004; Marko and Poirier, 2003; Poirier and Marko, 2002, 2003; Schiessel, 2006; van Holde and Zlatanova, 2006; Zlatanova, 2003; Zlatanova and Leuba, 2002, 2003, 2004a; Zlatanova and van Holde, 2006).

A great deal of research has investigated the occupancy of nucleosomes at specific loci on a DNA molecule or chromosome. Recently, researchers have systematically probed this occupancy in entire genomes (e.g., Ioshikhes et al., 2006; Lee et al., 2004; Segal et al., 2006; Yuan et al., 2005). However, the localization of a nucleosome at a specific site does not mean that it remains static. The nucleosome itself could open up. Perhaps, it is not surprising that while evolution has developed a packaging device to protect the DNA, it has also developed methods to open nucleosomes. For example, we understand that after replication, H3/H4 tetramers bind to the DNA followed by H2A/H2B dimers and linker histone (Worcel et al., 1978); transcription could be viewed as a reversal of that process (e.g., van Holde et al., 1992).

Originally, nucleosomes were considered solely as packaging devices to fit the 2 m of DNA within the confines of the cell nucleus, a compaction of $\sim 10\,000$ fold. The general working dogma was that they were immovable objects, i.e., like "rocks" that sequestered the DNA for packaging. With the discovery of nucleosome sliding (Beard, 1978; reviewed in van Holde, 1988), we understood that at least the DNA portion is mobile. The structure of the nucleosome (Luger et al., 1997) supported that notion because the B-factors, which indicate thermal fluctuations in the crystal structure, were significantly higher for the DNA portion than for the core histone fold region. The histone core portion was still like a "rock." Pioneering work of Prunell (Hamiche et al., 1996) demonstrated internal rearrangements of the histone fold domain core upon torsion applied by the DNA molecule.

Nucleosome fluctuations has lead to the observations of multiple states other than the canonical 147 bp nucleosomal core particle (Table 5.1). The motivation for the creation of this table was to describe some of the research that has led to the understanding of nucleosomes as mobile entities. The table is by no means exhaustive, but it does exhibit a wide display of work. Single molecule work is displayed alongside ensemble studies. We have attempted to describe the data in the table in terms of two states; however, not all of the papers can be easily described that way. One can see that the study of nucleosomes has expanded from the comparisons of nucleosomes with their naked DNA moieties at high ionic conditions to the study of internal reversible dynamics occurring in individual nucleosomes in real time.

A couple of recent examples of two or more states occurring in a single nucleosome are displayed in Figures 5.6 and 5.7. Using a nucleosome containing a donor and an acceptor

Table 5.1: Some examples of chromatin and of nucleosomes in multiple states

Experimental System	State 1	State 2	Comment	Reference(s)
Mononucleosomes isolated from mouse L929 cells were incubated under various salt conditions and temperature, sedimented via sucrose gradients, and fractions examined for sedimentation coefficient (in Svedbergs (S)).	Naked DNA (5S species)	Mononucleosomes (11S species)	Mononucleosomes after dilution to 2 μg/ml in 0.25 M NaCl, 10 mM Tris, 1 mM EDTA, and incubation at 37% for 2 h prior to loading into a sucrose gradient were found to consist of naked DNA (5S) and mononucleosomes (11S).	Cotton and Hamkalo (1981)
Sedimentation velocity measurements by analytical ultracentrifugation of 190±15 bp chicken erythrocyte mononucleosomes in buffer with various ionic concentrations.	Naked DNA (5.5S species)	Mononucleosomes	Treating mononucleosomes with 0.5 M NaCl leads to 80% naked mononucleosomal DNA. Nucleosomes in 0.5 M NaCl have up to 10% naked DNA. Dissociation is reversible.	Yager and van Holde (1984)
208 bp 5S rDNA mononucleosomal positioning DNA with additional ligated naked DNA to generate mononucleosomes with up to 414 bp in length were reconstituted and examined by nondenaturing two-dimensional gel electrophoresis.	Multiple histone octamer sliding events were observed.		Different applied temperatures affect the magnitude of these sliding events.	Meersseman et al. (1992)
Reconstitution of a single nucleosome on a 360 bp minicircle system.	In this torsionally constrained model mononucleosome system, nucleosomes have three states: (i) open without crossing over of the two linker DNA molecules (ii) negative crossover of the two arms (iii) a positive crossover.		The torsional constraints of this minicircle system affects the N-terminal tails of the two H2B and two H3 molecules, which align along the minor grooves of adjacent DNA gyres.	De Lucia et al. (1999) and Sivolob et al. (2003)

DNA templates of 150–220 bp with the *Xenopus borealis* 5S ribosomal RNA gene nucleosomal positioning sequence were end-labeled at opposite ends with a donor and an acceptor fluorescent dye pair. Reconstituted nucleosomes were analyzed by FRET.	No crossing over of the DNA arms is observed up to 220 bp mononucleosome particles.	Acetylation of all histones or of H3 specifically opens up the mononucleosome particle structure.	Addition of linker histone increases the FRET indicating a compaction of the DNA arms entering and exiting the mononucleosome. The addition of monovalent or divalent salts compact the mononucleosome.	Tóth et al. (2001, 2006)
λ-Phage DNA reconstituted with *Xenopus* egg extract and then disassembled using force measuring optical tweezers.	Naked DNA occurs after subjecting the chromatin fiber to 65 pN.	Individual disruptions have lengths of ~65 nm (191 bp).	15–40 pN is sufficient to disrupt individual nucleosomes (1 μm/s pulling rate). Stretch modulus of chromatin fiber determined to be 150 pN and the persistence length of a chromatin fiber, 30 nm.	Bennink et al. (2001)
17 repeats of 208 bp of the 5S sea urchin rRNA gene were reconstituted with core histones to form oligonucleosomes that were tethered between a surface and a microscopic bead and examined by optical tweezers.	Second transition is unwrapping of the final ~80 bp around the octamer. Keeping the exerted force to less than 60 pN allows some nucleosomes to refold.	Low force range (up to ~15 pN at 28 nm/s) removes the first wrap of DNA around the histone octamer.	Previous removal of all core histone tails reduces the nucleosome disruption force almost twofold. Incubation with p300, which acetylates core histones, reduces the nucleosome disruption force by 10%.	Brower-Toland et al. (2002, 2005)

(Continued)

Table 5.1: Continued

Experimental System	State 1	State 2	Comment	Reference(s)
147 bp DNA based on 601 nucleosome positioning template containing LexA binding site and end-labeled with donor dye. Acceptor dye at histone H3 V35C mutant. FRET should occur from the DNA end to the distal and the proximal acceptor dye, however, only the FRET from donor dye to the proximal acceptor should be sufficient for measuring.	Binding of LexA near the DNA end with the donor dye or increase in ionic concentration reduces FRET signal.	Canonical nucleosome has high FRET.	100 nM concentrations were used for fluorescence correlation spectroscopy (FCS); such concentrations ensured multiple molecules per measurement. FCS experiments in the absence of binding protein suggest that the ends (~30 bp) of nucleosomes spontaneously breathe once every 250 ms and then rewrap within ~10–50 ms.	Li and Widom (2004) and Li et al. (2005)
146 bp DNA molecule end labeled with dye near dyad. Either H4 T71C or H2B T112C labeled with other dye. Histone dye locations are far from dyad.	High salt: loss of FRET.	Low salt: canonical nucleosome.	At 0.55 M NaCl, H2A/H2B dimers dissociate from H3/H4 tetramers.	Park et al. (2004)
Theoretical study of changes occurring within a nucleosome due to external applied force.	The histone octamer is held by only one face of a DNA molecule.	The DNA is wrapped about 1.5 times (Zivanovic et al., 1988) around an octamer.	Nucleosome unwrapping due to external applied force involves an 180° rotation of the core histone octamer (Figure 5.8). Probably because of assistance of phosphate repulsion from adjacent DNA gyres, unwrapping of the first wrap requires less energy than the unwrapping the final wrap.	Kulic and Schiessel (2004) and Schiessel (2006)

154 bp DNA fragment with acceptor/donor at 47 and 122 bp sites. The two dyes are 75 bp (~25 nm) apart in the naked DNA template. Measured FRET suggests that they are about 4.4 nm apart in the assembled nucleosome.	Naked DNA FRET state; molecules at 0.13 E_{app}	High FRET state; molecules at 0.88 E_{app}	Reversible dynamic FRET was observed in individual time trajectories (Figure 5.6) indicating that a nucleosome can stochastically open and close. Both sites are as far away from the dyad as possible. Loss of FRET in the nucleosome requires detachment of almost a complete wrap of the DNA. Nucleosomes were tethered to a surface. Similar observations were made with cross-linked core histone octamers.	Tomschik et al. (2005)
2539 bp DNA fragment containing 10 5S nucleosomal positioning sequences reconstituted with chicken erythrocyte histones and stretched with an optical tweezers. Additionally, native chicken erythrocyte chromatin were examined.	Two disruption lengths are observed: the first at about 24 nm and the second at about 50 nm.		The authors note that single molecule chromatin fiber stretching experiments occurring at extremely low protein concentrations will tend to dissociate H2A/H2B dimers prior to removal of H3/H4 tetramers.	Claudet et al. (2005)
2582 bp DNA molecule containing the 601 nucleosome positioning sequence upon which a single nucleosome is reconstituted.	At ~3 pN, one wrap is removed from around the histone octamer.	At ~8–9 pN, the second wrap is removed.	In ranges of 2.1–2.9 pN applied force, nucleosomes fluctuate between two wraps and one wrap (Figure 5.7).	Mihardja et al. (2006)
Donor dye end-labeled (or 17 bp from end) 147 bp 601 nucleosome positioning sequence (Lowary and Widom, 1998) plus typically 20, 40, 60, or 78 bp flanking DNA on the other end. Acceptor dye at histone H2A 120C mutant.	Movement of the histone octamer away from the donor dye-labeled end reduces the overall FRET.	Unmoved histone octamer position gives high FRET in end-positioned mononucleosome.	Chromatin remodelers typically need at least 20–40 bp linker DNA to change the histone octamer position.	Yang et al. (2006)

(Continued)

Table 5.1: Continued

Experimental System	State 1	State 2	Comment	Reference(s)
λ-Phage DNA reconstituted with *Xenopus* egg extract and then disassembled using force measuring magnetic tweezers.	50 nm steps are observed, suggesting the removal of individual nucleosomes.		2 pN is sufficient to disassemble histones from the DNA. ATP hydrolysis assists with nucleosome rearrangement and disassembly.	Yan et al. (2007)
160 and 170 bp lengths of the 601 and 612 nucleosomal positioning sequences (Lowary and Widom, 1998).	601^{170} low FRET peak at 0.05 E_{app} 612^{160} low FRET peak at 0.05 E_{app}	601^{170} high FRET peak with range from 0.45 to 0.95 E_{app} 612^{160} high FRET peak at 0.05 E_{app} with range from 0.65 to 1.0 E_{app}	The distribution of the high-FRET signal of the 612^{160} nucleosome was much narrower than the distribution of the 601^{170} nucleosome in agreement with biochemical data that showed three histone octamer positions for the 601 sequence and only one position for the 612 sequence.	Gansen et al. (2007)
177 bp DNA of the 601 nucleosomal positioning (Lowary and Widom, 1998) with one dye at the pseudodyad (position 32) and the other dye 80 bp apart (position 32).	Low FRET state at 0.2 E_{app}	High FRET state at 0.87 E_{app}	3% intact nucleosomes display reversible FRET behavior in individual time trajectories. Nucleosomes were tethered to a surface.	Koopmans et al. (2007)

	Low-FRET state range from 0 to 0.2 E_{app}	High-FRET state range from 0.8 to 1.0 E_{app}		
160 bp DNA fragment with acceptor/donor dyes located at sites 40 and 120 bp. Nucleosome positioning sequences from sea urchin 5S rRNA gene, Gal10, and MMTVB were tested			The 5S nucleosome was the most stable of the three systems tested	Kelbauskas et al. (2007)
36 tandemly repeated 208 bp 5S nucleosome positioning sequences (~7500 bp) were reconstituted into chromatin fibers and torsionally manipulated using a magnetic tweezers	Nucleosomes have three states: (i) negative crossovers (ii) no crossovers of the linker DNA (iii) positive crossovers		As the path of DNA in nucleosomes follows a left-handed superhelix, the maximum extension of the chromatin fiber is observed with negative rotations. Disruption of individual nucleosomes observed with 7 pN. With sufficient positive torsional stress, the H2A/H2B dimers undock and nucleosomes undergo a chiral transition from a left-handed DNA superhelix to a right-handed superhelix	Bancaud et al. (2006, 2007)

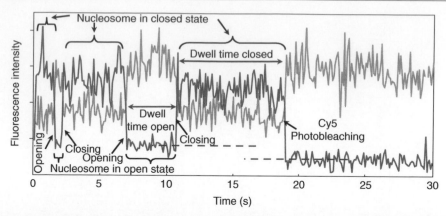

Figure 5.6: Two states of a nucleosome as observed by single-pair fluorescence resonance energy transfer (spFRET) time trajectory. Dark gray curve represents acceptor signal and light gray curve is donor dye signal. The anticorrelation of the two independent signals confirms that the two dyes are on the same DNA molecule. Reversible flipping between a high-FRET state and a low-FRET state occurs before acceptor dye photobleaching at ∼19 s. As the baseline of intensities of the photobleached state of the acceptor dye is lower than the low nucleosomal state intensities (see the two different dashed lines), the low nucleosomal state is not due to blinking of the acceptor dye. Adapted from Tomschik et al. (2005)

dye pair, we observed that an individual nucleosome can reversibly and dynamically fluctuate between the canonical closed state and a more open state where the dyes are too far apart for energy transfer (Figure 5.6) (Tomschik et al., 2005). Reversible fluctuations in a single nucleosome were later observed in optical tweezer experiments (Figure 5.7) (Mihardja et al., 2006). Using a long linear DNA molecule with a high affinity for binding one nucleosome at a single nucleosomal positioning sequence, researchers were able to reach forces where the nucleosome would "hop" back and forth between two states. At 2.1 pN applied force, the nucleosome would predominantly remain in the "canonical" state and only occasionally hop to a state with one unwrapping of the DNA gyres. At 2.9 pN, the nucleosome would predominantly remain with only one wrap of the DNA gyres and occasionally hop back into the fully wrapped canonical state.

An example of modeling of the unraveling of a single nucleosome due to applied force is displayed in Figure 5.8 (Kulic and Schiessel, 2004; Schiessel, 2006). One observes that the histone octamer undergoes a 180° rotation as the model nucleosome progresses from ~1.5 wraps of DNA around the octamer to a state where the octamer is only touching one surface of the DNA molecule. Since the eventual unwrapping of the nucleosome in

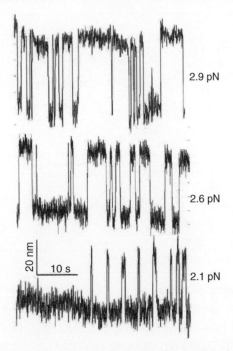

Figure 5.7: Two states of a nucleosome as observed using external force. For description, see text. Data from Mihardja et al. (2006)

Figure 5.8: Modeling of a nucleosome transitioning from a wrapped state (far left panel) to an unwrapped state with applied force (far right panel). For description, see text and Kulic and Schiessel (2004). From Schiessel (2006)

this model has been calculated due to the three-dimensional rotation of the octamer with respect to the axis of the DNA in the fiber, it would be interesting to know how this could occur in a fiber of nucleosomes and how adjacent nucleosomes may affect this process.

Figure 5.9 displays some of the various amino acid sites of the histone octamer within the nucleosome that have been converted into cysteines. Since cysteines can be chemically modified with the maleimide derivative of a fluorescent dye, these sites represent numerous locations for labeling. Because single-pair fluorescence resonance energy transfer (spFRET) is essentially a study of the "conversation" between two dyes and there

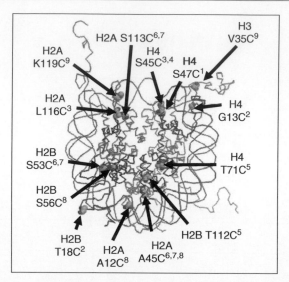

Figure 5.9: Locations of some of the amino acid residues in the nucleosome that have been converted to cysteines. Superscript numbers correspond to the following references: (1) Flaus et al. (1996), (2) Bruno et al. (2003), (3) Flaus et al. (2004), (4) Muthurajan et al. (2004), (5) Park et al. (2004), (6) Kassabov et al. (2003), (7) Kassabov and Bartholomew (2004), (8) Yang and Hayes (2004), and (9) Li and Widom (2004). Figure was prepared in PyMOL (DeLano, 2002) from pdb file 1kx5 (Davey et al., 2002)

are numerous locations to place dyes, the nucleosome presents a wealth of opportunities for future studies.

Conclusion

Contrary to questions about the sperm genome, investigations in somatic chromatin seek to uncover how the dynamics within and without nucleosomes leads to access of the underlying DNA template to biological processes. It is generally understood that this access is dynamic and a point of regulation. How it occurs can be uncovered by single molecule approaches. Many open questions remain. For example, we know that core histone H3 variant H3.3 undergoes replication-independent deposition into chromatin in living cells (Ahmad and Henikoff, 2002). How does that occur?

Presently, there is a boom in chromatin research. Previous ideas have become "rediscovered" and the field is expanding dramatically. There is a need to understand mechanistically what is occurring at the DNA and protein levels and which single molecule approaches can provide us these novel understandings. We hope that the reader has received a sense of current research using single molecule approaches in sperm and somatic chromatin.

Acknowledgments

We thank Dr J. van Noort for sharing experimental results prior to publication. We thank the many people who have contributed to the work we have described, including Drs Rod Balhorn, Miroslav Tomschik, Ken van Holde, Haocheng Zheng and Jordanka Zlatanova, and we appreciate a critical reading by Drs Syam P. Anand and Richard Steinman. This work was supported by NIH grant GM077872 (SHL) and startup funds from Washington State University and the Center for Reproductive Biology (LRB).

References

Abbasi, A. A., Prasad, A. S., Rabbani, P., and DuMouchelle, E. (1980). Experimental zinc deficiency in man. Effect on testicular function. *J Lab Clin Med* 96, 544–550.

Ahmad, K. and Henikoff, S. (2002). The histone variant H3.3 marks active chromatin by replication-independent nucleosome assembly. *Mol Cell* 9, 1191–1200.

Allen, M. J., Lee, J. D. T., Lee, C., and Balhorn, R. (1996). Extent of sperm chromatin hydration determined by atomic force microscopy. *Mol Reprod Dev* 45, 87–92.

Bancaud, A., Conde e Silva, N., Barbi, M., Wagner, G., Allemand, J. F., Mozziconacci, J., Lavelle, C., Croquette, V., Victor, J. M., Prunell, A., and Viovy, J. L. (2006). Structural plasticity of single chromatin fibers revealed by torsional manipulation. *Nat Struct Mol Biol* 13, 444–450.

Bancaud, A., Wagner, G., Conde, E. S. N., Lavelle, C., Wong, H., Mozziconacci, J., Barbi, M., Sivolob, A., Le Cam, E., Mouawad, L., Viovy, J. L., Victor, J. M., and Prunell, A. (2007). Nucleosome chiral transition under positive torsional stress in single chromatin fibers. *Mol Cell* 27, 135–147.

Beard, P. (1978). Mobility of histones on the chromosome of simian virus 40. *Cell* 15, 955–967.

Bennink, M. L., Leuba, S. H., Leno, G. H., Zlatanova, J., de Grooth, B. G., and Greve, J. (2001). Unfolding individual nucleosomes by stretching single chromatin fibers with optical tweezers. *Nat Struct Biol* 8, 606–610.

Bianchi, F., Rousseaux-Prevost, R., Bailly, C., and Rousseaux, J. (1994). Interaction of human P1 and P2 protamines with DNA. *Biochem Biophys Res Commun* 201, 1197–1204.

Bianchi, F., Rousseaux-Prevost, R., Sautiere, P., and Rousseaux, J. (1992). P2 protamines from human sperm are zinc-finger proteins with one CYS2/HIS2 motif. *Biochem Biophys Res Commun* 182, 540–547.

Brewer, L., Corzett, M., and Balhorn, R. (2002). Condensation of DNA by spermatid basic nuclear proteins. *J Biol Chem* 277, 38895–38900.

Brewer, L., Corzett, M., Lau, E. Y., and Balhorn, R. (2003). Dynamics of protamine 1 binding to single DNA molecules. *J Biol Chem* 278, 42403–42408.

Brewer, L. R., and Bianco, P. R. (2008). Laminar flow cells for single molecule studies. *Nat Methods* 5, 517–525.

Brewer, L. R., Corzett, M., and Balhorn, R. (1999). Protamine-induced condensation and decondensation of the same DNA molecule. *Science* 286, 120–123.

Brower-Toland, B., Wacker, D. A., Fulbright, R. M., Lis, J. T., Kraus, W. L., and Wang, M. D. (2005). Specific contributions of histone tails and their acetylation to the mechanical stability of nucleosomes. *J Mol Biol* 346, 135–146.

Brower-Toland, B. D., Smith, C. L., Yeh, R. C., Lis, J. T., Peterson, C. L., and Wang, M. D. (2002). Mechanical disruption of individual nucleosomes reveals a reversible multistage release of DNA. *Proc Natl Acad Sci USA* 99, 1960–1965.

Bruno, M., Flaus, A., Stockdale, C., Rencurel, C., Ferreira, H., and Owen-Hughes, T. (2003). Histone H2A/H2B dimer exchange by ATP-dependent chromatin remodeling activities. *Mol Cell* 12, 1599–1606.

Cairns, B. R. (2007). Chromatin remodeling: insights and intrigue from single molecule studies. *Nat Struct Mol Biol* 14, 989–996.

Caron, N., Veilleux, S., and Boissonneault, G. (2001). Stimulation of DNA repair by the spermatidal TP1 protein. *Mol Reprod Dev* 58, 437–443.

Cho, C., Jung-Ha, H., Willis, W. D., Goulding, E. H., Stein, P., Xu, Z., Schultz, R. M., Hecht, N. B., and Eddy, E. M. (2003). Protamine 2 deficiency leads to sperm DNA damage and embryo death in mice. *Biol Reprod* 69, 211–217.

Claudet, C., Angelov, D., Bouvet, P., Dimitrov, S., and Bednar, J. (2005). Histone octamer instability under single molecule experiment conditions. *J Biol Chem* 280, 19958–19965.

Claudet, C. and Bednar, J. (2006). Pulling the chromatin. *Eur Phys J E Soft Matter* 19, 331–337.

Cole, A., Meistrich, M. L., Cherry, L. M., and Trostle-Weige, P. K. (1988). Nuclear and manchette development in spermatids of normal and azh/azh mutant mice. *Biol Reprod* 38, 385–401.

Cotton, R. W. and Hamkalo, B. A. (1981). Nucleosome dissociation at physiological ionic strengths. *Nucleic Acids Res* 9, 445–457.

Davey, C. A., Sargent, D. F., Luger, K., Maeder, A. W., and Richmond, T. J. (2002). Solvent mediated interactions in the structure of the nucleosome core particle at 1.9 A resolution. *J Mol Biol* 319, 1097–113.

D'Occhio, M. J., Hengstberger, K. J., and Johnston, S. D. (2007). Biology of sperm chromatin structure and relationship to male fertility and embryonic survival. *Anim Reprod Sci* 101, 1–17.

DeLano, W. L. (2002). *The PyMOL Molecular Graphics System.* DeLano Scientific, San Carlos, CA.

De Lucia, F., Alilat, M., Sivolob, A., and Prunell, A. (1999). Nucleosome dynamics. III. Histone tail-dependent fluctuation of nucleosomes between open and closed DNA conformations. Implications for chromatin dynamics and the linking number paradox. A relaxation study of mononucleosomes on DNA minicircles. *J Mol Biol* 285, 1101–1119.

Dincer, S. L. and Oz, S. G. (1990). Zinc deficiency and male infertility. *Hosp Pract (Off Ed)* 25, 20.

Dooher, G. B. and Bennett, D. (1973). Fine structural observations on the development of the sperm head in the mouse. *Am J Anat* 136, 339–361.

Evenson, D. P., Larson, K. L., and Jost, L. K. (2002). Sperm chromatin structure assay: its clinical use for detecting sperm DNA fragmentation in male infertility and comparisons with other techniques. *J Androl* 23, 25–43.

Fawcett, D. W., Anderson, W. A., and Phillips, D. M. (1971). Morphogenetic factors influencing the shape of the sperm head. *Dev Biol* 26, 220–251.

Flaus, A., Luger, K., Tan, S., and Richmond, T. J. (1996). Mapping nucleosome position at single base-pair resolution by using site-directed hydroxyl radicals. *Proc Natl Acad Sci U S A* 93, 1370–1375.

Flaus, A., Rencurel, C., Ferreira, H., Wiechens, N., and Owen-Hughes, T. (2004). Sin mutations alter inherent nucleosome mobility. *EMBO J* 23, 343–353.

Gagnon, C. (1999). *The Male Gamete: From Basic Science to Clinical Applications.* Cache River Press, Vienna, IL.

Gansen, A., Hauger, F., Tóth, K., and Langowski, J. (2007). Single-pair fluorescence resonance energy transfer of nucleosomes in free diffusion: optimizing stability and resolution of subpopulations. *Anal Biochem* 368, 193–204.

Gatewood, J. M., Schroth, G. P., Schmid, C. W., and Bradbury, E. M. (1990). Zinc-induced secondary structure transitions in human sperm protamines. *J Biol Chem* 265, 20667–20672.

Green, G. R., Balhorn, R., Poccia, D. L., and Hecht, N. B. (1994). Synthesis and processing of mammalian protamines and transition proteins. *Mol Reprod Dev* 37, 255–263.

Hamiche, A., Carot, V., Alilat, M., De Lucia, F., O'Donohue, M. F., Revet, B., and Prunell, A. (1996). Interaction of the histone (H3–H4)$_2$ tetramer of the nucleosome with positively supercoiled DNA minicircles: potential flipping of the protein from a left- to a right-handed superhelical form. *Proc Natl Acad Sci USA* 93, 7588–7593.

Hud, N. V., Allen, M. J., Downing, K. H., Lee, J., and Balhorn, R. (1993). Identification of the elemental packing unit of DNA in mammalian sperm cells by atomic force microscopy. *Biochem Biophys Res Commun* 193, 1347–1354.

Hud, N. V. and Downing, K. H. (2001). Cryoelectron microscopy of lambda phage DNA condensates in vitreous ice: the fine structure of DNA toroids. *Proc Natl Acad Sci USA* 98, 14925–14930.

Hud, N. V., Downing, K. H., and Balhorn, R. (1995). A constant radius of curvature model for the organization of DNA in toroidal condensates. *Proc Natl Acad Sci USA* 92, 3581–3585.

Ioshikhes, I. P., Albert, I., Zanton, S. J., and Pugh, B. F. (2006). Nucleosome positions predicted through comparative genomics. *Nat Genet* 38, 1210–1215.

Kassabov, S. R. and Bartholomew, B. (2004). Site-directed histone–DNA contact mapping for analysis of nucleosome dynamics. *Methods Enzymol* 375, 193–210.

Kassabov, S. R., Zhang, B., Persinger, J., and Bartholomew, B. (2003). SWI/SNF unwraps, slides, and rewraps the nucleosome. *Mol Cell* 11, 391–403.

Kelbauskas, L., Chan, N., Bash, R., Debartolo, P., Sun, J., Woodbury, N., and Lohr, D. (2007). Sequence-dependent variations associated with H2A/H2B depletion of nucleosomes. *Biophys J* 94, 147–158.

Kierszenbaum, A. L., Rivkin, E., and Tres, L. L. (2003). Acroplaxome, an F-actin-keratin-containing plate, anchors the acrosome to the nucleus during shaping of the spermatid head. *Mol Biol Cell* 14, 4628–4640.

Kierszenbaum, A. L. and Tres, L. L. (2004). The acrosome–acroplaxome–manchette complex and the shaping of the spermatid head. *Arch Histol Cytol* 67, 271–284.

Koopmans, W. J., Brehm, A., Logie, C., Schmidt, T., and van Noort, J. (2007). Single-pair FRET microscopy reveals mononucleosome dynamics. *J Fluoresc* 17, 785–795.

Kulic, I. M. and Schiessel, H. (2004). DNA spools under tension. *Phys Rev Lett* 92, 228101.

Kundu, T. K. and Rao, M. R. (1995). DNA condensation by the rat spermatidal protein TP2 shows GC-rich sequence preference and is zinc dependent. *Biochemistry* 34, 5143–5150.

Kundu, T. K. and Rao, M. R. (1996). Zinc dependent recognition of a human CpG island sequence by the mammalian spermatidal protein TP2. *Biochemistry* 35, 15626–15632.

Langowski, J. and Schiessel, H. (2004). Theory and computational modeling of the 30 nm chromatin fiber. In: *Chromatin Structure and Dynamics: State-of-the-Art* (J. Zlatanova, and S. H. Leuba, Eds.), Vol. 39, pp. 397–420. Elsevier, Amsterdam.

Lee, C. K., Shibata, Y., Rao, B., Strahl, B. D., and Lieb, J. D. (2004). Evidence for nucleosome depletion at active regulatory regions genome-wide. *Nat Genet* 36, 900–905.

Leuba, S. H., Bennink, M. L., and Zlatanova, J. (2004). Single molecule analysis of chromatin. *Methods Enzymol* 376, 73–105.

Leuba, S. H. and Zlatanova, J. (Eds.) (2001). Biology at the Single molecule Level. Pergamon, Amsterdam.

Leuba, S. H. and Zlatanova, J. (2002). Single molecule studies of chromatin fibers: a personal report. *Arch Histol Cytol* 65, 391–403.

Levesque, D., Veilleux, S., Caron, N., and Boissonneault, G. (1998). Architectural DNA-binding properties of the spermatidal transition proteins 1 and 2. *Biochem Biophys Res Commun* 252, 602–609.

Lewis, J. D., Song, Y., de Jong, M. E., Bagha, S. M., and Ausió, J. (2003). A walk though vertebrate and invertebrate protamines. *Chromosoma* 111, 473–482.

Li, G., Levitus, M., Bustamante, C., and Widom, J. (2005). Rapid spontaneous accessibility of nucleosomal DNA. *Nat Struct Mol Biol* 12, 46–53.

Li, G. and Widom, J. (2004). Nucleosomes facilitate their own invasion. *Nat Struct Mol Biol* 11, 763–769.

Lowary, P. T. and Widom, J. (1998). New DNA sequence rules for high affinity binding to histone octamer and sequence-directed nucleosome positioning. *J Mol Biol* 276, 19–42.

Luger, K., Mäder, A. W., Richmond, R. K., Sargent, D. F., and Richmond, T. J. (1997). Crystal structure of the nucleosome core particle at 2.8 ℓÅ resolution. *Nature* 389, 251–260.

Marko, J. F. and Poirier, M. G. (2003). Micromechanics of chromatin and chromosomes. *Biochem Cell Biol* 81, 209–220.

Meersseman, G., Pennings, S., and Bradbury, E. M. (1992). Mobile nucleosomes – a general behavior. *EMBO J* 11, 2951–2959.

Meetei, A. R., Ullas, K. S., and Rao, M. R. (2000). Identification of two novel zinc finger modules and nuclear localization signal in rat spermatidal protein TP2 by site-directed mutagenesis. *J Biol Chem* 275, 38500–38507.

Meetei, A. R., Ullas, K. S., Vasupradha, V., and Rao, M. R. (2002). Involvement of protein kinase A in the phosphorylation of spermatidal protein TP2 and its effect on DNA condensation. *Biochemistry* 41, 185–195.

Meistrich, M. L., Trostle-Weige, P. K., and Russell, L. D. (1990). Abnormal manchette development in spermatids of azh/azh mutant mice. *Am J Anat* 188, 74–86.

Mihardja, S., Spakowitz, A. J., Zhang, Y., and Bustamante, C. (2006). Effect of force on mononucleosomal dynamics. *Proc Natl Acad Sci USA* 103, 15871–15876.

Muthurajan, U. M., Bao, Y., Forsberg, L. J., Edayathumangalam, R. S., Dyer, P. N., White, C. L., and Luger, K. (2004). Crystal structures of histone Sin mutant nucleosomes reveal altered protein–DNA interactions. *EMBO J* 23, 260–271.

Oko, R. J., Jando, V., Wagner, C. L., Kistler, W. S., and Hermo, L. S. (1996). Chromatin reorganization in rat spermatids during the disappearance of testis-specific histone, H1t, and the appearance of transition proteins TP1 and TP2. *Biol Reprod* 54, 1141–1157.

Oliva, R. (2006). Protamines and male infertility. *Hum Reprod Update* 12, 417–435.

Park, Y. J., Dyer, P. N., Tremethick, D. J., and Luger, K. (2004). A new fluorescence resonance energy transfer approach demonstrates that the histone variant H2AZ stabilizes the histone octamer within the nucleosome. *J Biol Chem* 279, 24274–24282.

Poirier, M. G. and Marko, J. F. (2002). Micromechanical studies of mitotic chromosomes. *J Muscle Res Cell Motil* 23, 409–431.

Poirier, M. G. and Marko, J. F. (2003). Micromechanical studies of mitotic chromosomes. *Curr Top Dev Biol* 55, 75–141.

Purohit, P. K., Inamdar, M. M., Grayson, P. D., Squires, T. M., Kondev, J., and Phillips, R. (2005). Forces during bacteriophage DNA packaging and ejection. *Biophys J* 88, 851–866.

Raukas, E. and Mikelsaar, R. H. (1999). Are there molecules of nucleoprotamine? *Bioessays* 21, 440–448.

Ruiz-Lara, S. A., Cornudella, L., and Rodriguez-Campos, A. (1996). Dissociation of protamine–DNA complexes by *Xenopus* nucleoplasmin and minichromosome assembly in vitro. *Eur J Biochem* 240, 186–194.

Russell, L. D., Russell, J. A., MacGregor, G. R., and Meistrich, M. L. (1991). Linkage of manchette microtubules to the nuclear envelope and observations of the role of the manchette in nuclear shaping during spermiogenesis in rodents. *Am J Anat* 192, 97–120.

Sato, H., Akama, K., Kojima, S., Miura, K., Sekine, A., and Nakano, M. (1999). Expression of a zinc-binding domain of boar spermatidal transition protein 2 in *Escherichia coli*. *Protein Expr Purif* 16, 454–462.

Schiessel, H. (2006). The nucleosome: a transparent, slippery, sticky and yet stable DNA–protein complex. *Eur Phys J E Soft Matter* 19, 251–262.

Segal, E., Fondufe-Mittendorf, Y., Chen, L., Thastrom, A., Field, Y., Moore, I. K., Wang, J. P., and Widom, J. (2006). A genomic code for nucleosome positioning. *Nature* 442, 772–778.

Sivolob, A., Lavelle, C., and Prunell, A. (2003). Sequence-dependent nucleosome structural and dynamic polymorphism. Potential involvement of histone H2B N-terminal tail proximal domain. *J Mol Biol* 326, 49–63.

Tomschik, M., Zheng, H., van Holde, K., Zlatanova, J., and Leuba, S. H. (2005). Fast, long-range, reversible conformational fluctuations in nucleosomes revealed by single-pair fluorescence resonance energy transfer. *Proc Natl Acad Sci USA* 102, 3278–3283.

Torregrosa, N., Dominguez-Fandos, D., Camejo, M. I., Shirley, C. R., Meistrich, M. L., Ballesca, J. L., and Oliva, R. (2006). Protamine 2 precursors, protamine 1/protamine 2 ratio, DNA integrity and other sperm parameters in infertile patients. *Hum Reprod* 21, 2084–2089.

Tóth, K., Brun, N., and Langowski, J. (2001). Trajectory of nucleosomal linker DNA studied by fluorescence resonance energy transfer. *Biochemistry* 40, 6921–6928.

Tóth, K., Brun, N., and Langowski, J. (2006). Chromatin compaction at the mononucleosome level. *Biochemistry* 45, 1591–1598.

Tsanev, R., Russev, G., Pashev, I., and Zlatanova, J. (1992). *Replication and Transcription of Chromatin*. CRC Press, Boca Raton.

Turner, B. M. (2002). *Chromatin and Gene Regulation: Mechanisms in Epigenetics*. Blackwell Science, Oxford.

van Holde, K. and Zlatanova, J. (2006). Scanning chromatin: a new paradigm?. *J Biol Chem* 281, 12197–12200.

van Holde, K. E. (1988). *Chromatin*. Springer-Verlag, New York.

van Holde, K. E., Lohr, D. E., and Robert, C. (1992). What happens to nucleosomes during transcription?. *J Biol Chem* 267, 2837–2840.

Wilkins, M. H. and Randall, J. T. (1953). Crystallinity in sperm heads: molecular structure of nucleoprotein in vivo. *Biochim Biophys Acta* 10, 192–193.

Wolffe, A. P. (1998). *Chromatin: Structure and Function*. Academic Press, New York.

Worcel, A., Han, S., and Wong, M. L. (1978). Assembly of newly replicated chromatin. *Cell* 15, 969–977.

Yager, T. D. and van Holde, K. E. (1984). Dynamics and equilibria of nucleosomes at elevated ionic strength. *J Biol Chem* 259, 4212–4222.

Yan, J., Maresca, T. J., Skoko, D., Adams, C. D., Xiao, B., Christensen, M. O., Heald, R., and Marko, J. F. (2007). Micromanipulation studies of chromatin fibers in *Xenopus* egg extracts reveal ATP-dependent chromatin assembly dynamics. *Mol Biol Cell* 18, 464–474.

Yang, J. G., Madrid, T. S., Sevastopoulos, E., and Narlikar, G. J. (2006). The chromatin-remodeling enzyme ACF is an ATP-dependent DNA length sensor that regulates nucleosome spacing. *Nat Struct Mol Biol* 13, 1078–1083.

Yang, Z. and Hayes, J. J. (2004). Large scale preparation of nucleosomes containing site-specifically chemically modified histones lacking the core histone tail domains. *Methods* 33, 25–32.

Yuan, G. C., Liu, Y. J., Dion, M. F., Slack, M. D., Wu, L. F., Altschuler, S. J., and Rando, O. J. (2005). Genome-scale identification of nucleosome positions in *S. cerevisiae*. *Science* 309, 626–630.

Zhao, M., Shirley, C. R., Hayashi, S., Marcon, L., Mohapatra, B., Suganuma, R., Behringer, R. R., Boissonneault, G., Yanagimachi, R., and Meistrich, M. L. (2004a). Transition nuclear proteins are required for normal chromatin condensation and functional sperm development. *Genesis* 38, 200–213.

Zhao, M., Shirley, C. R., Mounsey, S., and Meistrich, M. L. (2004b). Nucleoprotein transitions during spermiogenesis in mice with transition nuclear protein Tnp1 and Tnp2 mutations. *Biol Reprod* 71, 1016–1025.

Zhao, M., Shirley, C. R., Yu, Y. E., Mohapatra, B., Zhang, Y., Unni, E., Deng, J. M., Arango, N. A., Terry, N. H., Weil, M. M., Russell, L. D., Behringer, R. R., and

Meistrich, M. L. (2001). Targeted disruption of the transition protein 2 gene affects sperm chromatin structure and reduces fertility in mice. *Mol Cell Biol* 21, 7243–7255.

Zivanovic, Y., Goulet, I., Revet, B., Le Bret, M., and Prunell, A. (1988). Chromatin reconstitution on small DNA rings. II. DNA supercoiling on the nucleosome. *J Mol Biol* 200, 267–290.

Zlatanova, J. (2003). Forcing chromatin. *J Biol Chem* 278, 23213–23216.

Zlatanova, J. and Leuba, S. H. (2002). Stretching and imaging single DNA molecules and chromatin. *J Muscle Res Cell Motil* 23, 377–395.

Zlatanova, J. and Leuba, S. H. (2003). Chromatin fibers, one-at-a-time. *J Mol Biol* 331, 1–19.

Zlatanova, J. and Leuba, S. H. (2004a). Chromatin structure and dynamics: lessons from single molecule approaches. In: *Chromatin Structure and Dynamics: State-of-the-Art* (J. Zlatanova, and S. H. Leuba, Eds.), Vol. 39, pp. 369–396. Elsevier, Amsterdam.

Zlatanova, J. and Leuba, S. H. (Eds.) (2004b). Chromatin Structure and Dynamics: State-of-the-Art. Elsevier, Amsterdam.

Zlatanova, J. and van Holde, K. (2006). Single molecule biology: what is it and how does it work? *Mol Cell* 24, 317–329.

CHAPTER 6

Single Molecule Studies of Nucleic Acid Enzymes

Samir M. Hamdan

Department of Biological Chemistry and Molecular Pharmacology, Harvard Medical School, 240 Longwood Avenue, Boston, MA 02115, USA

Antoine M. van Oijen

Department of Biological Chemistry and Molecular Pharmacology, Harvard Medical School, 240 Longwood Avenue, Boston, MA 02115, USA

Summary

The replication, transcription, recombination, and repair of DNA are essential processes in the propagation, utilization, and upkeep of genetic information. Numerous different enzymes have evolved to catalyze the synthesis, digestion, unwinding, and unlinking of nucleic acids that are central to genomic maintenance. The development of single molecule techniques has allowed researchers to study the activity of nucleic acid enzymes at an unprecedented level of detail. This chapter will review the various experimental methods used and will discuss the type of information obtained by the study of nucleic acid enzymes at the single molecule level.

Key Words

Optical tweezers; magnetic tweezers; flow stretching; fluorescence resonance energy transfer; DNA elasticity; DNA polymerase; RNA polymerase; exonuclease; helicase; topoisomerase; replication.

Introduction

Understanding the complexity of biomolecules has been a goal for many decades. A wide variety of structural techniques, such as X-ray crystallography, NMR spectroscopy, and electron microscopy, combined with biochemical and molecular biological studies, have led to a quantitative understanding of the molecular properties and mechanisms underlying the processes of life. Yet, the static and ensemble-averaged view obtained with these techniques limits us in understanding the dynamic behavior of biomolecules and the nature of their interactions. Recent advances in imaging and molecular manipulation techniques have made it possible to observe individual molecules and record "molecular movies" that provide insight into their dynamics and reaction mechanisms. As is discussed elsewhere in this volume, these developments have led to a greater understanding of a large number of biological processes. A field of significant activity within the single molecule community is the study of processes that involve DNA as a substrate. The robustness of DNA, the often high processivity of nucleic acid enzymes, and the development of techniques that enable the manipulation of individual DNA molecules (reviewed in Bustamante et al., 2000a,b, 2003; Seidel and Dekker, 2007; van Oijen, 2007a,b) and the visualization of the proteins acting on them (reviewed in Ha, 2001a, 2004) have contributed to many exciting developments in the field of nucleic acid enzymology. This chapter provides an overview of the various techniques in use and will discuss a number of nucleic acid enzyme systems where single molecule approaches have proven to be particularly powerful in unraveling the molecular details of enzymatic activity.

Methods

Mechanical Manipulation of Individual DNA Molecules

The recent development of several methods to mechanically manipulate objects of microscopic scale has allowed researchers to stretch and twist individual DNA molecules. This ability not only allows for a detailed characterization of the mechanical properties of DNA and RNA itself (Bustamante et al., 2000b, 2003), but also makes it possible to study the activity of nucleic acid processing enzymes at the single molecule level.

Individual molecules, or particles attached to them, can be mechanically manipulated by introducing an external field that will exert a force on the molecule or particle. By attaching one end of the DNA to a solid surface and applying a laminar flow of aqueous buffer, a drag force can be exerted on DNA molecules themselves or on particles attached directly to the DNA or DNA-bound protein (Figure 6.1A) (Blainey et al., 2006; Kim et al., 2007; Lee et al., 2006; van Oijen et al., 2003). Combining the flow stretching of surface-anchored DNA molecules with wide-field microscopy allows for the simultaneous observation of many individual DNA molecules, thus significantly increasing data throughput.

Magnetic fields can be used to manipulate and apply forces on DNA or proteins that are tethered to magnetic particles, with the unique advantage that the presence of a preferred

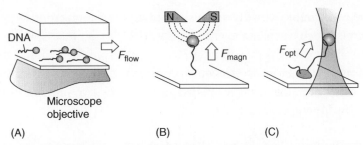

Figure 6.1: Schematic depiction of methods to mechanically stretch individual DNA molecules. (A) Flow stretching. The Stokes drag force exerted by a laminar flow on a bead at the free end of surface-attached DNA will cause the molecules to be stretched close to and parallel to the surface of a microscope coverslip. (B) Magnetic tweezers. The bead attached to the free end of the DNA is paramagnetic and can be manipulated by a set of magnets directly above the sample. The presence of a magnetic dipole moment in the bead allows torque to be introduced into the DNA. (C) Optical trapping. A tightly focused laser beam traps the bead attached to the free end of the DNA, allowing the DNA to be stretched. In the case illustrated here, the DNA is held to the surface by a surface-immobilized nucleic acid processing enzyme. Figure panel (A) reproduced from van Oijen, 2007b with permission. Panel (C) adapted from Bustamante et al. (2000) with permission

magnetization axis in magnetic beads allows them to be aligned with the direction of the applied magnetic field (Figure 6.1B) (Strick et al., 1996, 1998, 2000). This way, individual DNA molecules can be over- or underwound, thus allowing supercoils to be introduced.

Optical tweezers, or optical traps, exploit the fact that light exerts force on matter. Dielectric particles, such as polystyrene beads or bacteria, are attracted to the center of a tightly focused laser beam and can be trapped there (Figure 6.1C) (Ashkin, 1997; Ashkin and Dziedzic, 1987). A clear advantage of the optical trapping techniques is the large range in forces that can be applied (100 fN to 100 pN) and the subnanometer spatial resolution.

Fluorescence Detection of Individual DNA-Binding Proteins

Advances in fluorescence spectroscopy and microscopy have made it possible to detect the fluorescence from a single fluorophore under biological conditions (Moerner and Fromm, 2003). By labeling proteins with dye molecules at appropriate positions, their movement can be monitored by single molecule imaging (Toprak and Selvin, 2007). By making use of spectroscopic properties such as polarization and energy transfer, the protein's conformational dynamics can be followed at the microscopic level. Distance changes on a length scale comparable with the dimensions of biological macromolecules can be measured by fluorescence resonance energy transfer (FRET), a process where the excitation energy of a donor fluorophore is transferred to an acceptor dye via an induced dipole–dipole interaction. The strong distance dependence of this interaction allows FRET to be used as a molecular ruler to probe small changes in distance between two proteins or two sites on a protein (Ha, 2001b; Myong et al., 2006).

Single Molecule Studies of Nucleic Acid Enzymes

RNA Polymerases

RNA polymerase is the enzyme that reads out a DNA sequence and transcribes it into RNA. Found in all forms of life, it functions by locally melting the double-stranded DNA template and translocating along the DNA while polymerizing ribonucleotides in a template-directed fashion. The molecular mechanisms underlying transcription elongation have been studied for many decades, but numerous details remain unclear. The *Escherichia coli* RNA polymerase represents one of the first nucleic acid enzymes whose activity was studied at the single molecule level. Early experiments relied on the observation of the position of a small bead attached to the enzyme with the enzyme bound to a surface-tethered DNA substrate. From the amplitude of the Brownian motion of the bead, the length of the tether (i.e., the distance between the particle and the anchor point of the DNA on the surface) can be determined with an accuracy of 100 bp (Dohoney and Gelles, 2001; Schafer et al., 1991). An analysis of the particle motion allows for a direct observation of the position of the enzyme along the substrate while it synthesizes an RNA copy of the DNA template. A main advantage of this so-called tethered particle motion technique lies in its multiplexed nature. By combining appropriate immobilization densities with digital image processing, several dozens of individual tethers can be observed (Tolic-Norrelykke et al., 2004).

It was shown early on that individual RNA polymerase molecules transcribe at a constant rate over more than 1000 bp of template DNA. However, different molecules in the population appear to move at different rates (Schafer et al., 1991; Tolic-Norrelykke et al., 2004; Yin et al., 1994). The observation that a purified, seemingly homogeneous, population of enzyme molecules displayed drastically different turnover rates has been repeated for a number of different systems (Craig et al., 1996; Lu et al., 1998; Maier et al., 2000; Xue and Yeung, 1995; van Oijen et al., 2003) and will be discussed later in this chapter.

Early bulk-phase biochemical studies demonstrated that the RNA polymerase pauses frequently during the synthesis of an RNA transcript (Landick, 2006). This transcriptional pausing is believed to play an important role in the regulation of gene expression, and has been studied extensively by single molecule methods (Davenport et al., 2000; Forde et al., 2002; Herbert et al., 2006; Neuman et al., 2003). Using optical tweezers, transcriptional elongation traces of individual *E. coli* RNA polymerases could be obtained with high spatial resolution, allowing for a detailed study of the statistics of transcriptional pausing. These studies revealed a two-tiered pause mechanism in which particular sequences in the DNA promote entry into a paused state, followed by a secondary mechanism, such as RNA hairpin formation or backtracking, stabilizing the pause (Herbert et al., 2006).

Further technical improvements of optical trapping techniques allowed observation of movements of individual RNA polymerases along their template DNA with a resolution better than a single base pair (Abbondanzieri et al., 2005; Greenleaf and Block, 2006). These ultrasensitive methods were used to demonstrate that the unitary step size of RNA polymerase is a single base pair. An investigation of the elongation rate at different forces and ribonucleotide concentrations demonstrated that the translocation mechanism of RNA polymerases is mediated by a "Brownian ratchet" mechanism, where diffusion of the RNA polymerase between its pre- and posttranslocated states is directionally rectified by the binding of an incoming ribonucleotide (Abbondanzieri et al., 2005).

DNA Polymerases

DNA polymerases catalyze the synthesis of a new DNA strand on a single-stranded DNA template. Both prokaryotic and eukaryotic cells contain multiple types of DNA polymerases that are involved in replication, recombination, and repair. Among these are the replicative DNA polymerases, which faithfully copy long stretches of single-stranded DNA template by catalyzing the addition of the complementary nucleotides to a nucleic acid primer. Two metal ions in the active site bind and align the 3′-OH group of the primer and the incoming nucleotide to facilitate a nucleophilic reaction that results in a covalent addition of a nucleoside to the growing chain. Structural studies of DNA polymerases revealed an open and closed conformation of the active site corresponding to the absence and presence of a primer–template and an incoming nucleotide (Rothwell and Waksman, 2005). The two conformations are proposed to mediate the fidelity of nucleotide incorporation. In the closed conformation, the active site makes extensive contacts with both the growing primer–template strands and the incoming nucleotide to assure the fulfillment of the Watson–Crick base-pairing structure of double-stranded DNA prior to polymerization (Brautigam and Steitz, 1998; Johnson et al., 2003). The open conformation, on the other hand, is proposed to exchange the nucleotide in the active site prior to its incorporation if the nucleotide is incorrect and does not form sufficient hydrogen-bonding and base-stacking interactions. If a noncomplementary nucleotide is mistakenly incorporated into the growing chain, the polymerase undergoes a conformational change that transfers the primer–template strands to a spatially separated exonuclease active site for cleavage, a process commonly referred to as proofreading activity.

The ability to observe a single DNA polymerase moving along DNA while incorporating nucleotides allows for a more extensive study of the relation between enzyme–template interactions and kinetics. Single molecule techniques have been employed to study DNA synthesis by the replicative DNA polymerase of the bacteriophage T7 (Maier et al., 2000; Wuite et al., 2000). The T7 DNA polymerase is a 1:1 complex of the bacteriophage T7 gene 5 protein, containing polymerase and exonuclease activities, and the *E. coli* thioredoxin processivity factor (Tabor et al., 1987). The rate of a single T7 DNA polymerase synthesizing DNA on a single-strand DNA template was measured by utilizing the different elastic properties of single-stranded versus double-stranded DNA. The large difference in length between single-stranded and double-stranded DNA at

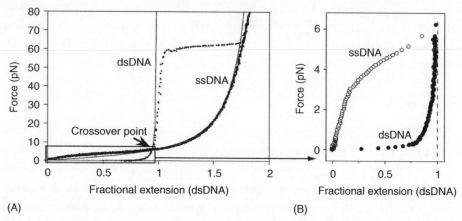

(A) (B)

Figure 6.2: Force-extension data of single- and double-stranded DNA (ssDNA and dsDNA). (A) At forces above the crossover point (~6 pN), single-stranded DNA is longer than double-stranded DNA. At ~65 pN, the double-stranded DNA undergoes a reversible and highly cooperative transition into a form that is considerably longer than B-form DNA. The solid line going through the ssDNA data indicates a fit of the force-extension curves with the worm-like chain (WLC) model, which describes the DNA polymer as a flexible rod with a well-defined persistence length (Bustamante et al., 2003, 2000b). Base pairing and electrostatic self-avoiding effects make the WLC model break down for single-stranded DNA at lower forces (Dessinges et al., 2002). (B) Below the crossover point, single-stranded DNA is shorter than the duplex DNA. Enzymatically catalyzed conversions between the two states can be visualized by changes in DNA length. Small differences between the two data sets in (A) and (B) can be attributed to different buffer conditions. Adapted from Wuite et al., 2000 and Lee et al., 2006 with permission

certain stretching forces (Figure 6.2) can be used to measure the activity of enzymes that convert one form of the DNA into the other (van Oijen, 2007a). The conversion of single-stranded DNA into double-stranded DNA by DNA polymerase activity can be detected as an increase in total DNA length, at stretching forces below ~6 pN, or as a decrease when the force is higher than this "crossover" force (Figure 6.2). The catalytic rates of the single enzymes showed large heterogeneity within the population of molecules. Furthermore, a single enzyme was shown to display fluctuations in its enzymatic rate over time (Maier et al., 2000). These time-dependent rate fluctuations were interpreted as small conformational changes in the protein modulating the geometry of the active site.

By means of optical (Wuite et al., 2000) or magnetic tweezers (Maier et al., 2000), the force can be changed in a controllable way and its effect on the enzyme's polymerization kinetics investigated. For a number of different DNA polymerases, these studies demonstrated that forces higher than 30–40 pN stall the enzyme (Maier et al., 2000; Wuite et al., 2000). Single molecule studies on the T7 DNA polymerase showed that even higher forces stimulate the exonuclease activity by several orders of magnitude (Wuite et al., 2000). The tension dependence of the polymerization rate was explained by a change in the conformation of the enzyme during the rate-limiting step. This conformational change may be related to a transient interaction of the enzyme with multiple bases of the single-stranded DNA (Goel et al., 2001, 2001; Maier et al., 2000; Wuite et al., 2000).

Local information on conformational changes occurring during polymerization can be obtained by the use of single molecule fluorescence. A recent study made use of the variations in fluorescent intensity of a Cy3-labeled primer–template caused by local changes in the environment (Luo et al., 2007). This assay provided a direct measure for conformational changes during polymerization. A rate-limiting conformational change within the active site was observed and found to be crucial during polymerization. The two conformers as observed in the single molecule experiments were attributed to the closed and open conformation of the active site as shown in crystal structures (Brautigam and Steitz, 1998; Johnson et al., 2003). The single molecule study revealed that the rate of this conformational change is much faster than the polymerization rate. When a noncomplementary nucleotide is inserted in the active site, the rate of this conformational transition drastically reduces, suggesting a key role in the kinetic proofreading mechanism described above.

Topoisomerases

Topoisomerases act to regulate the state of DNA supercoiling by cutting and rejoining one or both strands of the duplex DNA. Of this enzyme class, topoisomerase type I enzymes cleave only one strand and resolve coils by swiveling the DNA around the remaining single phosphodiester backbone bond, while type II enzymes cut both strands to relax supercoiling. The cutting, swiveling, and rejoining of the DNA by topoisomerases is likely to involve several biochemical steps that are associated with conformational changes. For decades, the different steps of topoisomerases activity were studied by analysis of DNA products that have been trapped at intermediate states. The static nature of the information that can be extracted from these experiments makes it challenging to study the dynamic properties of the biological processes that modify supercoiling. Advances in single molecule techniques made it possible to investigate these processes in real time. Using magnetic tweezers, a paramagnetic bead attached to the DNA can be rotated, hence introducing supercoiling (Figures 6.1B and 6.3). The ability of topoisomerases to add or remove supercoils in the DNA can be observed when

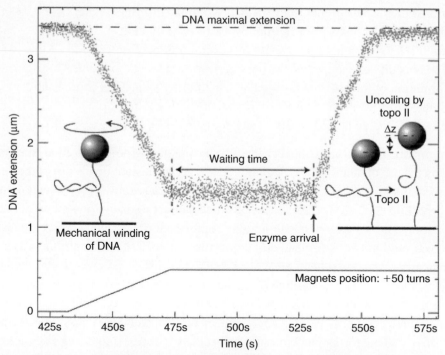

Figure 6.3: Detection of topoisomerase activity at the single molecule level. An individual DNA molecule tethered between a solid surface and a paramagnetic bead is stretched by a magnetic force in the upward direction. Rotation of the magnet in the horizontal plane results in rotation of the bead, which in turn causes the creation of supercoils in the DNA and a shortening of the DNA. Removal of supercoils by topoisomerase activity results in a detectable lengthening of the DNA. Reproduced from Charvin et al., 2005b with permission

the DNA is supercoiled to the point where it buckles and forms plectonemes, structures in which DNA wraps around itself to relieve supercoiling (Figure 6.3). The addition or removal of a supercoil then results in observable length changes of the DNA (Figure 6.3). Strick et al. (2000) applied this method to observe individual turnovers of type II topoisomerases, which were shown to relax the torsion in supercoiled DNA by the relaxation of two turns per enzymatic cycle. By applying different external forces and measuring the kinetics of DNA relaxation, it was possible to demonstrate that the re-ligation of the cut DNA represents the rate-limiting step in topoisomerase II activity. The direct measurement of the binding kinetics of topoisomerases to supercoiled DNA in the absence of the catalytic cofactor ATP allowed for a characterization of the ability of the topoisomerase to locate and bind to plectonemes. Topoisomerase II was shown to form crossover intermediates with DNA in a mechanism that involves clamping of the DNA. Although bulk assays demonstrated the existence of a long-lived intermediate between topoisomerases and plectonemes in the absence of ATP, analysis of the kinetics of clamp release using single molecule methods revealed a distinct fast and a slow step underlying this process. These observations suggested the existence of at least two different configurations of DNA–protein complexes in the absence of ATP. The unlinking of two braided DNA duplexes by prokaryotic and eukaryotic type II topoisomerases was studied by single molecule techniques, allowing a characterization of the specificity of the enzymes for certain topologies of supercoils (Charvin et al., 2003). Magnetic trapping techniques were also employed to study topoisomerase I (Koster et al., 2005) and its interaction with the antitumor drug toptecan (Koster et al., 2007), as well as the uncoiling mechanism of the *E. coli* enzymes topoisomerase IV (Charvin et al., 2005a; Crisona et al., 2000) and gyrase (Gore et al., 2006; Nollmann et al., 2007). The power of FRET techniques to report on local conformation enabled recent studies of topoisomerase II during cleavage of DNA (Smiley et al., 2007). Using a DNA substrate containing a donor and acceptor dye at the two ends of the topoisomerase cleavage sequence, two states of DNA binding (open and closed) were demonstrated and their dynamic transitions characterized.

Exonucleases

Several classes of nucleases involved in repair and recombination processively degrade duplex DNA in a unidirectional fashion to convert it into single-stranded DNA. Differences in elasticity between single-stranded and double-stranded DNA (Figure 6.2) were utilized to study the processive degradation of one strand of a duplex DNA by an individual λ exonuclease, an enzyme required for recombination in bacteriophage λ. Statistical analyses of the single molecule trajectories revealed that the catalytic rate is dependent on the local base content of the substrate DNA. By relating single molecule kinetics to the free energies of hydrogen bonding and base stacking, the authors established that the disruption

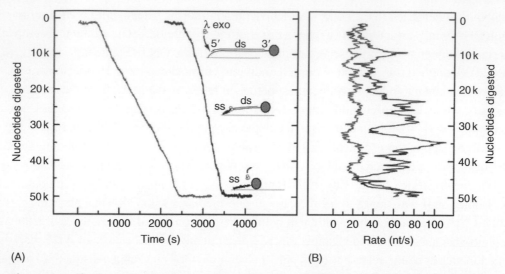

Figure 6.4: Single molecule observation of λ exonuclease-catalyzed digestion of DNA.
(A) Two single molecule traces displaying complete enzymatic digestion of one strand of the
48.5 kb long duplex substrate. Enzymatic conversion of double-stranded (ds) into single-
stranded (ss) DNA was observed as a shortening of the DNA (inset). (B) Time derivatives of
the two traces in (A) as a function of enzyme position on the substrate DNA. Adapted from
van Oijen et al. (2003) with permission

of hydrogen bonding and base stacking between the base that is cleaved and the DNA
substrate is the rate-limiting step in the catalytic cycle (Figure 6.4) (van Oijen et al., 2003).
The catalytic rate also exhibited large fluctuations independent of the sequence, which
were attributed to conformational changes of the enzyme–DNA complex. In another single
molecule study that monitored the catalytic activity of surface-immobilized λ exonuclease,
sequence-dependent pausing of the enzyme was observed (Perkins et al., 2003). A
combination of single molecule trajectories and ensemble gel electrophoresis data was used
to identify the exact sequence of one of the strong pause sites.

The observation of heterogeneity in turnover rates, as observed in single molecule
studies of λ exonuclease, T7 DNA polymerase, and *E. coli* RNA polymerase, point to
a deeper question whether this is a general phenomenon and all enzymes display this
substantial variation of activity within a population of seemingly identical molecules.
This phenomenon is not limited to nucleic acid enzymes; similar behavior was observed
for cholesterol oxidase (Lu et al., 1998), lactate dehydrogenase (Xue and Yeung, 1995),

and alkaline phosphatase (Craig et al., 1996). Such variability in catalytic rates might be rationalized by means of a rugged energy landscape, resulting in multiple possible macromolecular conformational substates (Frauenfelder et al., 1991). However, other enzymes display no significant molecule-to-molecule variation in their turnover rate. Kinesin (Schnitzer and Block, 1997) and *E. coli* topoisomerase II (Strick et al., 2000) are two examples of systems that have been studied extensively in single molecule experiments and display highly uniform catalytic rates across large sample sizes.

Helicases

DNA helicases are enzymes capable of unwinding duplex DNA to provide the single-stranded DNA templates that are required in processes such as replication and recombination. All helicases separate the strands of a double helix using the energy derived from nucleotide hydrolysis. Single molecule techniques are well suited to monitor rates and processivities of unwinding and study their dependence on the applied load.

Using the different elastic properties of single-stranded and double-stranded DNA (Figure 6.2), the unwinding activity of UvrD, a prokaryotic helicase involved in a number of repair pathways (Friedberg et al., 2006), has been studied at the single molecule level (Dessinges et al., 2004). The helicase was observed to switch strands during unwinding, leading to a reversed translocation of the protein away from the fork and a gradual reannealing of the two single-stranded DNA products. A Fourier analysis of the noise associated with the measurement of the extension of the DNA allowed the determination of the enzyme's step size.

Another illustrative example of a helicase that has been intensively investigated by various single molecule techniques is RecBCD, an enzyme with both helicase and nuclease activities that plays an important role in the repair of chromosomal DNA through homologous recombination (Friedberg et al., 2006). Unwinding and nucleolytic activity was studied by monitoring the displacement of intercalating fluorescent dye from a flow-stretched DNA substrate (Bianco et al., 2001). Both this experiment and a previous study (Dohoney and Gelles, 2001) pointed to a wide distribution of enzymatic rates. Encounter of the enzyme with the χ sequence, an eight-base sequence that regulates recombination in *E. coli* (Eggleston and West, 1997), was shown to change its nuclease activity (Bianco et al., 2001). Using optical tweezers to apply tension to a DNA template that was threaded through a surface-immobilized RecBCD enzyme, Perkins et al. (2004) observed continuous DNA unwinding with a precision of 2 nm, implying a unitary step size of the enzyme below 6 bp. Interestingly, they observed that the velocity of the enzyme remained constant for hundreds to thousands of base pairs, before suddenly switching to a different rate, pausing, or even occasionally sliding back along the DNA template.

Ha et al. (2002) pioneered the use of single molecule FRET in the study of the molecular mechanisms of helicase activity. Association of *E. coli* Rep helicase with a duplex DNA substrate and subsequent unwinding could be observed by a strategic labeling of the DNA with the donor molecule on one strand and the acceptor dye on the other (Figure 6.5) (Ha et al., 2002). Later experiments involved fluorescent labeling of the protein itself and led to the determination of the orientation of a Rep monomer bound to a single-stranded/double-stranded hybrid DNA template, as well as the relative orientation of one of its subdomains (Rasnik et al., 2004). Furthermore, the authors demonstrated that an interaction between the helicase and the 3′-terminus of one of the unwound DNA strands prevents the protein from dissociating from the DNA (Myong et al., 2005).

The monomeric hepatitis C nonstructural protein 3 helicase (NS3) is unusual in that it unwinds both DNA and RNA substrates. Using optical tweezers to simultaneously trap two beads that are attached to the ends of the DNA/RNA strands that contain an RNA duplex

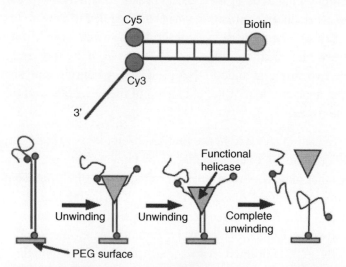

Figure 6.5: Schematic depiction of single molecule fluorescence resonance energy transfer (FRET) experiment to probe helicase-mediated unwinding of double-stranded DNA. The transfer of excitation energy from the donor fluorophore Cy3 to the acceptor fluorophore Cy5 is extremely sensitive to the physical distance between the two probes. Unwinding of the duplex will cause this distance to increase, allowing for detailed measurements of unwinding with subnanometer resolution. Adapted from van Ha et al. (2002) with permission

hairpin in the middle, unwinding activity at a spatial resolution of 2 bp could be obtained (Dumont et al., 2006). The study revealed that ATP coordinates the cyclic movement of NS3 in discrete steps of 11 bp. The unwinding itself, however, occurs in much more rapid and smaller steps of 3.6 bp. A model emerging from these studies describes the unwinding of RNA by the NS3 helicase in an inchworm-like fashion using two RNA-binding sites. NS3 is known to consist of three domains, two of which are RecA-like, in which the ATP-binding pockets play a role in coordinating their translocation and unwinding. In this model, one of the two domains is bound 11 bp away from the first domain, that is bound at the junction of the fork where unwinding takes place. In this case, the first domain at the fork will unwind the duplex RNA in steps of 2–5 bases until it physically contacts the third domain to start a new cycle of unwinding. Myong et al. (2007) used a single molecule FRET assay by attaching a donor and an acceptor dye at the junction of the DNA to map the steps involved in NS3 unwinding. Their study revealed the existence of three hidden steps within the 3 bp steps observed in DNA unwinding. During these three substeps, hydrolysis of one ATP molecule leads to building up of tension in the protein–DNA complex that is relieved via "burst" unwinding of 3 bp.

Many of the helicases that unwind DNA in the context of the replication machinery have not yet been studied at the single molecule level. Well-known examples of prokaryotic replicative helicases are the *E. coli* DnaB, the bacteriophage T7 gp4, and T4 gp41, all hexameric, donut-shaped proteins belonging to the DnaB superfamily. Extensive biochemical and structural characterization resulted in the picture that these proteins encircle a single strand of DNA and are able to translocate in the 5′ to 3′ direction (Patel and Picha, 2000). Upon encountering a single-stranded/double-stranded DNA junction, the complementary strand is displaced and the double-stranded DNA is unwound. It is not understood how the single-stranded DNA translocation activity is coupled to the unwinding, however. Two different models are typically considered: an active or passive coupling (Betterton and Julicher, 2005). A passive helicase acts as a Brownian ratchet: it waits for the thermal fluctuation that transiently melts the first few base pairs of the double-stranded DNA, and then moves forward and binds to the newly available single-stranded DNA. The active model describes a helicase that employs an irreversible power stroke to disrupt the double-stranded DNA. In this case, the hydrolysis of nucleotides is tightly coupled to the destabilization of the duplex, leading the helicase to unwind the double-stranded DNA without being significantly slowed down by the single-stranded/double-stranded DNA barrier.

Johnson et al. (2007) employed force to study the kinetics of T7 gp4 unwinding of double-stranded DNA, and translocation on single-stranded DNA, at the single molecule level. By tethering the 3′-terminus of one strand of the DNA to a surface and the

5′-terminus of the same strand to an optically trapped bead, forces can be applied that assist the unwinding of the double-stranded DNA. The rate of DNA translocation was measured by sensing the force change when the helicase, translocating on single-stranded DNA, approaches a replication fork, and the rate of unwinding was measured using feedback control of force to follow the movement of the fork. To unravel whether the helicase acts in a passive or active way, the rate of DNA unwinding was monitored at different unzipping forces. The relation between the unwinding rate and the unzipping force provided direct evidence that the gp4 helicases unwind DNA in active mode. Further single molecule studies on other hexameric, replicative helicases are needed to determine the generality of this observation.

The Replisome

An important direction in single molecule biophysical research is the study of larger, more complex multiprotein systems. A recent example of such studies is the single molecule observation of the activity of the multienzyme machinery that is responsible for DNA replication – the replisome. Lee et al. (2006) used the difference in elastic properties between single-stranded and double-stranded DNA to study how the T7 helicase and the T7 DNA polymerase cooperatively unwind and copy duplex DNA. In this experiment, a 48.5 kb long λ-phage duplex DNA is attached to a glass flow cell via the 5′-end of one strand whose 3′-end is linked to a bead (Figure 6.6). A constant laminar flow is applied such that the resultant drag on the bead stretches the DNA molecule. One end of the DNA contains a short stretch of noncomplementary DNA, forming a branched replication fork that allows the assembly of the helicases and DNA polymerases. Upon unwinding by the helicase and synthesis by the DNA polymerase, one single strand is converted to double-stranded DNA, but the strand that is attached to the glass surface remains in the single-stranded form. As a net result, parental double-stranded DNA was converted into single-stranded DNA, resulting in a gradual reduction of the length of the DNA between the bead and the surface. The rate of length reduction is related to the enzymatic rate at which the DNA is unwound and copied upon. Interestingly, the processivity of this strand-displacement synthesis is a factor of 20 higher than the processivity of the T7 DNA polymerase alone. This effect can be attributed to stabilizing protein–protein interactions between the helicase and the DNA polymerase, and to the absence of any processivity-reducing secondary structure in the single-stranded DNA template in the presence of a helicase. Recent work demonstrated that during strand-displacement synthesis, the polymerase dissociates transiently from the DNA but remains tethered to the helicase via a unique binding mode, allowing it to quickly locate the primer–template strand and continue DNA synthesis. Single molecule techniques

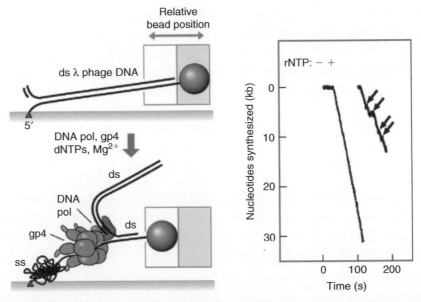

Figure 6.6: Single molecule observation of DNA replication. In the absence of lagging-strand synthesis, leading-strand synthesis causes the 5′ tail of the DNA to be converted to the single-stranded form. Attachment of the 5′-end to the surface allows the monitoring of this conversion as a change in total length of the DNA. Single molecule trajectories (right) show reduction of DNA length as leading-strand synthesis progresses. Above the traces is indicated whether ribonucleotides were present in the reaction mixture, without which no primase activity can occur. Activation of the primase activity results in a transient pausing of the replication fork (pauses denoted by arrows). Adapted from van Lee et al. (2006) with permission

suggested that during the transient dissociation of the polymerase, it rotates around the helicase to search for the primer–template strand (Hamdan et al., 2007), providing unique mechanistic insight into the processivity of DNA replication machineries.

The T7 gp4 is a bifunctional protein that is not only able to unwind the double-stranded DNA at the fork, but also capable of catalyzing the synthesis of the short RNA primers needed to initiate DNA polymerase activity on the lagging strand (Benkovic et al., 2001). Lee et al. (2006) studied the effect of the primase activity on leading-strand synthesis by introducing ribonucleotides to the single molecule replication reaction. In the presence of ribonucleotides, short pauses occurred in the leading-strand synthesis traces (Figure 6.6, trace 2; pauses are indicated by grey arrows). These pauses, with an average duration of several seconds, were demonstrated to result from primer synthesis by repeating these experiments with a gp4 defective in primase activity. The observation that leading-strand synthesis momentarily stops during primer synthesis explains how the slow enzymatic events on the lagging strand take place without leading-strand synthesis progressing too far ahead of the lagging-strand synthesis. The fleeting nature of the pauses in fork elongation and the difficulty of detecting such short-lived events in bulk-phase biochemical assays make this system a good example of how single molecule approaches can improve our understanding of how complex nucleic acid enzyme systems perform their tasks on DNA.

Conclusion

The ability to observe the activity of DNA-binding proteins on the single molecule level provides tremendous opportunities in the field of nucleic acid enzymology. The observation of single molecules allows the detection of transient intermediates and fluctuating enzymatic rates, all aspects that are hidden under the ensemble averaging in bulk experiments.

Whereas the development and use of mechanical manipulation and force in single molecule experiments has resulted in truly revolutionary advances in the field, the combination of the mechanical manipulation with the power of fluorescent probes to extract true microscopic information on the length scale of the protein motions relevant in the enzymatic mechanisms will likely lead to even more exciting developments and deeper understanding.

References

Abbondanzieri, E. A., Greenleaf, W. J., Shaevitz, J. W., Landick, R., and Block, S. M. (2005). Direct observation of base-pair stepping by RNA polymerase. *Nature* 438, 460–465.

Ashkin, A. (1997). Optical trapping and manipulation of neutral particles using lasers. *Proc Natl Acad Sci U S A* 94, 4853–4860.

Ashkin, A. and Dziedzic, J. M. (1987). Optical trapping and manipulation of viruses and bacteria. *Science* 235, 1517–1520.

Benkovic, S. J., Valentine, A. M., and Salinas, F. (2001). Replisome-mediated DNA replication. *Annu Rev Biochem* 70, 181–208.

Betterton, M. D. and Julicher, F. (2005). Opening of nucleic-acid double strands by helicases: active versus passive opening. *Phys Rev E Stat Nonlin Soft Matter Phys* 71, 011904.

Bianco, P. R., Brewer, L. R., Corzett, M., Balhorn, R., Yeh, Y., Kowalczykowski, S. C., and Baskin, R. J. (2001). Processive translocation and DNA unwinding by individual RecBCD enzyme molecules. *Nature* 409, 374–378.

Blainey, P. C., van Oijen, A. M., Banerjee, A., Verdine, G. L., and Xie, X. S. (2006). A base-excision DNA-repair protein finds intrahelical lesion bases by fast sliding in contact with DNA. *Proc Natl Acad Sci U S A* 103, 5752–5757.

Brautigam, C. A. and Steitz, T. A. (1998). Structural and functional insights provided by crystal structures of DNA polymerases and their substrate complexes. *Curr Opin Struct Biol* 8, 54–63.

Bustamante, C., Bryant, Z., and Smith, S. B. (2003). Ten years of tension: single molecule DNA mechanics. *Nature* 421, 423–427.

Bustamante, C., Macosko, J. C., and Wuite, G. J. L. (2000a). Grabbing the cat by the tail: manipulating molecules one by one. *Nat Rev Mol Cell Biol* 1, 130–136.

Bustamante, C., Smith, S. B., Liphardt, J., and Smith, D. (2000b). Single molecule studies of DNA mechanics. *Curr Opin Struct Biol* 10, 279–285.

Charvin, G., Bensimon, D., and Croquette, V. (2003). Single molecule study of DNA unlinking by eukaryotic and prokaryotic type-II topoisomerases. *Proc Natl Acad Sci U S A* 100, 9820–9825.

Charvin, G., Strick, T. R., Bensimon, D., and Croquette, V. (2005a). Topoisomerase IV bends and overtwists DNA upon binding. *Biophys J* 89, 384–392.

Charvin, G., Strick, T. R., Bensimon, D., and Croquette, V. (2005b). Tracking topoisomerase activity at the single molecule level. *Annu Rev Biophys Biomol Struct* 34, 201–219.

Craig, D. B., Arriaga, E. A., Wong, J. C. Y., Lu, H., and Dovichi, N. J. (1996). Studies on single alkaline phosphatase molecules: reaction rate and activation energy of a reaction catalyzed by a single molecule and the effect of thermal denaturation – the death of an enzyme. *J Am Chem Soc* 118, 5245–5253.

Crisona, N. J., Strick, T. R., Bensimon, D., Croquette, V., and Cozzarelli, N. R. (2000). Preferential relaxation of positively supercoiled DNA by *E. coli* topoisomerase IV in single molecule and ensemble measurements. *Genes Dev* 14, 2881–2892.

Davenport, R. J., Wuite, G. J., Landick, R., and Bustamante, C. (2000). Single molecule study of transcriptional pausing and arrest by *E. coli* RNA polymerase. *Science* 287, 2497–2500.

Dessinges, M. N., Lionnet, T., Xi, X. G., Bensimon, D., and Croquette, V. (2004). Single molecule assay reveals strand switching and enhanced processivity of UvrD. *Proc Natl Acad Sci U S A* 101, 6439–6444.

Dessinges, M. N., Maier, B., Zhang, Y., Peliti, M., Bensimon, D., and Croquette, V. (2002). Stretching single stranded DNA, a model polyelectrolyte. *Phys Rev Lett* 89, 248102.

Dohoney, K. M. and Gelles, J. (2001). Chi-sequence recognition and DNA translocation by single RecBCD helicase/nuclease molecules. *Nature* 409, 370–374.

Dumont, S., Cheng, W., Serebrov, V., Beran, R. K., Tinoco Jr., I., Pyle, A. M., and Bustamante, C. (2006). RNA translocation and unwinding mechanism of HCV NS3 helicase and its coordination by ATP. *Nature* 439, 105–108.

Eggleston, A. K. and West, S. C. (1997). Recombination initiation: easy as A, B, C, D … chi?. *Curr Biol* 7, R745–R749.

Forde, N. R., Izhaky, D., Woodcock, G. R., Wuite, G. J., and Bustamante, C. (2002). Using mechanical force to probe the mechanism of pausing and arrest during continuous elongation by *Escherichia coli* RNA polymerase. *Proc Natl Acad Sci U S A* 99, 11682–11687.

Frauenfelder, H., Sligar, S. G., and Wolynes, P. G. (1991). The energy landscapes and motions of proteins. *Science* 254, 1598–1603.

Friedberg, E. C., Walker, G. C., Siede, W., Wood, R. D., Schultz, R. A., and Ellenberger, T. (2006). *DNA Repair and Mutagenesis.* ASM Press, Washington, DC.

Goel, A., Astumian, R. D., and Herschbach, D. (2003). Tuning and switching a DNA polymerase motor with mechanical tension. *Proc Natl Acad Sci U S A* 100, 9699–9704.

Goel, A., Frank-Kamenetskii, M. D., Ellenberger, T., and Herschbach, D. (2001). Tuning DNA \"strings\": modulating the rate of DNA replication with mechanical tension. *Proc Natl Acad Sci U S A* 98, 8485–8489.

Gore, J., Bryant, Z., Stone, M. D., Nollmann, M., Cozzarelli, N. R., and Bustamante, C. (2006). Mechanochemical analysis of DNA gyrase using rotor bead tracking. *Nature* 439, 100–104.

Greenleaf, W. J. and Block, S. M. (2006). Single molecule, motion-based DNA sequencing using RNA polymerase. *Science* 313, 801.

Ha, T. (2001a). Single molecule fluorescence methods for the study of nucleic acids. *Curr Opin Struct Biol* 11, 287–292.

Ha, T. (2001b). Single molecule fluorescence resonance energy transfer. *Methods* 25, 78–86.

Ha, T. (2004). Structural dynamics and processing of nucleic acids revealed by single molecule spectroscopy. *Biochemistry* 43, 4055–4063.

Ha, T., Rasnik, I., Cheng, W., Babcock, H. P., Gauss, G. H., Lohman, T. M., and Chu, S. (2002). Initiation and re-initiation of DNA unwinding by the *Escherichia coli* Rep helicase. *Nature* 419, 638–641.

Hamdan, S. M., Johanson, D. J., Tanner, N., Lee, J. B., Qimron, U., Tabor, S., van Oijen, A. M., and Richards, C. C. (2007). Dynamic DNA helicase–DNA polymerase interactions assure processive replication fork movement. *Mol Cell* 27, 539–549. doi:10.1016/j.molcel.2007.06.020.

Herbert, K. M., La Porta, A., Wong, B. J., Mooney, R. A., Neuman, K. C., Landick, R., and Block, S. M. (2006). Sequence-resolved detection of pausing by single RNA polymerase molecules. *Cell* 125, 1083–1094.

Johnson, D. S., Bai, L., Smith, B. Y., Patel, S. S., and Wang, M. D. (2007). Single molecule studies reveal dynamics of DNA unwinding by the ring-shaped T7 helicase. *Cell* 129, 1299–1309.

Johnson, S. J., Taylor, J. S., and Beese, L. S. (2003). Processive DNA synthesis observed in a polymerase crystal suggests a mechanism for the prevention of frameshift mutations. *Proc Natl Acad Sci U S A* 100, 3895–3900.

Kim, S., Blainey, P. C., Schroeder, C. M., and Xie, X. S. (2007). Multiplexed single molecule assay for enzymatic activity on flow-stretched DNA. *Nat Methods* 4, 397–399.

Koster, D. A., Croquette, V., Dekker, C., Shuman, S., and Dekker, N. H. (2005). Friction and torque govern the relaxation of DNA supercoils by eukaryotic topoisomerase IB. *Nature* 434, 671–674.

Koster, D. A., Palle, K., Bot, E. S., Bjornsti, M. A., and Dekker, N. H. (2007). Antitumour drugs impede DNA uncoiling by topoisomerase I. *Nature* 448, 213–217.

Landick, R. (2006). The regulatory roles and mechanism of transcriptional pausing. *Biochem Soc Trans* 34, 1062–1066.

Lee, J.-B., Hite, R. K., Hamdan, S. M., Xie, X. S., Richardson, C. C., and van Oijen, A. M. (2006). DNA primase acts as a molecular brake during DNA replication. *Nature* 439, 621–624.

Lu, H. P., Xun, L., and Xie, X. S. (1998). Single molecule enzymic dynamics. *Science* 282, 1877–1882.

Luo, G., Wang, M., Konigsberg, W. H., and Xie, X. S. (2007). Single molecule and ensemble fluorescence assays for a functionally important conformational change in T7 DNA polymerase. *Proc Natl Acad Sci U S A* 104, 12610–12615.

Maier, B., Bensimon, D., and Croquette, V. (2000). Replication by a single DNA polymerase of a stretched single-stranded DNA. *Proc Natl Acad Sci U S A* 97, 12002–12007.

Moerner, W. E. and Fromm, D. P. (2003). Methods of single molecule fluorescence spectroscopy and microscopy. *Rev Sci Instrum* 74, 3597–3619.

Myong, S., Bruno, M. M., Pyle, A. M., and Ha, T. (2007). Spring-loaded mechanism of DNA unwinding by hepatitis C virus NS3 helicase. *Science* 317, 513–516.

Myong, S., Rasnik, I., Joo, C., Lohman, T. M., and Ha, T. (2005). Repetitive shuttling of a motor protein on DNA. *Nature* 437, 1321–1325.

Myong, S., Stevens, B. C., and Ha, T. (2006). Bridging conformational dynamics and function using single molecule spectroscopy. *Structure* 14, 633–643.

Neuman, K. C., Abbondanzieri, E. A., Landick, R., Gelles, J., and Block, S. M. (2003). Ubiquitous transcriptional pausing is independent of RNA polymerase backtracking. *Cell* 115, 437–447.

Nollmann, M., Stone, M. D., Bryant, Z., Gore, J., Crisona, N. J., Hong, S. C., Mitelheiser, S., Maxwell, A., Bustamante, C., and Cozzarelli, N. R. (2007). Multiple modes of *Escherichia coli* DNA gyrase activity revealed by force and torque. *Nat Struct Mol Biol* 14, 264–271.

Patel, S. S. and Picha, K. M. (2000). Structure and function of hexameric helicases. *Annu Rev Biochem* 69, 651–697.

Perkins, T. T., Dalal, R. V., Mitsis, P. G., and Block, S. M. (2003). Sequence-dependent pausing of single Lambda exonuclease molecules. *Science* 301, 1914–1918.

Perkins, T. T., Li, H. W., Dalal, R. V., Gelles, J., and Block, S. M. (2004). Forward and reverse motion of single RecBCD molecules on DNA. *Biophys J* 86, 1640–1648.

Rasnik, I., Myong, S., Cheng, W., Lohman, T. M., and Ha, T. (2004). DNA-binding orientation and domain conformation of the *E. coli* Rep helicase monomer bound to a partial duplex junction: single molecule studies of fluorescently labeled enzymes. *J Mol Biol* 336, 395–408.

Rothwell, P. J. and Waksman, G. (2005). Structure and mechanism of DNA polymerases. *Adv Protein Chem* 71, 401–440.

Schafer, D. A., Gelles, J., Sheetz, M. P., and Landick, R. (1991). Transcription by single molecules of RNA polymerase observed by light microscopy. *Nature* 352, 444–448.

Schnitzer, M. J. and Block, S. M. (1997). Kinesin hydrolyses one ATP per 8-nm step. *Nature* 388, 386–390.

Seidel, R. and Dekker, C. (2007). Single molecule studies of nucleic acid motors. *Curr Opin Struct Biol* 17, 80–86.

Smiley, R. D., Collins, T. R., Hammes, G. G., and Hsieh, T. S. (2007). Single molecule measurements of the opening and closing of the DNA gate by eukaryotic topoisomerase II. *Proc Natl Acad Sci U S A* 104, 4840–4845.

Strick, T. R., Allemand, J. F., Bensimon, D., Bensimon, A., and Croquette, V. (1996). The elasticity of a single supercoiled DNA molecule. *Science* 271, 1835–1837.

Strick, T. R., Allemand, J. F., Bensimon, D., and Croquette, V. (1998). Behavior of supercoiled DNA. *Biophys J* 74, 2016–2028.

Strick, T. R., Croquette, V., and Bensimon, D. (2000). Single molecule analysis of DNA uncoiling by a type II topoisomerase. *Nature* 404, 901–904.

Tabor, S., Huber, H. E., and Richardson, C. C. (1987). *Escherichia coli* thioredoxin confers processivity on the DNA polymerase activity of the gene 5 protein of bacteriophage T7. *J Biol Chem* 262, 16212–16223.

Tolic-Norrelykke, S. F., Engh, A. M., Landick, R., and Gelles, J. (2004). Diversity in the rates of transcript elongation by single RNA polymerase molecules. *J Biol Chem* 279, 3292–3299.

Toprak, E. and Selvin, P. R. (2007). New fluorescent tools for watching nanometer-scale conformational changes of single molecules. *Annu Rev Biophys Biomol Struct* 36, 349–369.

van Oijen, A. M. (2007a). Honey, I shrunk the DNA: DNA length as a probe for nucleic-acid enzyme activity. *Biopolymers* 85, 144–153.

van Oijen, A. M. (2007b). Single molecule studies of complex systems: the replisome. *Mol Biosyst* 3, 117–125.

van Oijen, A. M., Blainey, P. C., Crampton, D. J., Richardson, C. C., Ellenberger, T., and Xie, X. S. (2003). Single molecule kinetics of λ exonuclease reveal base dependence and dynamic disorder. *Science* 301, 1235–1239.

Wuite, G. J. L., Smith, S. B., Young, M., Keller, D., and Bustamante, C. (2000). Single molecule studies of the effect of template tension on T7 DNA polymerase activity. *Nature* 404, 103–106.

Xue, Q. and Yeung, E. S. (1995). Differences in the chemical reactivity of individual molecules of an enzyme. *Nature* 373, 681–683.

Yin, H., Landick, R., and Gelles, J. (1994). Tethered particle motion method for studying transcript elongation by a single RNA polymerase molecule. *Biophys J* 67, 2468–2478.

CHAPTER 7

Single Molecule Studies of Prokaryotic Translation

Colin Echeverría Aitken
Biophysics Program, Stanford University School of Medicine, Stanford, CA 94305, USA

R. Andrew Marshall
Department of Chemistry, Stanford University, Stanford, CA 94305, USA

Magdalena Dorywalska
Department of Structural Biology, Stanford University School of Medicine, Stanford, CA 94305, USA

Joseph D. Puglisi
Department of Structural Biology, Stanford University School of Medicine, Stanford, CA 94305, USA
Stanford Magnetic Resonance Laboratory, Stanford University School of Medicine, Stanford, CA 94305, USA

Summary

Translation is a multi-step, multi-component process requiring intricate communication to achieve speed, accuracy, and regulation. The next step in understanding this process is to reveal the functional significance of large-scale motions and dynamic interactions implied by static ribosome structures. This requires determination of the trajectories, time scales, forces, and biochemical signals that underlie these conformational changes and binding events. Single molecule methods have emerged as powerful tools for probing translation. Here, we chronicle the key discoveries in this nascent field, which have shed light on the dynamic nature of protein biosynthesis.

Key Words

translation; ribosome; protein synthesis; messenger RNA; Fluorescence resonance energy transfer.

Introduction

Protein expression is perhaps the most fundamental of biological processes. The simplest prokaryotic organisms are estimated to commit nearly half the dry weight of the cell and greater than 80% of its energy to this process (Maaloe, 1979). The first step of protein expression, transcription, creates a portable copy of the genetic information stored in the nucleus. The second step, translation, reads this copied information to synthesize the proteins responsible for executing the myriad tasks necessary to life.

The ribosome, a megadalton-scale molecular machine, interprets the copied genetic information temporarily stored in messenger RNA, catalyzing the synthesis of amino acids into proteins. The ribosome is composed of two subunits, large and small, which join on mRNA templates up to thousands of nucleotides long. Both subunits are composed of structured RNA and protein elements, and are highly conserved in all kingdoms of life (Green and Noller, 1997; Noller, 1984; Puglisi et al., 2000; Ramakrishnan, 2002; Ramakrishnan and Moore, 2001). The ribosome reads the mRNA template by providing transfer RNA (tRNA) molecules access to mRNA codons at three distinct sites on the ribosome: aminoacyl (A site), peptidyl (P site), and exit (E site). Aminoacyl-tRNA molecules are selected at the A site. Peptide bond formation occurs in the P site, whereas deacylated tRNA dissociates from the E site.

To meet the biological demands of the cell, translation must be both rapid and accurate. Indeed, protein synthesis is estimated to proceed at an in vivo rate of up to 20 peptide bonds per second (Bremer and Dennis, 1987; Kurland, 1992). Despite this speed, misincorporation occurs for less than 1 in 1000 amino acids. The ribosome achieves this degree of processivity and accuracy by interacting with a host of protein factors, a subset of which is highly conserved across prokaryotes and eukaryotes. These translation factors interact directly with RNA and protein components of the ribosome to modulate the energetics of key translation events. Several translation factors harness the energy from ATP or GTP hydrolysis to drive conformational rearrangements thought to underlie macromolecular motion.

Movement is at the heart of translation. The process begins with the factor-guided assembly of the small and large ribosomal subunits on the mRNA template. Once assembled, the ribosome repeatedly shuttles selected tRNAs through the A, P, and E sites, all the while catalyzing peptide bond formation and directionally stepping along the mRNA message. Throughout this process, the ribosome and several of its factors perform significant conformational rearrangements. The mechanistic role of these nanometer-scale motions is central to an understanding of the translation mechanism. However, the complexity of the translation cycle makes direct observation of these motions in bulk systems difficult.

Directly Observing Translation

Single molecule techniques provide powerful tools to investigate multistep, multicomponent processes such as translation. These techniques allow for the observation of distance changes on the sub-nanometer scale and the measurement of molecular forces in the piconewton range (Greenleaf et al., 2007; Weiss, 1999). Perhaps more importantly, single molecule data can be manipulated on a per molecule basis, eliminating temporal and population averaging effects. This permits direct characterization of transient and rare events, parallel reaction pathways, and asynchronous or heterogeneous processes, all of which present a challenge to bulk methods. The study of translation by single molecule methods promises to provide new insight into the workings of protein synthesis. To date, single molecule investigation of translation has employed both fluorescence and force-based methods.

Single molecule fluorescence methods provide time-resolved information. In particular, total internal reflection-based fluorescence resonance energy transfer (TIR-FRET) reports on distance and orientation changes in biological systems (Michalet et al., 2003; Moerner and Fromm, 2003; Weiss, 1999). FRET between donor and acceptor dyes is exquisitely sensitive to the distance between the two dipoles (R), with a dependence of $1/R^6$. Single molecule FRET measurements are thus particularly useful in the characterization of dynamic conformational changes. The power of these measurements is amplified when high-resolution structures are available to guide site-specific labeling with donor and acceptor fluorophores. Current methodology allows for millisecond-resolved observation of nanometer-scale distance changes. This regime is perfectly suited to the study of many key events in translation.

Direct manipulation of single molecules provides thermodynamic information (Gosse and Croquette, 2002; Greenleaf et al., 2007; Hormeno and Arias-Gonzalez, 2006; Muller et al., 2002). Optical trapping methods have recently been applied in the study of translation, yielding insight into the origins of ribosomal motion. These methods employ a tightly focused laser beam to capture small dielectric beads by radiation pressure. Attachment of the bead to a molecule of interest allows for manipulation. Immobilization of the molecule by either surface attachment or a second optical trap allows for the controlled application of force. Optical trap manipulation can directly assay the stability of macromolecular interactions as well as the magnitude and nature of molecular motions.

Single molecule force measurements promise to complement single molecule fluorescence measurements, providing a more detailed understanding of the translation mechanism.

The Translation Cycle

The current mechanistic model of translation (Figure 7.1) was born of decades of genetic, biochemical, and biophysical experimentation. This model is composed of four component steps: initiation, elongation, termination, and recycling. Each of these steps contains a series of precisely timed molecular events involving multiple molecular components.

Initiation assembles the translation apparatus on the mRNA template in preparation for protein synthesis. This requires identification of the appropriate start site to ensure that the correct protein is synthesized. Not surprisingly, then, initiation is thought to be the most highly regulated and, consequently, rate-limiting step in translation (Gualerzi and Pon, 1990; Kozak, 1999; Laursen et al., 2005; Marintchev and Wagner, 2004). In prokaryotes, where transcription and translation processes are coupled, the small (30S) ribosomal subunit binds the mRNA template through interactions between RNA elements of the 30S subunit and the Shine–Dalgarno (SD) sequence located in the 5′ end of the mRNA (Shine and Dalgarno, 1974). The assembled 30S:mRNA complex then accommodates an initiator tRNA^{fMet} in the P site, an event controlled by the three initiation factors IF1, IF2, and IF3. These factors tune the P site's affinity for the initiator tRNA, thus ensuring proper assembly of the initiation complex (Antoun et al., 2006a,b). The entire process, which culminates with the joining of the large (50S) subunit, has been estimated to occur at a rate of $2.8\,\mu M^{-1}s^{-1}$ in vivo. This corresponds to a delay of 2 s between initiation events on one mRNA molecule (Underwood et al., 2005). Observed initiation rates in vitro are several orders of magnitude slower.

Initiation is followed by elongation, a cyclic process that culminates in the synthesis of the polypeptide specified by the mRNA template (Nilsson and Nissen, 2005; Noller et al., 2002; Ogle and Ramakrishnan, 2005; Rodnina et al., 2006; Rodnina and Wintermeyer, 2001; Wintermeyer et al., 2004). Each round of elongation begins with the selection of the appropriate aminoacyl-tRNA. More specifically, the ribosome must discriminate between cognate tRNA, which base-pair with all three bases of the mRNA codon, and near- and non-cognate tRNA, which do not correspond to the mRNA codon. This selection occurs in two steps: initial selection and proofreading. During initial selection, a ternary complex of aminoacyl-tRNA, the ribosome-activated GTPase EF-Tu (elongation factor Tu), and GTP binds the ribosome reversibly. Binding of cognate tRNA results in activation of EF-Tu for GTP hydrolysis. Upon hydrolysis, EF-Tu dissociates, freeing the ribosome to accommodate the aminoacyl-tRNA into the A site and prompting peptide bond formation.

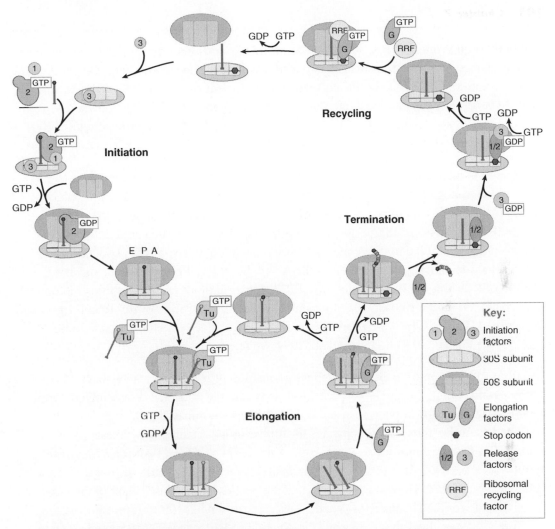

Figure 7.1: The prokaryotic translation cycle. Shown is the current model, which has been derived from biochemical and biophysical studies. Initiation, mediated by initiation factors 1,2, and 3, culminates in the joining of 30S and 50S subunits on the mRNA message primed with initiator tRNA in the P site. This complex, aided by the elongation factors Tu and G, subsequently undergoes multiple rounds of elongation. Termination, under the control of release factors 1, 2, and 3, frees the newly synthesized polypeptide upon recognition of the stop codon. Ribosomal recycling factor and elongation factor G then prepare the translational machinery for subsequent initiation events. Abbreviations: A, ribosomal A site; E, ribosomal E site; G, elongation factor G; P, ribosomal P site; RRF, ribosome recycling factor; Tu, elongation factor Tu. Reprinted, with permission, from the Annual Review of Biochemistry, Volume 77 ©2008 by Annual Reviews www.annualreviews.org

The individual events of tRNA selection occur on the millisecond timescale, as measured both in vivo and in vitro (Gromadski and Rodnina, 2004; Pape et al., 1998, 1999). Peptide bond formation is followed by the coordinated movement of A- and P-site tRNAs, and their associated mRNA codons, into the P and E sites, respectively. This movement, termed translocation, is catalyzed by EF-G (Noller et al., 2002; Wintermeyer et al., 2001, 2004). After translocating, the ribosome is ready to decode the next mRNA codon.

Elongation is the central event in translation; several antibiotic classes target the translational machinery during elongation. These antibiotics, which are approximately 10 000 times smaller than the ribosome, bind to functional regions of the ribosome and disrupt tRNA selection and translocation. The aminoglycosides, such as streptomycin and paromomycin, bind at the A site and cause misreading of the mRNA codons during tRNA selection. Tetracycline also binds the A site; however, it completely blocks the binding of tRNA at the A site (Cundliffe, 1980). Other antibiotics, such as thiostrepton and kirromycin, target the ribosomal components involved in activation of EF-Tu and EF-G and impair the GTP hydrolysis step during each round of peptide bond formation (Modolell et al., 1971; Pestka, 1970; Rodnina et al., 1994; Vogeley et al., 2001; Wolf et al., 1977).

The ribosome continues to cycle through the elongation phase until it encounters a stop codon. At this point, the nascent peptide is released, and the ribosome disassembled. Stop codon recognition and peptide release together comprise termination, which prepares the components of the translational apparatus for further rounds of protein synthesis (Bertram et al., 2001; Kisselev and Buckingham, 2000; Kisselev et al., 2003; Poole and Tate, 2000; Wimberly et al., 2000). In prokaryotes, termination is facilitated by the codon-specific release factors RF1 and RF2. However, it appears that the peptidyl transferase center of the ribosome is responsible for the chemistry of peptide release (Arkov et al., 1998). A third factor, the GTPase RF3, stimulates the activity and dissociation of RF1 and RF2; upon binding the ribosome, RF3 exchanges GDP for GTP (Freistroffer et al., 1997; Grentzmann et al., 1995). Release is followed by recycling, the complete dissociation of the remaining translational machinery – ribosomal subunits, deacylated tRNA, and mRNA. This is facilitated by the ribosome recycling factor, or RRF, as well as EF-G and IF3 (Karimi et al., 1999).

Translation at Atomic Resolution

High-resolution structures of the ribosome have added atomic detail to existing models of translation. X-ray crystallography has yielded structures of the ribosome alone, and in complex with several components of the translation apparatus (Moore and Steitz, 2003; Noller, 2005; Noller and Baucom, 2002; Ogle and Ramakrishnan, 2005; Ogle et al.,

2003; Puglisi et al., 2000; Ramakrishnan, 2002; Ramakrishnan and Moore, 2001; Wilson et al., 2002; Yonath and Bashan, 2004). Cryo-electron microscopy (cryo-EM) techniques have also contributed structural insights into the workings of the ribosome and its ligands (Frank, 2003; Mitra and Frank, 2006). Taken together, these structures provide a structural framework within which to interpret existing genetic, biochemical, and biophysical data. In addition to highlighting the role of ribosomal RNA (rRNA), high-resolution structures of the translation machinery reveal its sheer scale and suggest large conformational rearrangements of mechanistic importance.

The ribosome is a massive, multisubunit macromolecule. Twenty-five nanometers in diameter, the ribosome contains active sites responsible for decoding mRNA, activating ribosomal GTPases, and catalyzing peptide bond formation; each site is at least 5 nm from its closest neighbor. The binding sites for A-, P-, and E-site tRNAs are likewise separated by 4 nm (Yusupov et al., 2001). The ribosome not only facilitates communication between these distinct regions, but also orchestrates the motion of tRNAs and the mRNA template, themselves large macromolecules. Structures of the ribosome in complex with various ligands reveal an intricate web of RNA–RNA, RNA–protein, and protein–protein interactions. These atomic-level interactions serve as direct physical links to the mechanisms of tRNA selection, GTPase activation, and peptide bond formation.

These structures further imply that large-scale molecular motions underlie these mechanisms. Direct comparison of free and tRNA-bound ribosome structures reveals that significant remodeling of local and global architecture of the small subunit occurs upon tRNA binding (Berk et al., 2006; Ogle et al., 2001, 2002; Schuwirth et al., 2005). Cryo-EM structures of free and EF-G-bound pre-translocation complexes have identified a reorientation of the small subunit with respect to the large subunit, suggesting a ratchet motion involved in translocation (Frank and Agrawal, 2000). Crystal structures of GTP- and GDP-bound EF-Tu show large, domain-specific conformational rearrangements upon GTP hydrolysis (Kjeldgaard et al., 1993; Nissen et al., 1995). High-resolution structural methods have provided insight into the molecular working of the translation apparatus as well as hinting at various molecular motions of potential mechanistic importance. However, the function and dynamic nature of these motions remain unknown.

Single Molecule Translation

Single molecule measurements have provided direct evidence of the dynamic events of translation. Single molecule fluorescence studies have explored the mechanism of tRNA selection in real time. Force-based techniques have been employed to probe the physical origins of ribosomal movement. Interpreted within the context of existing biochemical and high-resolution structural data, these results have yielded new insights into the workings of translation in prokaryotes.

tRNA Selection in Real Time

An elongating ribosome must repeatedly select the appropriate aminoacyl-tRNA from a pool of over 40 tRNAs and catalyze peptide bond formation. The events prior to peptide bond formation determine the fidelity of peptidyl transfer, while the events that follow ensure reading frame maintenance. Characterizing tRNA movement on the ribosome during tRNA selection and translocation is imperative to understanding how these events are regulated.

The free energy differences between codon–anticodon base-pairing for cognate, near-cognate, and non-cognate tRNAs cannot explain the fidelity of tRNA selection. Therefore, a kinetic scheme has been proposed in which irreversible phosphate bond hydrolysis separates two distinct stages: initial selection and proofreading (Hopfield, 1974; Ninio, 1975). A model for this scheme, highlighting the role of induced fit in tRNA selection, is built on numerous biochemical and biophysical studies (Ogle and Ramakrishnan, 2005; Ogle et al., 2003; Rodnina et al., 2005, 2001).

Initial selection begins with ternary complex binding to the large and small ribosomal subunits. The interaction is mediated by proteins L7 and L12 of the large subunit and rRNA, protein, and mRNA of the small subunit A site (Hamel et al., 1972; Yoshizawa et al., 1999). At this stage, recognition of correct codon–anticodon pairing signals a conformational change in the 30S subunit, which increases stability of cognate tRNA 350-fold (Gromadski and Rodnina, 2004; Ogle et al., 2003). Codon-dependent discrimination between non-cognate, near-cognate, and cognate tRNAs at this stage is known as initial selection. Recognition of the correct codon–anticodon pair on the 30S subunit then signals the GTP hydrolysis by EF-Tu, which is bound over 8 nm away on the 50S subunit. This signaling, known as GTPase activation, accelerates EF-Tu-catalyzed GTP hydrolysis 50 000-fold compared to near-cognate tRNA (Rodnina and Wintermeyer, 2001). The allostery linking GTPase activation and codon recognition is not fully understood. Following GTP hydrolysis, a structural rearrangement of EF-Tu leads to its dissociation from the ribosome. tRNA then undergoes a second codon-dependent discrimination step, known as proofreading. Proofreading commits the tRNA to engage in peptidyl transfer. As in initial selection, proofreading is likely aided by conformational changes that accelerate both stable binding of cognate tRNA and dissociation of near-cognate tRNA (Ogle and Ramakrishnan, 2005; Rodnina and Wintermeyer, 2001).

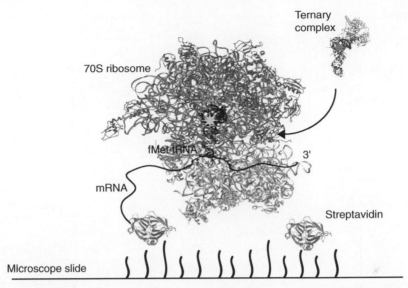

Figure 7.2: Monitoring tRNA selection in real time. A schematic representation of the ternary complex delivery to surface-immobilized ribosomes. In these experiments, 70S ribosomes containing a Cy3-labeled fMet-tRNA^fMet were immobilized to a quartz microscope slide using the high-affinity interaction between a 5′-biotinylated mRNA and streptavidin. To prevent nonspecific adhesion of biomolecules to the quartz surface, a commercially available amino-functionalized silane was deposited on the surface, which could then coupled to amine-reactive polyethylene glycol, a fraction of which were biotinylated. This created a biocompatible surface that can be further derivatized with streptavidin and allowed for the spatial resolution of ribosome complexes. Stopped-flow delivery of Cy5-labeled ternary complex permitted the real-time observation of tRNA selection on single ribosomes

Using a single molecule fluorescence assay, Blanchard et al. (2004a,b) observed tRNA selection in real time. In these experiments, fully-assembled 70S ribosome complexes initiated with Cy3-labeled fMet-tRNA^fMet were specifically immobilized via a 5′-biotinylated mRNA to a quartz microscope side (Figure 7.2). Stopped-flow delivery of ternary complex with Cy5-labeled Phe-tRNA^Phe resulted in FRET between ribosome-bound fMet-tRNA^fMet and incoming Phe-tRNA^Phe. The time evolution of this FRET signal reports directly on the conformation of ribosome-bound tRNA during a single round of tRNA selection. Postsynchronizing the FRET trajectories of several hundred individual molecules to the arrival of inter-tRNA FRET revealed that tRNA selection progresses through three states characterized by low ($\sim$0.3), mid ($\sim$0.5), or high ($\sim$0.75) FRET (Figure 7.3).

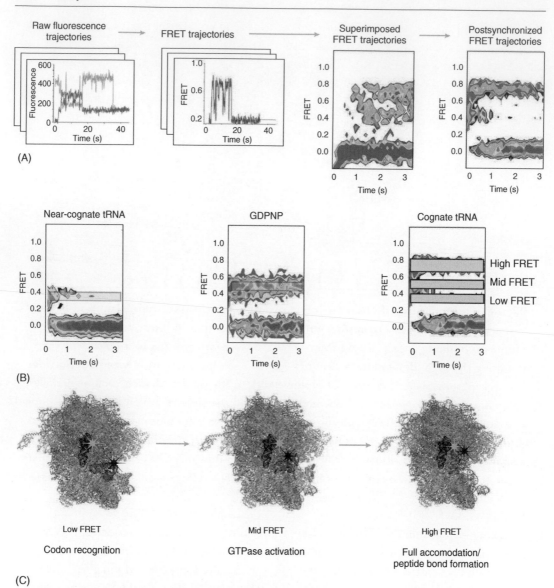

Figure 7.3: Single molecule tRNA dynamics. (A) Donor and acceptor fluorescence intensity traces were collected from hundreds of individual ribosome complexes. Calculated FRET traces were subsequently superimposed to visualize ensemble behavior. Postsynchronization to the initial appearance of FRET revealed the chronological trajectory of FRET changes. (B) Postsynchronized color overlays under various conditions identified three distinct FRET states, which were assigned with the aid of existing biochemical and structural data. Analysis of

High-resolution structures and biochemical controls aided the assignment of these FRET states to initial tRNA binding, GTPase activation, and fully accommodated tRNA, respectively. Transit from the low-FRET state to high-FRET state occurs in ~ 100 ms, which agrees with the rate of tRNA selection determined biochemically (Rodnina and Wintermeyer, 2001). When viewed in the context of the current model for tRNA selection, these results provide unique insight into the role of tRNA dynamics in discrimination.

During real-time tRNA delivery to cognate and near-cognate mRNA codons, Blanchard et al. (2004a) observed short-lived transitions to low FRET with a median lifetime, or $\tau \sim 50$ ms. Interestingly, when these delivery experiments were repeated to non-cognate codons, transitions to the low-FRET state were completely eliminated. Initial tRNA binding to L7 and L12 is not observable because the distance between P-site tRNA and the incoming tRNA exceeds the FRET range of Cy3 and Cy5. Therefore, the transient excursions to low-FRET during tRNA selection represent a reversible, codon-dependent tRNA-binding step. Prior to single molecule observation, this step had not been identified.

Cognate tRNA sampling of the low-FRET state is rare (0.24 samples per ribosome), with the majority of molecules rapidly transitioning to mid and then high FRET (Blanchard et al., 2004a). The nonhydrolyzable GTP analogue, GDPNP, eliminates transitions to the high-FRET state. Instead, a stable ribosomal complex ($\tau \sim 8$ s) is formed at mid FRET. Therefore, the transition from low to mid FRET occurs prior to GTP hydrolysis. This, along with the observation that near-cognate tRNAs are rejected before reaching mid FRET, suggests that initial selection occurs before the low to mid FRET transition. By comparing the probability that near-cognate and cognate tRNAs reach the mid-FRET state, it was determined that during initial section, cognate tRNAs are sixfold more likely to proceed through initial selection. These results agree with prior measurements of selection efficiency (Gromadski and Rodnina, 2004; Pape et al., 1999).

Real-time observation of tRNA delivery in the presence of GDPNP also reveals the dynamics of GTPase activation on the ribosome. Lee et al. (2007) characterized both long- and short-lived transitions ($\tau \sim 8$ and < 0.5 s, respectively) to mid FRET in the presence of GDPNP. The authors interpreted transitions to long-lived mid-FRET states as successful GTPase activation, whereas transitions to short-lived mid-FRET states were

individual state lifetimes, as well as the ensemble progression through these states, provides new insight into the nature of tRNA selection. (C) Interpretation of FRET trajectories in the context of high-resolution structural data provides a dynamic model for tRNA selection.

regarded as unsuccessful attempts at GTPase activation (Lee et al., 2007). Analyzing these fluctuations reveals that cognate tRNAs preferentially transit to mid FRET threefold faster than near-cognate tRNAs. Successful and unsuccessful attempts at GTPase activation were not observed prior to single molecule fluorescence approaches.

In response to initial tRNA selection, the ribosome stimulates GTP hydrolysis by EF-Tu through interactions with the 50S subunit. The sarcin–ricin loop (SRL), a universally conserved region of 23S rRNA, interacts with incoming ternary complex and regulates GTP hydrolysis by EF-Tu (Bilgin and Ehrenberg, 1994; Rodnina et al., 2000; Valle et al., 2003a). Cleavage of the SRL by toxins – sarcin, ricin, and restrictocin – inhibits tRNA delivery and translocation (Hausner et al., 1987). In single molecule tRNA delivery assays, incoming tRNAs stall at the mid-FRET state on ribosome complexes treated with restrictocin (Blanchard et al., 2004a). tRNA progress is unimpeded prior to reaching mid FRET; however, transit to the high-FRET state is 13-fold slower as compared to unmodified ribosomes (Figure 7.4). These results confirm the participation of the SRL in GTPase activation.

During uninhibited tRNA delivery, Blanchard et al. (2004a) observed rapid progression from mid to high FRET. The conformational change represented by this transition requires GTP hydrolysis and brings the labeled tRNA within ∼4 nm, as estimated by the FRET efficiency of ∼0.75. X-ray crystallography demonstrates that A- and P-site tRNAs are separated by roughly 4.5 nm (Yusupov et al., 2001). The transition from mid to high FRET represents proofreading, the second codon-dependent tRNA discrimination step. Overall error frequency for single ribosomes was determined to be $\sim 7 \times 10^{-3}$. To account for the difference between the observed error frequency and the discrimination during initial selection, the efficiency ratio for proofreading (the discrimination step following GTP hydrolysis by EF-Tu) was determined to be 24:1 according to Blanchard et al. (2004a).

Single molecule fluorescence techniques have provided direct evidence for the dynamic motion of tRNAs during selection, and highlighted the kinetics regulating this process. During initial selection, cognate tRNAs transition from low to mid FRET at a threefold faster rate than near-cognate tRNAs (Blanchard et al., 2004a; Lee et al., 2007). X-ray crystallography reveals that local and global changes in small-subunit architecture accompany cognate codon–anticodon interaction (Ogle et al., 2001; Yoshizawa et al., 1999). In response to initial selection, ribosomal rearrangements decrease the barrier to tRNA movement. Near-cognate tRNAs do not cause similar changes upon 30S binding (Ogle et al., 2002). Additionally, the GTPase domain of the 50S subunit displays a high degree of conformational flexibility that is believed to position the incoming ternary complex for rapid GTP hydrolysis (Valle et al., 2002, 2003a). Other conformational

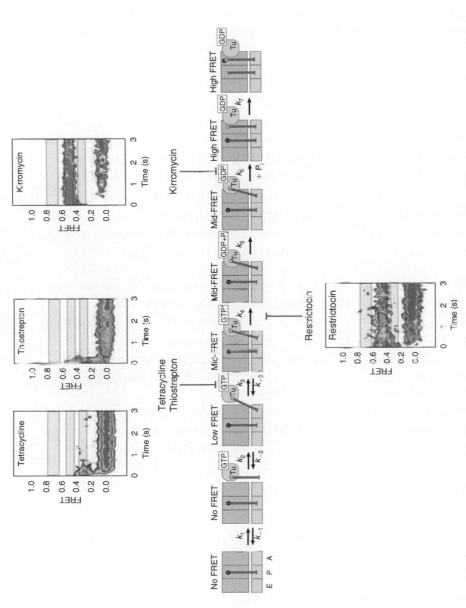

Figure 7.4: Inhibitors of tRNA selection. The current model for tRNA selection, based on bulk biochemical and biophysical studies, as well as recent single molecule results. Single molecule studies have revealed the mode of action of several inhibitors of tRNA selection. Both tetracycline and thiostrepton were shown to inhibit initial tRNA selection. Tetracycline inhibits progression to the GTPase-activated state likely via steric interactions, which hinder tRNA movement. Thiostrepton may bind at the GTPase activation center in competition with ternary complex. Restrictocin, an enzyme shown to cleave the Sarcin–Ricin loop, slows GTP hydrolysis. Kirromycin interferes with tRNA accommodation likely by inhibiting conformational changes in EF-Tu. Reprinted, with permission, from the Annual Review of Biochemistry, Volume 77 ©2008 by Annual Reviews www.annualreviews.org

changes in the 30S and 50S subunits are also believed to modulate tRNA dynamics (Ogle et al., 2003; Ogle and Ramakrishnan, 2005; Rodnina et al., 2005, 2001). These conformational rearrangements suggest that the ribosome actively accommodates incoming tRNA molecules. This induced-fit mechanism is at the origin of the kinetic tRNA selection observed by single molecule techniques.

Consistent with this interpretation, conformational flexibility of the ribosome affects tRNA dynamics. Single molecule fluorescence reveals that ribosome-targeting antibiotics modulate tRNA dynamics during selection (Figure 7.4). In the presence of tetracycline, FRET trajectories are characterized by stochastic fluctuations to low FRET (Blanchard et al., 2004a; Cundliffe, 1980). The sampling rate of the low-FRET state is consistent with the rate of tRNA arrival determined in the absence of antibiotic, and is not changed in the presence of near-cognate tRNA (Blanchard et al., 2004a). This suggests that tetracycline does not affect initial tRNA binding, but instead impedes the conformational change required for GTPase activation. This observation is consistent with structural and biochemical studies of tetracycline mechanism (Brodersen et al., 2000; Gordon, 1969).

The aminoglycoside antibiotic paromomycin is a potent miscoding agent that causes structural rearrangements in the small subunit that mimic those seen during cognate tRNA recognition (Carter et al., 2000; Fourmy et al., 1996, 1998; Ogle et al., 2001). In the presence of paromomycin, near-cognate tRNAs transition to the high-FRET state with increased frequency (S.C. Blanchard, S. Chu, and J.D. Puglisi, unpublished observations). This result also points to the importance of induced fit in initial selection of tRNA.

Like tetracycline, the cyclic peptide antibiotic thiostrepton hinders tRNA movements required for GTPase activation. However, thiostrepton does not block the transition to mid FRET. Instead, it dramatically decreases the lifetime of the mid-FRET state (Gonzalez et al., 2007). Therefore, thiostrepton likely inhibits stable association of ternary complex at the GTPase center during tRNA selection (Gonzalez et al., 2007; Mazumder, 1973; Modolell et al., 1971; Pestka, 1970; Rodnina et al., 1999).

In ternary complex delivery assays, kirromycin stalls tRNA delivery at the mid-FRET state (Blanchard et al., 2004a). However, kirromycin does not bind the ribosome. Kirromycin instead binds to EF-Tu inhibiting the conformational rearrangements necessary for its dissociation from the ribosome; bound EF-Tu likely restricts tRNA movement (Rodnina et al., 1994; Vogeley et al., 2001; Wolf et al., 1977). The mechanisms of these antibiotics highlight the relationship between ribosomal rearrangements and tRNA dynamics, and their role in tRNA discrimination.

tRNA Dynamics of Translocation

As a result of peptide bond formation, the growing nascent peptide chain is transferred to the A-site tRNA. In response to peptidyl transfer, and to allow for further rounds of elongation, the ribosome translocates the A- and P-site tRNAs and accompanying mRNA codons into the P and E sites, respectively. This concerted, large-scale movement requires the ribosome to disrupt specific interactions previously employed to orient these molecules for efficient peptidyl transfer and reading frame maintenance (Korostelev et al., 2006; Namy et al., 2006; Selmer et al., 2006). The link between peptidyl transfer and translocation remains unclear.

Immediately prior to peptide bond formation, tRNAs in the A and P sites are positioned vertically, with their anticodon and acceptor stems occupying their corresponding sites on the large and small subunits. This tRNA arrangement can be observed as a high-FRET state in tRNA delivery experiments. In addition to this classical positioning of tRNA, biochemical studies have identified a "hybrid" arrangement occurring as an intermediate state following peptide bond formation (Moazed and Noller, 1989). In the hybrid state, interactions between the acceptor stems of A- and P-site tRNAs and the large subunit are replaced by new interactions with the P and E sites, respectively; the position of the tRNA anticodon stems in the A and P sites remains unchanged. As compared to the classical tRNA arrangement, hybrid state tRNAs are oriented diagonally with respect to the mRNA template. The acylation state of A-site tRNA, to which hybrid state formation is sensitive, also affects the rate and fidelity of translocation, buttressing interpretations of the hybrid state as a translocation intermediate (Semenkov et al., 2000; Sharma et al., 2004). In fact, the hybrid state has recently been directly identified as an intermediate along the pathway of EF-G-catalyzed translocation (Dorner et al., 2006). Notably, formation of the hybrid state does not depend on translation factors; the ribosome itself senses peptidyl transfer and subsequently triggers translocation. Perhaps more importantly, the transient hybrid state has not been captured in high-resolution structures.

Single molecule methods, however, have succeeded in directly observing the hybrid state. Upon successful tRNA accommodation following ternary complex delivery, evidenced by appearance of the high-FRET state, a reversible transition to an intermediate, observed as a second mid-FRET state, occurs (Blanchard et al., 2004b; Kim et al., 2007). Replacing tRNA^{fMet} with deacylated tRNA^{fMet} in a delivered ternary complex – preventing peptide bond formation – blocks this reversible transition. Put differently, peptide bond formation precedes formation of an observed intermediate tRNA configuration: the hybrid state. Single molecule observation of ribosome complexes with different A-site tRNAs buttresses this interpretation. Kim and coworkers observed hybrid state dynamics for ribosomes with Cy3-labeled P-site tRNA^{fMet} and either Cy5 Phe-tRNA^{Phe} or *N*-acetylated Cy5 Phe-tRNA^{Phe}, a peptidyl tRNA analogue. *N*-acetylated tRNA favored the hybrid state configuration, consistent with bulk and single molecule studies demonstrating preferential hybrid state formation for A-site bound peptidyl tRNA. Interestingly, the classical hybrid equilibrium was modulated by magnesium ion concentration. At [Mg^{2+}] exceeding 10 mM, the classical state is heavily favored, regardless of A-site tRNA identity; low [Mg^{2+}] reveals a hybrid state preference in the presence of A-site peptidyl tRNA.

Munro et al. (2007) recently identified a novel hybrid state intermediate. This metastable state was detected by employing improved time resolution as well as statistical modeling of FRET trajectories. Mutation of the A loop, which base-pairs with the acceptor stem of A-site tRNA, resulted in preferential formation of the conventional hybrid state; mutation of the P loop, responsible for base-pairing with P-site tRNA, favored formation of the novel hybrid state. Both hybrid states were sensitive to deletion of the ribosomal protein L1, implicated in E-site remodeling (Subramanian and Dabbs, 1980). These observations suggest the novel hybrid state is a configuration in which only the P-site tRNA adopts a diagonal, hybrid position.

In their application to the translocation event, single molecule methods have demonstrated their power to provide insight into the workings of the translational machinery. Single molecule observation of tRNA dynamics has provided direct observation of early translocation events, revealing several key intermediate states. In fact, recent measurements further suggest a role for ribosome dynamics in hybrid state formation. Viomycin, which inhibits translocation by modulating intersubunit dynamics, decreases the fluctuation between classical and hybrid states (Ermolenko et al., 2007). As with tRNA discrimination, the ribosome actively modulates tRNA dynamics during translocation.

Origins of Ribosome Movement on mRNA

Protein synthesis requires the directional movement of the ribosome along an mRNA. Between each round of peptide bond formation, the ribosome steps three nucleotides down an mRNA in a 5′ to 3′ direction. This process is known as translocation. Each step the ribosome takes during elongation requires the breaking and reforming of the interactions holding the mRNA and the ribosome together. Precise regulation of these association and dissociation events is essential to maintain the appropriate reading frame during translation and avoid synthesis of aberrant proteins.

mRNAs make both sequence-specific and non-specific interactions with the ribosome. Most bacterial mRNAs contain a purine-rich SD sequence, which forms Watson–Crick base pairs with nucleotides in the 3′ end of small ribosomal subunit RNA (Shine and Dalgarno, 1974). These interactions guide the selection of the start site during translation initiation. Additionally, the small ribosomal subunit provides a substantial platform for mRNA binding. Structural and biochemical studies have revealed that approximately 30 nucleotides of mRNA can be accommodated on the ribosome, and that both rRNA and ribosome proteins make favorable interactions with ribosome-bound mRNA (Huttenhofer and Noller, 1994; Shatsky et al., 1991; Steitz, 1969; Yusupova et al., 2001, 2006). It remains unknown how these interactions change during translation to allow for ribosome movement.

By performing optical tweezer-based force measurements, Uemura et al. (2007) directly measured the strength of the mRNA–ribosome interaction at various stages during translation. In this study, ribosomal complexes containing a P-site tRNA and a 5′-biotinylated mRNA were immobilized on a streptavidin-coated surface (Figure 7.5). To probe the mRNA–ribosome complex, the ribosome was attached to a polystyrene bead held within an optical trap, while the mRNA-tethered surface was moved in single direction. Movement of the surface displaces the ribosome-attached bead from the center of the optical trap. Rupture of the mRNA–ribosome interaction occurs when the force pulling the bead into the laser trap exceeds that which holds mRNA and ribosome together. The measured rupture force is equivalent to the strength of mRNA–ribosome interaction (Uemura et al., 2007).

In the absence of tRNA, ribosome complexes containing mRNA without SD sequence could not be immobilized, demonstrating that the SD sequence stabilizes mRNA on the ribosome. Ribosome complexes containing mRNA with a strong SD sequence displayed a rupture force of 10.6 pN. When this experiment was repeated with ribosome complexes containing deacylated P-site tRNA^{fMet} or both deacylated P-site tRNA^{fMet} and A-site Phe-tRNA^{Phe}, the rupture force increased by ~5 pN and ~10 pN, respectively. Ribosome complexes formed with a SD-deficient mRNA and deacylated P-site tRNA^{fMet} displayed a rupture force of 4.8 pN, and further addition of A-site Phe-tRNA^{Phe} increased the rupture force to 14.8 pN. Therefore, the interaction of tRNA with mRNA on the ribosome increases the stability of the mRNA–ribosome interaction.

To simulate the ribosome following peptide bond formation, force measurements were repeated with either SD-containing or SD-deficient ribosomal complexes containing deacylated tRNA^{fMet} in the P-site and A-site-bound *N*-acetyl-Phe-tRNA^{Phe}. The result, an observed rupture force of ~12 pN for both complexes, demonstrates that following peptide bond formation mRNA–ribosome contacts are disrupted (Uemura et al., 2007). Furthermore, because both SD-containing and SD-deficient ribosomal complexes display the same rupture force, the ribosome–SD interaction is broken upon peptide bond formation. In the context

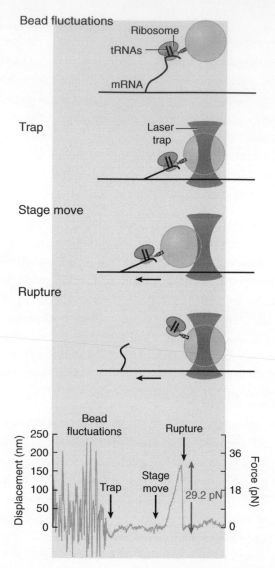

Figure 7.5: Single molecule measurements of mRNA–ribosome interaction. The ribosome–bead complex, tethered to the surface via mRNA, shows random fluctuations around the point of attachment. The fluctuations become suppressed once the bead is trapped with optical tweezers. Next, movement of the microscope stage results in gradual displacement of the bead from the trap center position, which can be correlated to an increasing force exerted on the ribosomal complex. Once the external force exceeds the intermolecular forces within the complex, the complex dissociates, or ruptures, and the bead returns to the trap center position. Measurements are repeated multiple times under each experimental condition to obtain a population distribution of rupture forces. Reprinted, with permission, from the Annual Review of Biochemistry, Volume 77 ©2008 by Annual Reviews www.annualreviews.org

of previous biochemical and structural studies, these single molecule force measurements provide a model for ribosome movement along mRNA. Peptide bond formation is coupled to destabilization of the mRNA–ribosome complex, allowing the ribosome to step to the next codon (Fredrick and Noller, 2002; Lill et al., 1989; Semenkov et al., 2000; Valle et al., 2003b; Zavialov and Ehrenberg, 2003). The precise mechanism for this allosteric signaling remains unknown. However, propagation of 50S conformational changes through tRNA or the subunit interface has been proposed (Uemura et al., 2007).

Real-Time Observation of Gene Expression at Single Molecule Level

The regulation of gene expression requires the concerted effort of the cell's transcriptional and translational machinery. The multistep, multicomponent pathways that tune mRNA and protein synthesis are required for the cell to respond in a rapid and controlled manner to external stimuli. Current methods for the characterization of protein expression are limited to measuring average expression levels over cell populations. Observation of protein synthesis in real time at the single-cell level eliminates the population and temporal averaging, which veil the timing and detail of this process.

Recently, Yu and colleagues demonstrated that single molecule fluorescence can probe in vivo protein synthesis in real time (Cai et al., 2006; Yu et al., 2006). By engineering a strain of *Escherichia coli* with a single copy of a chimeric gene containing a fast maturing yellow fluorescent protein (Venus) and membrane protein (Tsr), production of individual Tsr-Venus could be monitored in vivo. Transcription of the Tsr-Venus gene was under control of the *lac* promoter, which in these experiments was kept at highly repressed state. Under these conditions, gene expression was observed to proceed via random bursts during which multiple copies of the Tsr-Venus fusion protein are formed (Yu et al., 2006). These bursts in gene expression result from rare and transient dissociations of the *lac* repressor, which allow transcription of the Tsr-Venus mRNA. Rapid translation of the message results in the arrival of fluorescent fusion proteins at the cell surface. Further analysis reveals that on average each burst event results from the synthesis of one mRNA, and this single mRNA is translated by the ribosome approximately four times. This method represents a novel tool for observing the stochastic nature of protein expression, and heralds the possibility of real-time observation of cellular responses to environmental changes and other stimuli (Cai et al., 2006).

Conclusions

Single molecule observation of the translational machinery in vitro and in vivo has provided a window into the dynamic nature of protein synthesis. Using single molecule fluorescence to monitor the dynamic conformational rearrangements between P-site tRNA and incoming aminoacyl-tRNA has made possible the observation of codon–anticodon sampling and unsuccessful attempts at GTPase activation during initial tRNA selection. Antibiotics have been shown to disrupt tRNA dynamics during selection, providing insight into their modes of action. Single molecule fluorescence has also provided a unique perspective on the transient hybrid tRNA state. For the first time, the dynamic nature of ribosome-bound tRNA has been directly observed and characterized. Peptide bond formation, ribosome–tRNA interactions, and intersubunit dynamics have been shown to govern the conformational dynamics of tRNA. Optical tweezer-based force measurements have determined the strength of mRNA–ribosome complex at various states during translation. These observations have linked peptide bond formation with the weakening of the mRNA–ribosome interaction, which may be necessary for rapid and accurate ribosome movement along the mRNA during translocation. Finally, observing the synthesis of single proteins in vivo has highlighted directly the stochastic nature of gene expression.

Single molecule observation has proved to be a powerful tool in illuminating the highly regulated, multicomponent pathways of translation. These approaches, in concert with decades of biochemical data and high-resolution structural information, have helped to transform stop-animation models of mechanism into directly observed, detailed, and dynamic movies of biochemical function.

References

Antoun, A., Pavlov, M. Y., Lovmar, M., and Ehrenberg, M. (2006a). How initiation factors maximize the accuracy of tRNA selection in initiation of bacterial protein synthesis. *Mol Cell* 23, 183–193.

Antoun, A., Pavlov, M. Y., Lovmar, M., and Ehrenberg, M. (2006b). How initiation factors tune the rate of initiation of protein synthesis in bacteria. *EMBO J* 25, 2539–2550.

Arkov, A. L., Freistroffer, D. V., Ehrenberg, M., and Murgola, E. J. (1998). Mutations in RNAs of both ribosomal subunits cause defects in translation termination. *EMBO J* 17, 1507–1514.

Berk, V., Zhang, W., Pai, R. D., and Cate, J. H. (2006). Structural basis for mRNA and tRNA positioning on the ribosome. *Proc Natl Acad Sci U S A* 103, 15830–15834.

Bertram, G., Innes, S., Minella, O., Richardson, J., and Stansfield, I. (2001). Endless possibilities: translation termination and stop codon recognition. *Microbiology* 147, 255–269.

Bilgin, N. and Ehrenberg, M. (1994). Mutations in 23 S ribosomal RNA perturb transfer RNA selection and can lead to streptomycin dependence. *J Mol Biol* 235, 813.

Blanchard, S. C., Gonzalez, R. L., Kim, H. D., Chu, S., and Puglisi, J. D. (2004a). tRNA selection and kinetic proofreading in translation. *Nat Struct Mol Biol* 11, 1008–1014.

Blanchard, S. C., Kim, H. D., Gonzalez Jr., R. L., Puglisi, J. D., and Chu, S. (2004b). tRNA dynamics on the ribosome during translation. *Proc Natl Acad Sci U S A* 101, 12893–12898.

Bremer, H. and Dennis, P. P. (1987). Modulation of chemical composition and other parameters of the cell by growth rate. *In* Escherichia Coli and Salmonella Typhimurium: Cellular and Molecular Biology (Neidhardt, F. C., Ingranham, J. L., Low, K. B., Magasanik, B., Schaechter, M., and Umbarger, H. E., Eds.), Vol. 2, pp. 1527–1542. American Society for Microbiology, Washington, DC.

Brodersen, D. E., Clemons Jr., W. M., Carter, A. P., Morgan-Warren, R. J., Wimberly, B. T., and Ramakrishnan, V. (2000). The structural basis for the action of the antibiotics tetracycline, pactamycin, and hygromycin B on the 30S ribosomal subunit. *Cell* 103, 1143–1154.

Cai, L., Friedman, N., and Xie, X. S. (2006). Stochastic protein expression in individual cells at the single molecule level. *Nature* 440, 358–362.

Carter, A. P., Clemons, W. M., Brodersen, D. E., Morgan-Warren, R. J., Wimberly, B. T., and Ramakrishnan, V. (2000). Functional insights from the structure of the 30S ribosomal subunit and its interactions with antibiotics. *Nature* 407, 340–348.

Cundliffe, E. (1980). *In* Ribosomes, Structure, Function, and Genetics (Chambliss, G. H., Ed.), pp. 555–581. University Park Press, Baltimore.

Dorner, S., Brunelle, J. L., Sharma, D., and Green, R. (2006). The hybrid state of tRNA binding is an authentic translation elongation intermediate. *Nat Struct Mol Biol* 13, 234–241.

Ermolenko, D. N., Spiegel, P. C., Majumdar, Z. K., Hickerson, R. P., Clegg, R. M., and Noller, H. F. (2007). The antibiotic viomycin traps the ribosome in an intermediate state of translocation. *Nat Struct Mol Biol* 14, 493–497.

Fourmy, D., Recht, M. I., Blanchard, S. C., and Puglisi, J. D. (1996). Structure of the A site of *Escherichia coli* 16S ribosomal RNA complexed with an aminoglycoside antibiotic. *Science* 274, 1367–1371.

Fourmy, D., Yoshizawa, S., and Puglisi, J. D. (1998). Paromomycin binding induces a local conformational change in the A-site of 16 S rRNA. *J Mol Biol* 277, 333–345.

Frank, J. (2003). Electron microscopy of functional ribosome complexes. *Biopolymers* 68, 223–233.

Frank, J. and Agrawal, R. K. (2000). A ratchet-like inter-subunit reorganization of the ribosome during translocation. *Nature.* 406, 318–322.

Fredrick, K. and Noller, H. F. (2002). Accurate translocation of mRNA by the ribosome requires a peptidyl group or its analog on the tRNA moving into the 30S P site. *Mol Cell* 9, 1125–1131.

Freistroffer, D. V., Pavlov, M. Y., MacDougall, J., Buckingham, R. H., and Ehrenberg, M. (1997). Release factor RF3 in *E. coli* accelerates the dissociation of release factors RF1 and RF2 from the ribosome in a GTP-dependent manner. *EMBO J* 16, 4126–4133.

Gonzalez Jr., R. L., Chu, S., and Puglisi, J. D. (2007). *RNA* 13, 2091–2097.

Gordon, J. (1969). Hydrolysis of gaunosine 5′-triphosphate assoicated with binding of aminoacyl transfer ribonucleic acid to ribosomes. *J Biol Chem.* 244, 5680.

Gosse, C. and Croquette, V. (2002). Magnetic tweezers: micromanipulation and force measurement at the molecular level. *Biophys J* 82, 3314–3329.

Green, R. and Noller, H. F. (1997). Ribosomes and translation. *Annu Rev Biochem.* 66, 679.

Greenleaf, W. J., Woodside, M. T., and Block, S. M. (2007). High-resolution, single molecule measurements of biomolecular motion. *Annu Rev Biophys Biomol Struct* 36, 171–190.

Grentzmann, G., Brechemier-Baey, D., Heurgue-Hamard, V., and Buckingham, R. H. (1995). Function of polypeptide chain release factor RF-3 in *Escherichia coli*. RF-3 action in termination is predominantly at UGA-containing stop signals. *J Biol Chem* 270, 10595–10600.

Gromadski, K. B. and Rodnina, M. V. (2004). Kinetic determinants of high-fidelity tRNA discrimination on the ribosome. *Mol Cell* 13, 191.

Gualerzi, C. O. and Pon, C. L. (1990). Initiation of mRNA translation in prokaryotes. *Biochemistry* 29, 5881–5889.

Hamel, E., Koka, M., and Nakamoto, T. (1972). Requirement of an Eshcerichia-Coli 50S ribosomal-protein component for effective interaction of ribosome with T and G factors and with gaunosine triphosphate. *J Biol Chem* 247, 805.

Hausner, T. P., Atmadja, J., and Nierhaus, K. H. (1987). Evidence that the G2661 region of 23S ribosomal-RNA is located at the ribosomal-binding sites of both elongation-factors. *Biochimie* 69, 911.

Hopfield, J. J. (1974). Kinetic proofreading: a new mechanism for reducing errors in biosynthetic processes requiring high specificity. *Proc Natl Acad Sci U S A* 71, 4135–4139.

Hormeno, S. and Arias-Gonzalez, J. R. (2006). Exploring mechanochemical processes in the cell with optical tweezers. *Biol Cell* 98, 679–695.

Huttenhofer, A. and Noller, H. F. (1994). Footprinting mRNA–ribosome complexes with chemical probes. *EMBO J* 13, 3892–3901.

Karimi, R., Pavlov, M. Y., Buckingham, R. H., and Ehrenberg, M. (1999). Novel roles for classical factors at the interface between translation termination and initiation. *Mol Cell* 3, 601–609.

Kim, H. D., Puglisi, J., and Chu, S. (2007). Fluctuations of tRNAs between classical and hybrid states. *Biophys J* 93, 3575–3582.

Kisselev, L., Ehrenberg, M., and Frolova, L. (2003). Termination of translation: interplay of mRNA, rRNAs and release factors? *EMBO J* 22, 175–182.

Kisselev, L. L. and Buckingham, R. H. (2000). Translational termination comes of age. *Trends Biochem Sci* 25, 561–566.

Kjeldgaard, M., Nissen, P., Thirup, S., and Nyborg, J. (1993). The crystal structure of elongation factor EF-Tu from *Thermus aquaticus* in the GTP conformation. *Structure* 1, 35–50.

Korostelev, A., Trakhanov, S., Laurberg, M., and Noller, H. F. (2006). Crystal structure of a 70S ribosome–tRNA complex reveals functional interactions and rearrangements. *Cell* 126, 1065–1077.

Kozak, M. (1999). Initiation of translation in prokaryotes and eukaryotes. *Gene* 234, 187–208.

Kurland, C. G. (1992). Translational accuracy and the fitness of bacteria. *Annu Rev Genet* 26, 29–50.

Laursen, B. S., Sorensen, H. P., Mortensen, K. K., and Sperling-Petersen, H. U. (2005). Initiation of protein synthesis in bacteria. *Microbiol Mol Biol Rev* 69, 101–123.

Lee, T. H., Blanchard, S. C., Kim, H. D., Puglisi, J. D., and Chu, S. (2007). The role of fluctuations in tRNA selection by the ribosome. *Proc Natl Acad Sci U S A* 104, 13661–13665.

Lill, R., Robertson, J. M., and Wintermeyer, W. (1989). Binding of the 3′ terminus of tRNA to 23S rRNA in the ribosomal exit site actively promotes translocation. *EMBO J* 8, 3933–3938.

Maaloe, O. (1979). *In* Biological Regulation and Development (Goldberger, R. F., Ed.), Vol. 1, pp. 487–542. Plenum Press, New York.

Marintchev, A. and Wagner, G. (2004). Translation initiation: structures, mechanisms and evolution. *Q Rev Biophys* 37, 197–284.

Mazumder, R. (1973). Effect of thiostrepton on recycling of *Escherichia coli* initiation factor 2. *Proc Natl Acad Sci U S A* 70, 1939–1942.

Michalet, X., Kapanidis, A. N., Laurence, T., Pinaud, F., Doose, S., Pflughoefft, M., and Weiss, S. (2003). The power and prospects of fluorescence microscopies and spectroscopies. *Annu Rev Biophys Biomol Struct* 32, 161–182.

Mitra, K. and Frank, J. (2006). Ribosome dynamics: insights from atomic structure modeling into cryo-electron microscopy maps. *Annu Rev Biophys Biomol Struct* 35, 299–317.

Moazed, D. and Noller, H. F. (1989). Intermediate states in the movement of transfer RNA in the ribosome. *Nature* 342, 142–148.

Modolell, J., Cabrer, B., Parmeggiani, A., and Vazquez, D. (1971). Inhibition by siomycin and thiostrepton of both aminoacyl-tRNA and factor G binding to ribosomes. *Proc Natl Acad Sci U S A* 68, 1796–1800.

Moerner, W. E. and Fromm, D. P. (2003). Methods of single molecule fluorescence spectroscopy and microscopy. *Rev Sci Instrum* 74, 3597–3619.

Moore, P. B. and Steitz, T. A. (2003). The structural basis of large ribosomal subunit function. *Annu Rev Biochem* 72, 813–850.

Muller, D. J., Janovjak, H., Lehto, T., Kuerschner, L., and Anderson, K. (2002). Observing structure, function and assembly of single proteins by AFM. *Prog Biophys Mol Biol* 79, 1–43.

Munro, J. B., Altman, R. B., O'Connor, N., and Blanchard, S. C. (2007). Identification of two distinct hybrid state intermediates on the ribosome. *Mol Cell* 25, 505–517.

Namy, O., Moran, S. J., Stuart, D. I., Gilbert, R. J. C., and Brierley, I. (2006). A mechanical explanation of RNA pseudoknot function in programmed ribosomal frameshifting. *Nature* 441, 244.

Nilsson, J. and Nissen, P. (2005). Elongation factors on the ribosome. *Curr Opin Struct Biol* 15, 349–354.

Ninio, J. (1975). Kinetic amplification of enzyme discrimination. *Biochimie* 57, 587.

Nissen, P., Kjeldgaard, M., Thirup, S., Polekhina, G., Reshetnikova, L., Clark, B. F., and Nyborg, J. (1995). Crystal structure of the ternary complex of Phe-tRNA[Phe], EF-Tu, and a GTP analog. *Science* 270, 1464–1472.

Noller, H. F. (1984). Structure of ribosomal RNA. *Annu Rev Biochem* 53, 119–162.

Noller, H. F. (2005). RNA structure: reading the ribosome. *Science* 309, 1508–1514.

Noller, H. F. and Baucom, A. (2002). Structure of the 70 S ribosome: implications for movement. *Biochem Soc Trans* 30, 1159–1161.

Noller, H. F., Yusupov, M. M., Yusupova, G. Z., Baucom, A., and Cate, J. H. (2002). Translocation of tRNA during protein synthesis. *FEBS Lett* 514, 11–16.

Ogle, J. M., Brodersen, D. E., Clemons Jr., W. M., Tarry, M. J., Carter, A. P., and Ramakrishnan, V. (2001). Recognition of cognate transfer RNA by the 30S ribosomal subunit. *Science* 292, 897–902.

Ogle, J. M., Carter, A. P., and Ramakrishnan, V. (2003). Insights into the decoding mechanism from recent ribosome structures. *Trends Biochem Sci* 28, 259–266.

Ogle, J. M., Murphy, F. V., Tarry, M. J., and Ramakrishnan, V. (2002). Selection of tRNA by the ribosome requires a transition from an open to a closed form. *Cell* 111, 721–732.

Ogle, J. M. and Ramakrishnan, V. (2005). Structural insights into translational fidelity. *Annu Rev Biochem* 74, 129–177.

Pape, T., Wintermeyer, W., and Rodnina, M. V. (1998). Complete kinetic mechanism of elongation factor Tu-dependent binding of aminoacyl-tRNA to the A site of the *E. coli* ribosome. *EMBO J* 17, 7490–7497.

Pape, T., Wintermeyer, W., and Rodnina, M. (1999). Induced fit in initial selection and proofreading of aminoacyl-tRNA on the ribosome. *EMBO J* 18, 3800–3807.

Pestka, S. (1970). Thiostrepton: a ribosomal inhibitor of translocation. *Biochem Biophys Res Commun* 40, 667–674.

Poole, E. and Tate, W. (2000). Release factors and their role as decoding proteins: specificity and fidelity for termination of protein synthesis. *Biochim Biophys Acta* 1493, 1–11.

Puglisi, J. D., Blanchard, S. C., and Green, R. (2000). Approaching translation at atomic resolution. *Nat Struct Biol* 7, 855–861.

Ramakrishnan, V. (2002). Ribosome structure and the mechanism of translation. *Cell* 108, 557–572.

Ramakrishnan, V. and Moore, P. B. (2001). Atomic structures at last: the ribosome in 2000. *Curr Opin Struct Biol* 11, 144–154.

Rodnina, M. V., Beringer, M., and Wintermeyer, W. (2006). Mechanism of peptide bond formation on the ribosome. *Q Rev Biophys* 39, 203–225.

Rodnina, M. V., Fricke, R., and Wintermeyer, W. (1994). Transient conformational states of aminoacyl-transfer RNA during ribosome binding catalyzed by elongation-factor Tu. *Biochemistry* 33, 12267.

Rodnina, M. V., Gromadski, K. B., Kothe, U., and Wieden, H. J. (2005). Recognition and selection of tRNA in translation. *FEBS Lett* 579, 938–942.

Rodnina, M. V., Savelsbergh, A., Matassova, N. B., Katunin, V. I., Semenkov, Y. P., and Wintermeyer, W. (1999). Thiostrepton inhibits the turnover but not the GTPase of elongation factor G on the ribosome. *Proc Natl Acad Sci U S A* 96, 9586–9590.

Rodnina, M. V., Stark, H., Savelsbergh, A., Wieden, H. J., Mohr, D., Matassova, N. B., Peske, F., Daviter, T., Gualerzi, C. O., and Wintermeyer, W. (2000). GTPases mechanisms and functions of translation factors on the ribosome. *Biol Chem* 381, 377–387.

Rodnina, M. V. and Wintermeyer, W. (2001). Fidelity of aminoacyl-tRNA selection on the ribosome: kinetic and structural mechanisms. *Annu Rev Biochem* 70, 415–435.

Schuwirth, B. S., Borovinskaya, M. A., Hau, C. W., Zhang, W., Vila-Sanjurjo, A., Holton, J. M., and Cate, J. H. (2005). Structures of the bacterial ribosome at 3.5 Å resolution. *Science* 310, 827–834.

Selmer, M., Dunham, C. M., Murphy, F. V. T., Weixlbaumer, A., Petry, S., Kelley, A. C., Weir, J. R., and Ramakrishnan, V. (2006). Structure of the 70S ribosome complexed with mRNA and tRNA. *Science* 313, 1935–1942.

Semenkov, Y. P., Rodnina, M. V., and Wintermeyer, W. (2000). Energetic contribution of tRNA hybrid state formation to translocation catalysis on the ribosome. *Nat Struct Biol* 7, 1027–1031.

Sharma, D., Southworth, D. R., and Green, R. (2004). EF-G-independent reactivity of a pre-translocation-state ribosome complex with the aminoacyl tRNA substrate puromycin supports an intermediate (hybrid) state of tRNA binding. *RNA* 10, 102–113.

Shatsky, I. N., Bakin, A. B., Bogdanov, A., and Vasiliev, V. D. (1991). How does the mRNA pass through the ribosome?. *Biochimie* 73, 937–945.

Shine, J. and Dalgarno, L. (1974). The 3′-terminal sequence of *Escherichia coli* 16S ribosomal RNA: complementarity to nonsense triplets and ribosome binding sites. *Proc Natl Acad Sci U S A* 71, 1342–1346.

Steitz, J. A. (1969). Polypeptide chain initiation: nucleotide sequences of the three ribosomal binding sites in bacteriophage R17 RNA. *Nature* 224, 957–964.

Subramanian, A. R. and Dabbs, E. R. (1980). Functional studies of ribosomes lacking protein L-1 from mutant *Escherichia coli. Eur J Biochem* 112, 425.

Uemura, S., Dorywalska, M., Lee, T. H., Kim, H. D., Puglisi, J. D., and Chu, S. (2007). Peptide bond formation destabilizes Shine–Dalgarno interaction on the ribosome. *Nature* 446, 454–457.

Underwood, K. A., Swartz, J. R., and Puglisi, J. D. (2005). Quantitative polysome analysis identifies limitations in bacterial cell-free protein synthesis. *Biotechnol Bioeng* 91, 425.

Valle, M., Sengupta, J., Swami, N. K., Grassucci, R. A., Burkhardt, N., Nierhaus, K. H., Agrawal, R. K., and Frank, J. (2002). Cryo-EM reveals an active role for aminoacyl-tRNA in the accommodation process. *EMBO J* 21, 3557–3567.

Valle, M., Zavialov, A., Li, W., Stagg, S. M., Sengupta, J., Nielsen, R. C., Nissen, P., Harvey, S. C., Ehrenberg, M., and Frank, J. (2003a). Incorporation of aminoacyl-tRNA into the ribosome as seen by cryo-electron microscopy. *Nat Struct Biol* 10, 899–906.

Valle, M., Zavialov, A., Sengupta, J., Rawat, U., Ehrenberg, M., and Frank, J. (2003b). Locking and unlocking of ribosomal motions. *Cell* 114, 123–134.

Vogeley, L., Palm, G. J., Mesters, J. R., and Hilgenfeld, R. (2001). Conformational change of elongation factor Tu (EF-Tu) induced by antibiotic binding. Crystal structure of the complex between EF-Tu-GDP and aurodox. *J Biol Chem* 276, 17149.

Weiss, S. (1999). Fluorescence spectroscopy of single biomolecules. *Science* 283, 1676–1683.

Wilson, D. N., Blaha, G., Connell, S. R., Ivanov, P. V., Jenke, H., Stelzl, U., Teraoka, Y., and Nierhaus, K. H. (2002). Protein synthesis at atomic resolution: mechanistics of translation in the light of highly resolved structures for the ribosome. *Curr Protein Pept Sci* 3, 1.

Wimberly, B. T., Brodersen, D. E., Clemons Jr., W. M., Morgan-Warren, R. J., Carter, A. P., Vonrhein, C., Hartsch, T., and Ramakrishnan, V. (2000). Structure of the 30S ribosomal subunit. *Nature* 407, 327–339.

Wintermeyer, W., Peske, F., Beringer, M., Gromadski, K. B., Savelsbergh, A., and Rodnina, M. V. (2004). Mechanisms of elongation on the ribosome: dynamics of a macromolecular machine. *Biochem Soc Trans* 32, 733–737.

Wintermeyer, W., Savelsbergh, A., Semenkov, Y. P., Katunin, V. I., and Rodnina, M. V. (2001). Mechanism of elongation factor G function in tRNA translocation on the ribosome. *Cold Spring Harb Symp Quant Biol* 66, 449–458.

Wolf, H., Chinali, G., and Parmeggiani, A. (1977). Mechanisms of the inhibition of protein synthesis by kirromycin: role of elongation factor Tu and ribosomes. *Eur J Biochem* 75, 67.

Yonath, A. and Bashan, A. (2004). Ribosomal crystallography: initiation, peptide bond formation, and amino acid polymerization are hampered by antibiotics. *Annu Rev Microbiol* 58, 233–251.

Yoshizawa, S., Fourmy, D., and Puglisi, J. D. (1999). Recognition of the codon–anticodon helix by ribosomal RNA. *Science* 285, 1722–1725.

Yu, J., Xiao, J., Ren, X., Lao, K., and Xie, X. S. (2006). Probing gene expression in live cells, one protein molecule at a time. *Science* 311, 1600–1603.

Yusupov, M. M., Yusupova, G. Z., Baucom, A., Lieberman, K., Earnest, T. N., Cate, J. H. D., and Noller, H. F. (2001). Crystal structure of the ribosome at 5.5 Å resolution. *Science* 292, 883–896.

Yusupova, G., Jenner, L., Rees, B., Moras, D., and Yusupov, M. (2006). Structural basis for messenger RNA movement on the ribosome. *Nature* 444, 391–394.

Yusupova, G. Z., Yusupov, M. M., Cate, J. H., and Noller, H. F. (2001). The path of messenger RNA through the ribosome. *Cell* 106, 233–241.

Zavialov, A. V. and Ehrenberg, M. (2003). Peptidyl-tRNA regulates the GTPase activity of translation factors. *Cell* 114, 113–122.

CHAPTER 8

Single Ion Channels

David Colquhoun

Department of Pharmacology, University College London, Gower Street, London WC1E 6BT, UK

Remigijus Lape

Department of Pharmacology, University College London, Gower Street, London WC1E 6BT, UK

Lucia Sivilotti

Department of Pharmacology, University College London, Gower Street, London WC1E 6BT, UK

Summary

Moerner [(2007) *Proc Natl Acad Sci U S A* 104, 12596–12602] points out the very rapid increase since 1990 of papers with "single molecule" in the title. Had he searched for "single channel," or "single ion channel," he would have found that the rapid increase occurred in the mid-1970s.

By the time single molecule work was starting in biochemistry, the single ion channel field had already developed the tools that were needed for analysis and interpretation of observations on single molecules.

Single ion channels have the great advantage that it is possible to obtain long electrophysiological recordings (10^4–10^5 transitions between open and shut), with good signal-to-noise ratio and good time resolution (down to about $20\,\mu s$ for channels with a "medium-sized" conductance, i.e., 40–50 pS). For ion channels gated by a neurotransmitter, this can be done in the steady state or after fast concentration changes. In the case of the glycine channel, the information contained in steady-state experimental recordings has allowed up to 18 free parameters (rate constants) to be estimated [Burzomato et al. (2004) *J Neurosci* 24, 10924–10940; Colquhoun et al. (2003) *J Physiol (Lond)* 547, 699–728] for postulated mechanisms with 10 or more discrete states. Furthermore, this can be done with an exact allowance for missed events (events that are too short to be detected reliably because they are shorter than the experimental resolution).

Recently, there have been exciting advances in single molecule fluorescence methods. In the context of ligand-gated ion channels, such methods have enormous potential. It would be wonderful if it became possible to use fluorescence methods to measure, on a single molecule, the binding of the ligand (agonist) that activates the channel at the same time as measuring the opening and shutting of the channel. Such measurements could give an enormous amount of information about how the molecules work. Fluorescence records that resemble single-channel records have indeed been observed [e.g. Borisenko et al. (2003) *Biophys J* 84, 612–622; Sonnleitner et al. (2002) *Proc Natl Acad Sci U S A* 99, 12759–12764]. Up to now, these records are too short and have insufficient time resolution to achieve our aim, but at the present rate of progress our aim is not impossible.

In the simplest case, the fluorescence signal would change rapidly, in a stepwise fashion, when a ligand was bound, allowing measurement of the durations of times when a receptor was occupied and when it was vacant. In this case, no new theory would be needed because optimum fitting methods that can be used for any specified reaction mechanism already exist.

First, we shall review briefly some results obtained with single molecule fluorescence methods, with the aim of comparing them with single ion channel results. Next, we shall mention some of the stochastic theory that has been developed for the interpretation of single ion channel records and discuss how it might be applied to other sorts of single molecule observations. Finally, an example will be given of an analysis of an ion channel mechanism based on single molecule observations.

Comparison of Fluorescence Methods with Single-Channel Recording

Recordings of the current that flows through single ion channels show the alternation between the open and shut conformations of a single-channel molecule. A channel has several shut states, and usually several open states too. It would be very useful if methods could be devised to get information about transitions between states that are not distinguishable from one another in an electrical recording. To be useful, this information would need to have a resolution as good as that of the electrical record (10–40 μs resolution).

In recent years, structural information has emerged about some ion channels, though this information is still incomplete in most cases. But crystallographic structures are static. What we need now is dynamic information on the conformational changes in the channel before it opens.

Optical methods are obvious candidates for this. Fluorescence techniques have been applied to ensembles of ion channels for about a decade (Mannuzzu et al., 1996; reviewed in Gandhi and Isacoff, 2005; and Zheng, 2006). These methods rely on having a fluorophore in a specific position on a recombinant channel. This can be done either by attaching a sulfhydryl-reactive reagent to a cysteine residue or by producing an appropriate fusion protein of the channel protein with a genetically encoded label such as one of the forms of green fluorescent protein (GFP). A rearrangement in the channel structure can affect the fluorophore properties if it changes the environment around the probe (making it more or less hydrophobic) or the accessibility of the probe to quenching by small molecules or ions that can be added to the medium. In addition, if a pair of fluorophores can be attached to the protein, changes in their distance may result and be measurable as Förster resonance energy transfer (FRET).

All these measurements can be done at the same time as the electrophysiological recording in whole cells or in excised patches. Nevertheless, this work is usually carried out on ensembles of hundreds of channels in order to ensure a reliable signal-to-noise ratio.

Other than single-channel recording, there are relatively few single molecule experiments on ion channels, possibly because of the difficulty of carrying out single molecule experiments in live cells or at least in a fairly intact biological membrane preparation (for a review see Sonnleitner and Isacoff, 2003).

Single-Particle Tracking of Channels

The biggest group of studies is the application of single-particle tracking to channels, in which single channels are observed for the purpose of studying their lateral diffusion in the cell membrane and their interaction with the cytoskeleton and specific proteins such as gephyrin.

Receptors can be tracked by imaging a single fluorophore attached to the receptor by an antibody (see for AMPA receptors Tardin et al., 2003). The limitation of using an organic dye as a reporter is that photobleaching severely limits the amount of data that can be acquired. Choquet and coworkers (Tardin et al., 2003) report obtaining single-receptor trajectories lasting up to 4 s (on average 244 ms; 33 Hz acquisition rate and 30 ms integration) with Cy5 or Alexa-647 dyes. Single-particle tracks lasting a few seconds were achieved for $GABA_A$ receptors by labeling them with antibody-derivatized fluorescent nanospheres (0.1 µm in diameter) (Peran et al., 2001) and much longer times (tracks of 200 s) were achieved for glycine receptors by Triller and coworkers by using antibody-coated latex beads attached to an *myc* tag in the channel (Meier et al., 2001). Latex beads are said to allow recording times of up to a couple of minutes before they stick to membranes (Dahan et al., 2003). However, they are large (0.5 μm in diameter), which reduces spatial resolution, may limit the probe access to the protein of interest, or indeed affect the very movements to be tracked by producing "drag." A better compromise is that of using quantum dots, which are smaller (15–20 nm including the coating) and allow long recordings because of their resistance to photobleaching (reviewed in Bannai et al., 2006). However, they exhibit a property known as "blinking" in which their fluorescence output transiently blinks off and then on again. Although the blinking phenomenon can be exploited to confirm that a particular spot of light is produced by a single entity, the intermittent appearance and disappearance makes tracking and time series analysis more difficult. However, the excellent resistance of quantum dots to photobleaching means that recording periods are effectively limited simply by the cell survival during illumination. The time-lapse recording protocol described and discussed in Bannai et al. (2006) leads to a time resolution at 25 μs for a 5–10 nm point accuracy.

This group extended their previous work on glycine receptors in neurons in culture by using streptavidin-coated quantum dots attached to the receptor via a subunit-specific antibody and a biotinylated Fab fragment (Dahan et al., 2003). It was estimated that quantum dots allowed more than 20 min recording (vs. 5 s for Cy3-coupled antibody) and gave a signal-to-noise ratio an order of magnitude greater than Cy3 (Dahan et al., 2003). For extrasynaptic receptors, these measurements gave a diffusion coefficient value four times higher than the previous estimate from latex beads, suggesting that these had effectively "slowed" the receptor in the previous study. Whether this artifact always distorts results may depend both on the receptor and its effective diffusion rate and on the probe (see for instance Tardin et al., 2003). Even in these recent papers, the linker between the quantum dot and the receptor was still bulky, and further improvements may come from the adoption of quantum dots derivatized with peptide receptor ligands such as those that have been developed for G-protein–linked receptors (Zhou et al., 2007). For a discussion of the analysis of such results see Bouzigues and Dahan (2007) and Ghosh and Wirth (2007). Techniques for single-particle tracking have been used for a variety of channels, including AMPA, NMDA, $GABA_A$, and potassium channels (see for instance Borgdorff and Choquet, 2002; Bouzigues and Dahan, 2007; Groc et al., 2006; O'connell et al., 2006). For more on this subject, see also Chapter 9, this volume.

Stoichiometry

While photobleaching limits the duration of data collection from a fluorophore molecule, it has been turned to good use for the purpose of counting subunits in channels by

Ulbrich and Isacoff (2007). This is an elegant technique, which involves the imaging by total internal reflection fluorescence (TIRF) microscopy of recombinant channel molecules expressed in *Xenopus* oocytes at low expression density. The fluorescence signal comes from GFP tags inserted in the channel subunits, and the stoichiometry of a single channel is assessed by counting individual steps in the emission as the fluorophore is photobleached. Direct counting is not entirely straightforward because approximately 20% of GFP tags are not fluorescent, maybe because of defects in folding during assembly. Furthermore, two bleaching steps may appear as one in the record because of temporal proximity (the authors estimate that the dwell times for each level of fluorescence observed with their protocol imply that 10% of steps would be missed). Even so, the results of the technique appear to be robust and reliable if proper control experiments are carried out.

Imaging Ion Fluxes Through Single Ion Channels

Fluorescent calcium ion indicators have allowed the visualization of localized calcium ion transients (sparks) for a long time (Cheng et al., 1993), but it is only recently that technical refinements have allowed restricting the volume to be probed to that immediately adjacent to the plasma membrane. It is in this narrow volume that changes in calcium concentrations can be expected to follow closely the activity of single ion channels (Shuai and Parker, 2005).

An excellent example of what is now possible is given by the work of Demuro and Parker (2003, 2005) on voltage-gated calcium channels and muscle nicotinic acetylcholine receptors. The authors have used TIRF microscopy to restrict the excitation of fluo-4 dextran loaded in a *Xenopus* oocyte to the 100 nm thickness of the evanescent wave and obtained simultaneous recordings from many (100 or more) individual channels. Muscle nicotinic channels appear to be randomly distributed in the oocyte membrane and showed little lateral movement over tens of seconds. Only stage VI oocytes were used, as these have lost most of the membrane microvilli. It is not clear whether all the functional channels are imaged or whether the remaining microvilli restrict the proportion of imaged channels to those in favorable locations.

A highly sensitive CCD camera made a frame rate of $500\,s^{-1}$ possible with good signal-to-noise ratio, and the factors that limit the time resolution on these records become the affinity of the fluorescent dye and the volume of the intracellular space sampled in the process. Shuai and Parker (2005) estimated that 1–2 ms resolution is possible with the best "compromise" choices, in particular with a fluorescent dye affinity for calcium ions of between 1 and 5 μM (a higher affinity would slow equilibration and a lower one would reduce sensitivity). While this is much less than the resolution achievable with electrophysiological recording, the "optical patch" technique allows the activity of a particular single-channel molecule to be followed in time. This is not possible in normal patch clamping (unless you can be sure that the patch contains only one channel, which is something that cannot be easily achieved for ligand-gated channels). It is also remarkable that the channels in the optical patch remained accessible to drugs applied to the bulk bath solution, despite the fact that the TIRF conditions require the oocyte membrane to be resting on the coverslip of the microscope. Irrespective of the time resolution, this method may also allow counting of the number of functional channels in an area (assuming no channel is missed because of its position in a membrane invagination outside the optical "slice" that is excited). It would be very useful to be able to count the number of functional channels in a patch from which channels are being recorded by patch clamp, but the geometry of a normal patch (i.e., a dome on the tip of a glass pipette) makes TIRF imaging impossible. The only possibility would be if the recent development of planar patching devices (Klemic et al., 2002) could be extended to make it feasible to obtain electrophysiological recordings from planar patches held close ($\leqslant$100 nm) to the refractive interface needed for TIRF.

Detecting Conformational Changes in Single-Channel Molecules

Proof of principle that this can be achieved was provided in Shaker potassium channels by Sonnleitner et al. (2002), who labeled the S4 voltage-sensing domain of these channels with tetramethylrhodamine (at very low concentration, to ensure that the optical patch would have a low density of fluorophores and that each channel would bear one dye molecule at most on its four S4 domains). The position of the labeled residue was chosen among those that gave a large fluorescence change upon channel activation in ensemble recordings. Changes in the fluorophore emission in response to voltage steps applied by two-electrode voltage clamp were measured by TIRF microscopy of the oocyte membrane. Once again, TIRF techniques were essential to reduce background fluorescence and this required careful preparation of the oocyte, including removal of both the vitelline membrane and the extracellular matrix by enzyme treatment. It is likely that the imaging monitors only channels located at the tip of microvilli in the oocyte membrane. Because several channels were imaged at one time (by a CCD camera with a 25 ms frame time), the time resolution was less than the theoretical maximum of the technique. The same authors (Sonnleitner and Isacoff, 2003) estimate that this could be 10 μs, if only one channel is monitored at a time with an avalanche photodiode and if the fluorophore is excited at a near-saturating level. If that were possible, the resolution would become comparable with that of electrical recording. However, photobleaching means that there would be a trade-off between temporal resolution and the average duration of the optical recording. If a fluorophore molecule emits 10^5–10^6 photons before irreversible damage, and 100 photons per data point are needed to make the measurement signal-to-noise ratio acceptable, only 1000–10000 data points can be obtained (Ha, 2001) (Figures 8.1 and 8.2).

Simultaneous Electrophysiological and Optical Recording from Single Channels

The suggestion that single-channel recording and single-pair FRET could be advantageously combined was made as early as 1999 (Weiss, 1999), 3 years after the first demonstration of single-pair FRET (Ha et al., 1996), but to our knowledge there are only two cases in which this "holy grail" has been achieved. Both studies monitored the single-channel activity of gramicidin in artificial bilayers and have used single-pair FRET to determine the dimerization of gramicidin (Borisenko et al., 2003; Harms et al., 2003).

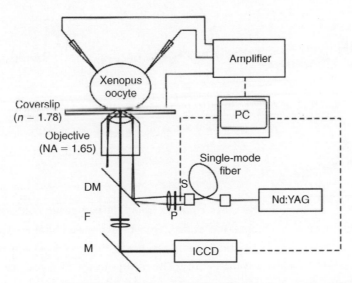

Figure 8.1: An experimental setup for total internal fluorescence microscopy (TIRF) on oocytes. The line drawing shows the arrangement of a Nd:YAG laser for fluorophore excitation through a single-mode optical fiber for spatial filtering and expansion, a shutter (S), a polarizer (P), and a dichroic mirror (DM). Detection is with an image-intensified cooled charge-coupled device (ICCD), via the appropriate filters (F) and a mirror (M). Channels are activated by voltage steps applied by a two-electrode voltage clamp. Reproduced with permission from Sonnleitner et al. (2002), where further details of the method can be found

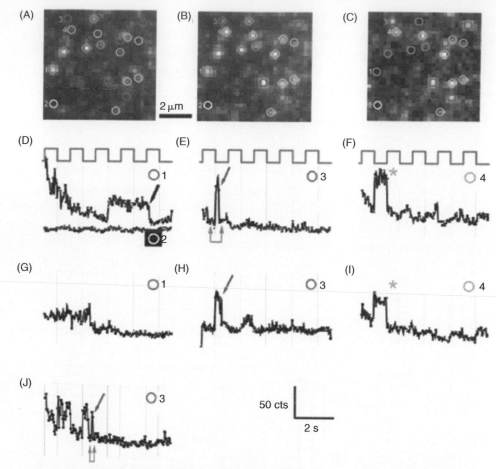

Figure 8.2: Fractional fluorescence changes upon voltage-gated conformational change in single potassium channel molecules. Oocyte-expressed Shaker-type potassium channels were labeled at a specific cysteine residue in their voltage-sensing domain with tetramethylrhodamine-5-maleimide. The frames in the first row are averages ($n = 2$, 30×30 pixel, corresponding to $6.75\,\mu m \times 6.75\,\mu m$) taken immediately before (A) or immediately after (B) the oocyte is repolarized from +20 to −100 mV. C is the difference between B and A. The intensity of fluorescence is measured from this sort of image at the circles over a period of 10 s, and examples of different trajectories are shown in the three rows below (D–J), together with the steps in the command voltage (from −100 to +20 mV, 1 s pulses at each voltage). Some of the trajectories (numbers) were obtained from the optical patch in A–C. Some spots (see traces, D and G) are bright throughout the voltage perturbation and are likely to be nonspecifically

The experimental details are somewhat different: Borisenko et al. (2003) added to planar bilayers Cy3 gramicidin (donor) and a conductance mutant form of Cy5 gramicidin (acceptor) in order to be able to distinguish in the electrophysiological recordings the heterodimers (which are the only channels that have a donor and an acceptor and therefore could give rise to FRET) from homodimer channels (i.e., containing two acceptor or two donor molecules).

In contrast, Harms et al. (2003) allowed only heterodimers to form, as the gramicidin tagged with the donor (tetramethylrhodamine) and that tagged with the acceptor dye (Cy5) were added from different sides of the bilayer, which was formed on the tip of a patch pipette. In both studies, images were taken with a 5 or 6 ms exposure, every 7–1000 ms. The temporal resolution is therefore far slower than can be attained with electrical measurements. Both groups report that some openings failed to be associated with detectable FRET events. This could be for a variety of reasons. For example, it could result from blinking or photobleaching of one of the dyes in the pair, incomplete imaging of the membrane (especially at the edges of the field), and the possibility that in some dimerized channels or in some states of the open channels the angle between the fluorophores is inefficient for FRET. Conversely, FRET events were observed in the absence of openings. Again it is impossible to be sure of the explanation for this, which could be that dimers can form in parts of the image where there is no bilayer as such, or that the dimers do not conduct because they represent intermediate conformations that precede opening.

This brief survey shows that the potential of fluorescence methods for contributing to improvement of channel kinetic measurements has yet to be fulfilled, but also that there is hope that it soon might be. This topic is discussed further in Chapter 9, this volume. Next, we shall consider the extent to which the existing single molecule stochastic theory, and existing methods for extracting information from experimental data, may be useful for optical measurements from single molecules.

---

bound rhodamine molecules. Other dim spots (white circle, D) also represent background signal. Channels labeled with a fluorophore in this position responded to depolarization with a decrease in fluorescence and are the bright spots in the difference image, C. After a short time, voltage pulses cease to elicit fluorescence changes, probably because of bleaching (arrows and asterisks). Recording at 20 frames per second, high-intensity illumination: 1.2 mW per 20 μm diameter spot. Reproduced with permission from Sonnleitner et al. (2002)

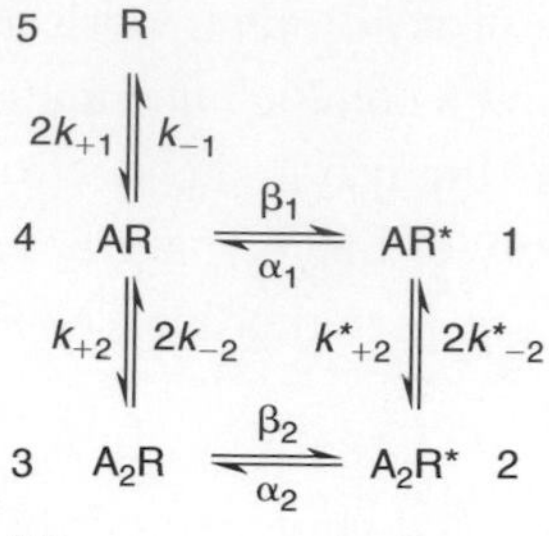

$$
\begin{array}{ccc}
& 5 \quad R & \\
& 2k_{+1} \big\Updownarrow k_{-1} & \\
4 \quad AR & \underset{\alpha_1}{\overset{\beta_1}{\rightleftharpoons}} & AR^* \quad 1 \\
k_{+2}\big\Updownarrow 2k_{-2} & & k^*_{+2}\big\Updownarrow 2k^*_{-2} \\
3 \quad A_2R & \underset{\alpha_2}{\overset{\beta_2}{\rightleftharpoons}} & A_2R^* \quad 2
\end{array}
$$

(A)

$$
Q = \left[
\begin{array}{ccc:ccc}
-(\alpha_1 + k^*_{+2}[A]) & k^*_{+2}[A] & 0 & \alpha_1 & 0 \\
2k^*_{-2} & -(\alpha_2 + 2k^*_{-2}) & \alpha_2 & 0 & 0 \\ \hdashline
0 & \beta_2 & -(\beta_2 + 2k_{-2}) & 2k_{-2} & 0 \\
\beta_1 & 0 & k_{+2}[A] & -(\beta_1 + k_{-1} + k_{+2}[A]) & k_{-1} \\
0 & 0 & 0 & 2k_{+1}[A] & -2k_{+1}[A]
\end{array}
\right]
$$

(B)

Figure 8.3: The simplest mechanism for a (muscle-type) nicotinic acetylcholine receptor (see text for a detailed description)

How to Get Information about Mechanisms from Single Molecule Measurements

For the purposes of quantitative analysis, virtually all mechanisms for receptors and for enzymes suppose that the protein can exist in one or another of several discrete states. An example is shown in Figure 8.3. For example, the states may differ because they have 0, 1, or more ligand molecules bound, or because the channel is in the open or shut conformation. The observed electrophysiological record distinguishes only between the open and the shut states.

In the interpretation of macroscopic recordings, that is, from populations of channels, it is supposed that the transitions between these discrete states obey the law of mass action. When dealing with a single molecule, the usual kinetic analyses have to be recast in the form of probabilities. This was done by Colquhoun and Hawkes (1977, 1982). In order to do this, it is supposed that the transitions between states are a Markov process[1] (this implies the law of mass action, though the converse is not necessarily true).

---

[1]   A Markov process is a random (stochastic) process that has no "memory," in the sense that the future evolution of the system depends only on its present state, not its past history. Put another way, events that occur in nonoverlapping time intervals are independent.

The only way to write the theory for anything but the very simplest mechanisms is to use matrix notation.[2] This has the great advantage that expressions look very simple, while at the same time being entirely general (e.g., Colquhoun and Hawkes, 1995a,b).

At the heart of every application of this sort of analysis lies a table that contains all of the rate constants for transitions between all possible pairs of states. The table, usually called the Q matrix, defines the mechanism completely.

The Q Matrix Defines the Mechanism

Consider, as an example, the mechanism shown in Figure 8.3A. The diagram shows the simplest mechanism for a (muscle-type) nicotinic acetylcholine receptor. The agonist (e.g., acetylcholine) is denoted A, and the receptor (which has two binding sites for the agonist) is denoted R (when shut) or R^* (when open). The mechanism has five discrete states, numbered 1–5, and the names of the rate constants for transitions between these states are written on the arrows. These rate constants are written down in a table (the Q matrix, Figure 8.3B), such that the entry in the ith row and the jth column is the transition rate from state i to state j. For example, the transition rate for opening of the doubly liganded receptor, β_2, is the transition from state 3 to state 2, and so appears in the third row and second column. Also notice that zero entries correspond to states that are not connected (for instance, states 5 and 1, see entries on fifth row and first column). The Q matrix defines the mechanism entirely. Incidentally, the transition rate from state 1 to state 1 means nothing; the elements on the diagonal are defined (simply for convenience) so that the rows add up to zero.

Macroscopic Experiments

The formulation is incredibly simple if we are recording from many molecules. We want to know how the fraction of molecules in each state varies with time after a perturbation (e.g., a concentration jump or a jump in membrane potential), and what equilibrium occupancies are eventually achieved. Call the occupancies of the five states $p_1(t), p_2(t), \ldots, p_5(t)$, at a time t, and write these in a table (a row vector) denoted $\boldsymbol{p}(t)$. Thus

$$\boldsymbol{p}(t) = \begin{bmatrix} p_1(t) & p_2(t) & p_3(t) & p_4(t) & p_5(t) \end{bmatrix} \tag{8.1}$$

---

[2]   Most biologists are not familiar with matrix notation, but we have found that it is quite feasible to get a working knowledge of the subject in a 5-day graduate school course, for example.

Whole books have been written about macroscopic kinetics, but they all boil down to special cases of the quite general solution to the problem

$$p(t) = p(0)\exp(Qt) \tag{8.2}$$

where $p(0)$ represents the occupancies at zero time (the time when the perturbation is applied). This solution holds as long as Q is constant, that is, the transition rates are all constants. In practice, this simple expression is most conveniently separated into $k - 1$ exponential components (where k is the number of states in the mechanism), plus an equilibrium component. The equilibrium occupancies, $p(\infty)$, are defined by

$$p(\infty)Q = 0 \tag{8.3}$$

In practice, it always turns out, except in the simplest cases, that macroscopic recordings do not contain enough information to define all the transition rates in the mechanism. Single molecule measurements contain more information and may be able to define the mechanism better.

From macroscopic kinetics, we now move to the more interesting topic of single molecules.

Single Molecule Experiments

Figure 8.4 shows part of an experimental single-channel recording.

The baseline of the record (top) represents the condition when the receptor is in one or other of the shut states. There is no way to tell which of the shut states it is in at any particular moment (there are three shut states in Figure 8.3A, for example). For that reason the process is described as an *aggregated Markov process*, or, sometimes, as a hidden Markov process, though the latter term is used in more than one sense, so is less desirable.

Similarly, the upward deflections show times when the channel is open, that is, in one or other of the open states (again we cannot tell which).

It is characteristic of a Markov process that the distribution (probability density function) for the length of time spent in any individual state is a simple exponential distribution

$$f(t) = \lambda \exp(-\lambda t) \tag{8.4}$$

where $\tau = 1/\lambda$ is the mean lifetime of the state in question. But for an aggregated process it gets slightly more complicated because what we observe (the duration of an opening or a shutting) represents the sojourn of the channel in a *set* of states (either open states

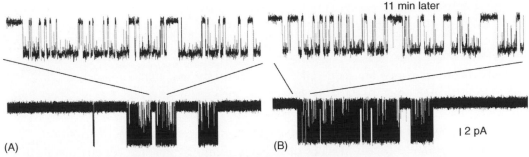

Figure 8.4: Currents produced by the activity of a single-channel molecule can be recorded for a relatively long time at high temporal resolution. An example of a recording of muscle nicotinic receptor single-channel currents activated by 10 μM acetylcholine at −100 mV transmembrane potential. Openings are downward. Bottom traces are stretches of 5 s taken at the beginning of the recording (A) or 11 min later (B). Top traces (continuous records of 0.5 s) are depicted on an expanded timescale to show the fine structure of activations. Note that noise level, amplitude and structure of the activations remain unchanged in long records. It is, however, impossible to know whether the channel activity in the different groups of openings (clusters) comes from the same channel molecule or not, as the number of channels in the patch is not known

or shut states) and this dwell time can be made up by many individual visits to different states in that set.

Define a set A that contains all the open states (states 1 and 2 in Figure 8.3) and a set F that contains all the shut states (states 3, 4, and 5 in the example).[3] The distribution of open times can then be written, quite generally, in the very simple form (Colquhoun and Hawkes, 1982)

$$f(t) = \varphi_A \exp(Q_{AA}t)(-Q_{AA})u_A \qquad (8.5)$$

In this expression, Q_{AA} is the subsection of the Q matrix that refers to transitions between open states. For example, for the mechanism in Figure 8.3A, the partition of Q into open and shut states is marked in Figure 8.3B and

---

[3]   The sets of open and closed states might, more obviously, be denoted O and C, but the notation A and F is widely used, for historical reasons, and emphasizes that the results are valid for any set of states that are distinguishable in the experimental record.

$$Q_{AA} = \begin{bmatrix} -(\alpha_1 + k^*_{+2}[A]) & k^*_{+2}[A] \\ 2k^*_{-2} & -(\alpha_2 + 2k^*_{-2}) \end{bmatrix}$$

(8.6)

The only other things that are needed are an initial vector, φ_A, that contains the probabilities that an opening starts in each of the open states (this can also be calculated from the Q matrix, see Colquhoun and Hawkes, 1982), and a vector u_A with elements all 1. This expression is directly analogous with the simple exponential distribution, Eq. (8.4), except that $-Q_{AA}$ is in place of λ, but it is entirely general, for any mechanism, however many states it may have.

The distribution of shut times is exactly analogous: all you have to do is to interchange A and F in Eq. (8.5).

Exactly the same expressions can be used for any subset of states that is distinguishable in the data. Imagine, for example, that a fluorescence record could be obtained that showed step changes when a binding site became occupied. Then, set A could be defined as the set of occupied states, and the distribution of the durations of occupancies could be calculated from Eq. (8.5).

Fitting Data

The problem in real life is to decide what mechanism describes your experimental results. In the early days of single-channel analysis, this was done by counting the number of components in the various types of dwell-time distributions and patching together the information into a plausible mechanism. This method produced the first dissection of agonist affinity and efficacy in a ligand-gated ion channel (Colquhoun and Sakmann, 1981, 1985), but it has several limitations. For example, analysis of distributions does not make use of all the information in the records; the appropriate number of components in the distributions can be ambiguous, estimating the underlying rate constants from observed time constants is indirect and difficult, and a proper correction for missed events is not possible until a mechanism is specified.

The modern strategy is to start by postulating a mechanism, obtain the values of the rate constants that fit your results best, and then see if the fit is adequate.

Several methods have been proposed for fitting, but only two have been used for real experiments. They are both based on maximizing the likelihood of the observed sequence of open and shut times (or durations of sojourns in A and not-A in general). The observed record must first be idealized (Colquhoun and Sigworth, 1995) to produce a list of open and shut times. The calculation of the likelihood of this observed sequence is not unambiguous.

Two programs are freely available for doing this: QUB (http://www.qub.buffalo.edu/) and HJCFIT. They use somewhat different methods for doing the calculations, and a comparison of the pros and cons for each method is available at http://www.ucl.ac.uk/Pharmacology/dcpr95.html#hjcfit. The only method for which the performance has been properly tested is HJCFIT (Burzomato et al., 2004; Colquhoun et al., 2003).

Other methods of fitting experimental data have been suggested for electrophysiological records from ion channels. In particular, hidden Markov methods (HMM) have been proposed (Qin et al., 2000; Venkataramanan et al., 1998a,b, 2000) for analysis of single ion channel data from electrophysiological recording. It has also been suggested as a method for fitting FRET records (McKinney et al., 2006). Whatever theoretical virtues HMM may have, they are far too slow for routine use on real data, as the probability of being in each state has to be calculated for every individual sample point in the raw data (3×10^7 points in a 10 min record at 50 kHz). In any case, HMM have never been tested by analysis of repeated simulated experiments, so it is not known whether they have any advantage or not.

The Number of Channels in a Patch

More information could be obtained from single-channel recordings if we knew how many channels were present. Occasionally that can be estimated, but usually it cannot. This has led to development of methods for the analysis of single-channel records that work even when the number of channels is not known. Part of the analysis consists of defining sequences of events in the experimental record that are likely to come from the same molecule. For instance, ligand-gated ion channels of the nicotinic or glutamate superfamily at low concentrations of agonists open in bursts, where a burst is essentially the events between the binding of the first molecule of ligand and the unbinding of the last molecule. Bursts of openings are separated by long shut times, and it is impossible to be sure that one burst originates from the same individual channel as the burst that preceded it. This means that the information contained in the shut times between the bursts cannot be used for fitting (because if there is more than one channel present, these shut times will be shorter than expected from the mechanism), and the likelihood must be calculated separately for each burst. A method has been devised for recovering some of the information lost in this way, by using "CHS vectors" for calculation of the likelihood (Colquhoun et al., 1996), and this has proved to be useful in practice (Colquhoun et al., 2003).

It would be still better to know the number of channels. In principle, one might imagine that a combination of fluorescence and patch-clamp methods might be able to tell one how many channels were present, but this has not been achieved yet.

The HJCFIT Method

The name of this program is derived from the papers by Hawkes, Jalali, and Colquhoun (Hawkes et al., 1990, 1992) that provided the exact solution to the missed events problem. In practice, it is not possible to detect very short events (in our case, very short openings or shuttings). In our single-channel recordings, the shortest resolvable event is usually 20–40 μs, although with suboptimal methods of idealization (or with channels with small conductances) it may be much longer. Since shut time components with mean durations in the 10–20 μs range seem to be common in many sorts of channel, many events may be missed and accurate allowance for this is essential. For example, the more the shuttings that are missed, the longer the openings will appear to be. What we need is the distribution of *apparent* open and shut times, as extended by missed shuttings and openings, respectively. Our way of calculating these gives what we refer to as HJC distributions. These describe what is actually observed under these circumstances, rather than the simple results such as Eq. (8.5), which would apply only with perfect resolution. The correction can be quite big. For example, the mean apparent open time may be about 20-fold longer than the true open time at high glycine concentrations (Burzomato et al., 2004; see Figure 8.6). If events are not missed, the form of the distribution of dwell times is a mixture of exponentials under our conditions, but HJC distributions have not got this form for the shortest events (below about three times the resolution).

Despite these complications, the practice of fitting a mechanism to the observations is reasonably straightforward. Records are obtained at a range of agonist concentrations and idealized, namely, turned into lists of openings and shuttings, including their duration. At this stage, the effective time resolution of the record is estimated. A preliminary analysis is carried out in order to identify which stretches of the record can be assumed to come from the same channel. At high agonist concentrations, these can be very large groups of openings (clusters) identified by the lack of double openings[4] despite relatively high open probability. At low agonist concentrations, only short sections of the record (bursts) can be assumed to come from the same individual channel.

Everything is now ready to be entered into the HJCFIT program. The program calculates the likelihood of the observations (typically several sets obtained at different agonist concentrations) from the postulated mechanism, the resolution, and a set of guesses for the rate constants (i.e., the elements of the Q matrix). The rate constant values are then changed until that likelihood is maximized. The QUB program also maximizes likelihood,

---

[4]   Indicated by double the normal channel current.

but it calculates the likelihood in a way somewhat different from ours. In order to judge how well the mechanism fits the data, the original observations are then displayed as dwell-time distributions, conditional distributions, and P_{open} (probability of being open), and predictions of the fit (HJC distributions) are *superimposed* on the observations. The plausibility of the mechanism fit is judged by the extent to which these predictions superimpose on the data.

In principle, the calculation of the likelihood is simple. We define a matrix, $G_{\text{AF}}(t)$, which contains (roughly speaking) the probability densities for staying within the subset of states A (e.g., open) for a time t and then leaving A for a state not in A (in subset F). Colquhoun and Hawkes (1982) show that this is given by

$$G_{\text{AF}}(t) = \exp(Q_{\text{AA}}t)Q_{\text{AF}} \tag{8.7}$$

where Q_{AF} is simply the top right subsection of the Q matrix, as illustrated in Figure 8.3B. $G_{\text{FA}}(t)$ is defined simply by interchanging A and F in Eq. (8.7).

The likelihood, l, of a whole sequence of observed (apparent) open and shut times can now be calculated, as described by Colquhoun et al. (1996).

$$l = \varphi_A G_{\text{AF}}(t_{o1})\;\; G_{\text{FA}}(t_{s1})\;\; G_{\text{AF}}(t_{o2})\;\; G_{\text{FA}}(t_{s2})\;\; G_{\text{AF}}(t_{o3}) \ldots u_F \tag{8.8}$$

where t_{o1}, t_{o2}, ... represent the first, second apparent open time, and t_{s1}, t_{s2}, ... first, second apparent shut time, etc. Note that openings and shuttings occur in this expression in the order in which they are observed. Thus, φ_A is a $1 \times k_A$ row vector giving the probabilities that the first opening starts in each of the open states. The first two factors, $\varphi_A G_{\text{AF}}(t_{o1})$, form a $1 \times k_F$ row vector the elements of which give the probabilities that the first shut time, t_{s1}, starts in each of the shut states, at the end of an open time of length t_{o1}. Then the first three terms, $\varphi_A G_{\text{AF}}(t_{o1}) G_{\text{FA}}(t_{s1})$, form a $1 \times k_A$ row vector the elements of which give the probabilities that the next open time, t_{o2}, starts in each of the open states, after an opening of length t_{o1} and a shutting of length t_{s1}. And so on up to the end of the observations, where the elements of the last row vector are summed over shut states to give the (scalar) value of the likelihood by (post)multiplying the row vector by u_F, the unit column vector (size $k_F \times 1$). The process of building up the product in Eq. (8.8) gives, at each stage, the joint density of the time intervals recorded thus far, and it can be regarded as progressive calculation of a vector that specifies probabilities for which state the next interval starts in, which will depend on the durations of those intervals. This process uses all the information in the record about correlations between intervals.

In real life, the calculation of $G_{AF}(t)$ and $G_{FA}(t)$ is more complicated than using the simple result in Eq. (8.7), because their definition must be changed to allow for missed events, as described by Colquhoun et al. (1996).

Fitting of 9465 resolved intervals with 9 free parameters takes less than 30 s on a standard PC. The fit produces estimates of all the rate constants, that is, of the Q matrix. It also produces estimates of their errors and their correlations (their covariance matrix). At this stage the predictions of the fit are compared with the data. In this method there is no need to fit separately distributions of apparent open times, shut times, burst lengths, etc. (and so no need to agonize about how many exponentials to fit). Rather, all the information in the record is taken into account simultaneously to get estimates of the Q matrix, and the predictions calculated from these estimates are superimposed on the data to see how successful the fit was. Examples are shown in the next section.

Some Recent Results from Single-Channel Recording

The high resolution of single-channel recording methods has allowed quite complicated reaction mechanisms, with up to 18 free parameters, to be estimated from experimental data, with reasonable precision (Burzomato et al., 2004). That is far more than can be defined by any macroscopic method, for which it is rare for even five free parameters to be estimatable. Furthermore, rate constants up to $130\,000\,\text{s}^{-1}$ can be found reliably, and that is much faster than can be found by other estimation methods. In particular, the rival QUB method does not seem to be capable of getting estimates of opening rate constants from low-concentration data.

The fitting process will be illustrated by an example from a study designed to cast light on the mechanism of the activation of the heteromeric glycine receptor by the neurotransmitter glycine and partial agonists.

After fitting over 30 different putative mechanisms, the choice was reduced to 2 (shown in Figure 8.5A and B). The results could be fitted by mechanisms with three binding sites (but not by those with only two). That is not surprising in view of the reported $\alpha 1_3 \beta_2$ subunit composition. With three binding sites, the minimum mechanism would have four shut states (with 0, 1, 2, or 3 glycine molecules bound) and three open states (the lower two rows in Figure 8.5A). But it was found that four shut states are not enough to give a good fit over the whole concentration range (analogous results have been found with the nicotinic receptor and the BK channel).

The simplest way to add extra shut states to the mechanism is to have just one connection between each additional closed state and the original scheme, as this puts the least

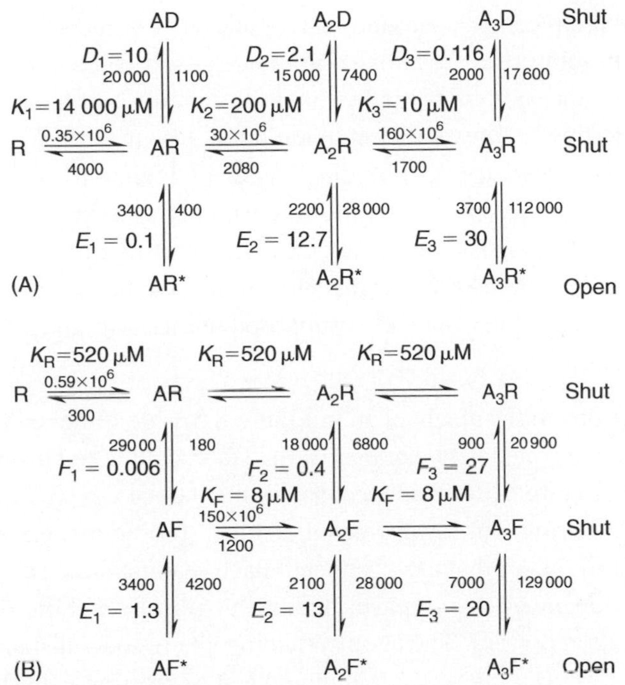

```
                AD              A₂D             A₃D      Shut
          D₁=10 ↕          D₂=2.1 ↕        D₃=0.116 ↕
            20000  ‖1100      15000  ‖7400     2000  ‖17600
      K₁=14000 µM ↕    K₂=200 µM ↕    K₃=10 µM ↕
        0.35×10⁶               30×10⁶            160×10⁶
     R ⇌ AR          A₂R ⇌         A₃R   Shut
          4000              2080              1700
           3400 ‖400      2200 ‖28000      3700 ‖112000
        E₁ = 0.1 ↕      E₂ = 12.7 ↕     E₃ = 30 ↕
     (A)       AR*             A₂R*            A₃R*     Open
```

$D_1=10$, $D_2=2.1$, $D_3=0.116$

$K_1=14\,000\ \mu M$, $K_2=200\ \mu M$, $K_3=10\ \mu M$

$E_1 = 0.1$, $E_2 = 12.7$, $E_3 = 30$

```
         K_R=520 µM     K_R=520 µM    K_R=520 µM
            0.59×10⁶
      R ⇌ AR ⇌ A₂R ⇌ A₃R   Shut
            300
          29000 ‖180     18000 ‖6800     900 ‖20900
        F₁ = 0.006 ↕    F₂ = 0.4 ↕      F₃ = 27 ↕
                       K_F = 8 µM     K_F = 8 µM
                        150×10⁶
          AF ⇌ A₂F ⇌ A₃F   Shut
                        1200
          3400 ‖4200     2100 ‖28000     7000 ‖129000
        E₁ = 1.3 ↕      E₂ = 13 ↕       E₃ = 20 ↕
     (B)      AF*            A₂F*            A₃F*    Open
```

$K_R=520\ \mu M$

$F_1 = 0.006$, $F_2 = 0.4$, $F_3 = 27$

$K_F = 8\ \mu M$

$E_1 = 1.3$, $E_2 = 13$, $E_3 = 20$

Figure 8.5: Activation mechanisms for the heteromeric glycine channel. Reproduced with permission from Burzomato et al. (2004), see text for a detailed description

constraints on the fit. These shut states could, for example, be distal to the open state (Hatton et al., 2003; Jones and Westbrook, 1995; Rothberg and Magleby, 1999; Salamone et al., 1999). An example of this approach is shown in Figure 8.5A where the scheme has three extra shut states, and is similar to that proposed for $GABA_A$ receptors by Jones and Westbrook (1995). The scheme has 18 free parameters, all of which are more or less well defined in a simultaneous fit to recordings at four different glycine concentrations (values of the rate constants are shown in the figure).

Although a good fit was obtained (Burzomato et al., 2004), the approach of simply adding unconstrained shut states is somewhat unsatisfactory, for several reasons.

First, the additional shut states are added arbitrarily to get a fit. Do these states correspond with physical reality? Although they are labeled D to suggest that they are desensitized states, they are all short-lived and do not account for the desensitization seen with macroscopic methods.

Second, for the glycine receptor we found that a good fit with the scheme in Figure 8.5A is achieved only if binding sites are allowed to interact to produce a strong increase in affinity as more ligand molecules bind (what is commonly, but unhelpfully, called "cooperativity of binding"). Thus, the first binding has low affinity ($K_1 = 14\,000\,\mu M$), but once one site is occupied, the next binding has much higher affinity ($K_2 = 200\,\mu M$), and when two are already occupied, the third has even higher affinity ($K_3 = 10\,\mu M$). The problem with this finding is that it is not clear how this interaction occurs, given the relatively long distance between the binding sites in the channel. How can one site detect when another is occupied, if the channel is supposed not to have undergone any major change of conformation?

Both of these objections to the mechanism in Figure 8.5A are removed if the three "extra" shut states are inserted between the resting state and the open state to give the "flip" mechanism in Figure 8.5B. This mechanism fits the data as well as that in Figure 8.5A, despite having four fewer free parameters. Fitting mechanisms of this sort is useful only if the states that are postulated have real physical existence. The flip mechanism has the advantage that it makes a plausible postulate about the physical nature of the channel activation process. The agonist binding site is some distance from the channel gate, and it has been suggested that there must be intermediate states between the initial binding and the final opening of the channel (e.g., Colquhoun, 2005; Grosman et al., 2000). The flip mechanism postulates that such an intermediate state lasts sufficiently long to be detected in the observations. The extra flip conformation states represent a (concerted) change in conformation that occurs after the agonist binds but *before* the channel opens.

One plausible speculation would be that this transition represents the "domain closure" seen in crystal structures of the extracellular domain of glutamate channels (Jin et al., 2003). In the "flip" mechanism, the apparent increase in affinity with increase in ligand formation is explained in the same way as was originally postulated for hemoglobin (Wyman and Allen, 1951). The binding sites behave independently and do not interact, but can exist in two conformations, resting (low affinity) and "flipped" (higher affinity). Within each conformation, the affinity of each binding step is the same. For example, for the glycine receptor we estimated that the affinity (as measured by the dissociation equilibrium constant) is $520\,\mu M$ for the first, second, and third binding to the resting conformation. Similarly, it is also postulated that the affinity is the same for each successive binding to the flipped conformation (F). But the affinity for the flipped conformation ($8\,\mu M$) is 65 times greater than for the resting conformation, so binding favors the higher affinity F state and hence activation of the receptor.

From the point of view of affinity and efficacy (Colquhoun, 1998), we see that there are now two different affinities, but only one of them tells us about the resting state of the receptor, so $K = 520\,\mu M$ is the "affinity" in the sense that Stephenson (1956) originally intended (but failed to measure). The efficacy now involves two different steps, flipping and opening. Although the gating constant still increases with the number of agonist molecules bound, it does not increase as much as the flipping constant (65-fold increase for each ligand bound). So, according to this interpretation, flipping (while shut) is more important in determining efficacy than the opening reaction itself.

How the Fit is Judged

Figure 8.6 shows how the results of a mechanism fit can be displayed. The data fitted were obtained at four glycine concentrations and fitted with the "flip" mechanism in panel A. Panels B and C show different ways of plotting the recorded data (P_{open} dose–response curve, open and shut time distributions, and conditional mean open time plots). The last of these is a measure of the (negative) correlation that is observed between the length of an opening and the length of the adjacent shut time. In each case, the actual data are shown together with the predictions for that distribution, calculated from the rate constant values obtained from fitting the mechanism to the idealized records. Note that, for example, the distribution of apparent open times has not been fitted directly, but the solid line that has been superimposed on the observed histogram is the HJC distribution of apparent open times that has been calculated from the optimum values of the rate constants found in the fit, which was carried out on a set of four idealized records together with the mechanism and the resolution ($30\,\mu s$ in this case).

In Figure 8.6, the dashed lines show the predictions of the fit for what would be seen with perfect resolution, as calculated from the fitted rate constants by the simple results such as Eq. (8.5). The extent to which the ideal (dashed line) distribution lies to the left of the HJC distribution (solid line) is a measure of the distorting effect of missed short events. This is particularly obvious for open times in Figure 8.6B. The apparent open time is roughly 20-fold more than the true open time at $1000\,\mu M$ glycine, because of the large number of missed brief shuttings. This emphasizes the importance of making accurate corrections for missed events. The conditional mean open time plots (bottom row) show the observations, which are the mean duration of those openings that are adjacent to shut times in a specified range, as filled circles joined by a solid line, and the predicted correlations as open symbols and dashed lines.

The purpose of these displays is to make it visible to the eye how well or how badly the mechanism describes the data.

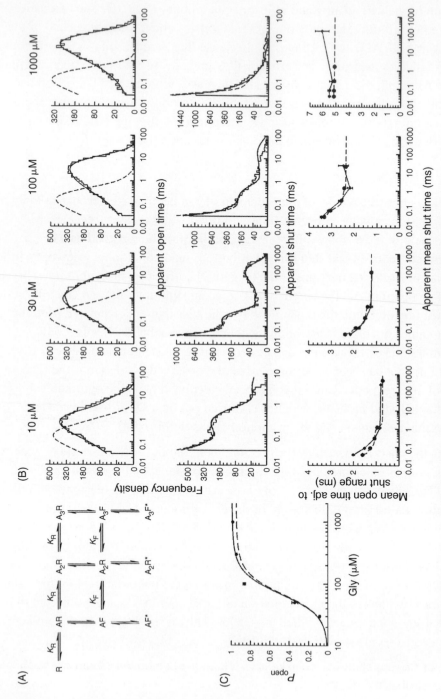

Figure 8.6: The "flip" mechanism provides an accurate description of the single-channel activity of the glycine heteromeric channel. Reproduced with permission from Burzomato et al. (2004), see text for a detailed description

Implications of the "Flip" Mechanism

In contrast to the heteromeric receptor that has just been described, the homomeric glycine receptor, when interpreted in terms of the flip mechanism, shows a smaller affinity difference between the resting and flipped conformations of the receptor (6-fold rather than 65-fold; Burzomato et al., 2004). This explains the smaller "apparent cooperativity" of the homomeric receptor. It also shows how unhelpful the term "cooperativity" is when discussing mechanisms.

A mutant glycine receptor that causes an inherited defect in the effectiveness of glycinergic synapses in mice, α1A52Sβ, shows hardly any difference in affinity between resting and flipped conformations (Plested et al., 2007). This means that on mutant channels the neurotransmitter is not as effective in producing the conformational changes that precede opening and provides a new way of interpreting the effects of mutations.

It has been supposed, ever since the work of del Castillo and Katz (1957), that partial agonists produce a small maximum response because of inefficiency of the gating step (the major conformation change that accompanies the opening and shutting of the channel). We now have evidence, for both nicotinic and glycine receptors, that the actual open-shut step is very similar for both full and partial agonists (Lape et al., 2008). What limits the maximum response for partial agonists is not gating but flipping.

According to this interpretation, it seems likely that the root of partial agonism lies before the gating step, and perhaps, therefore, nearer to the extracellular binding site region. This has interesting implications for structural studies. There seems to be little hope of relating structure and function if we have not got a description of function that describes the actual physical events that occur during channel activation.

Conclusions

The example described above shows how much has been achieved in understanding how ion channels work in the 30 years that have elapsed since the current signal produced by a single ion channel molecule was first recorded in a physiological membrane (Neher and Sakmann, 1976). Single molecule currents had been observed even earlier in lipid bilayers with gramicidin (Hladky and Haydon, 1970, 1972) and "excitability inducing material" (Latorre et al., 1972).

Other single molecule techniques are increasingly being applied to ion channels, which raises the hope that they soon will contribute information to complement and enrich that obtained from electrophysiology. It would, for example, be very useful to have a way to count the number of functional channels in the patch from which currents are recorded. More ambitiously, it would be very useful to have additional time-resolved information that would allow us to "disaggregate," at least in part, the different states in the sets of our Markov mechanisms. In the future, we expect that the single molecule optical methods described earlier will enable us to address some of these critical questions. An ability to make both electrical and optical measurements simultaneously would certainly advance our basic understanding. For any type of channel, this means finding a way to detect and report conformational changes short of opening at a time resolution similar to that achieved by electrical recordings. In addition to that, for ligand-gated ion channels it would be helpful to be able to detect whether one or more binding sites are occupied. The theoretical basis for analysis of single ion channel signals has been developed since 1977, and has now reached a state where mechanisms of any complexity can be fitted to data. Existing theory could cope with other single molecule measurements of the sort described above with little or no change.

References

Bannai, H., Levi, S., Schweizer, C., Dahan, M., and Triller, A. (2006). Imaging the lateral diffusion of membrane molecules with quantum dots. *Nat Protoc* 1, 2628–2634.

Borgdorff, A. J. and Choquet, D. (2002). Regulation of AMPA receptor lateral movements. *Nature* 417, 649–653.

Borisenko, V., Lougheed, T., Hesse, J., Fureder-Kitzmuller, E., Fertig, N., Behrends, J. C., Woolley, G. A., and Schütz, G. J. (2003). Simultaneous optical and electrical recording of single gramicidin channels. *Biophys J* 84, 612–622.

Bouzigues, C. and Dahan, M. (2007). Transient directed motions of $GABA_A$ receptors in growth cones detected by a speed correlation index. *Biophys J* 92, 654–660.

Burzomato, V., Bcato, M., Groot-Kormelink, P. J., Colquhoun, D., and Sivilotti, L. G. (2004). Single-channel behavior of heteromeric $\alpha 1\beta$ glycine receptors: an attempt to detect a conformational change before the channel opens. *J Neurosci* 24, 10924–10940.

Cheng, H., Lederer, W. J., and Cannell, M. B. (1993). Calcium sparks: elementary events underlying excitation-contraction coupling in heart muscle. *Science* 262, 740–744.

Colquhoun, D. (1998). Binding, gating, affinity and efficacy. The interpretation of structure-activity relationships for agonists and of the effects of mutating receptors. *Br J Pharmacol* 125, 923–948.

Colquhoun, D. (2005). From shut to open: what can we learn from linear free energy relationships? *Biophys J* 89, 3673–3675.

Colquhoun, D., Hatton, C. J., and Hawkes, A. G. (2003). The quality of maximum likelihood estimates of ion channel rate constants. *J Physiol (Lond)* 547, 699–728.

Colquhoun, D. and Hawkes, A. G. (1977). Relaxation and fluctuations of membrane currents that flow through drug-operated channels. *Proc R Soc Lond B Biol Sci* 199, 231–262.

Colquhoun, D. and Hawkes, A. G. (1982). On the stochastic properties of bursts of single ion channel openings and of clusters of bursts. *Philos Trans R Soc Lond B Biol Sci* 300, 1–59.

Colquhoun, D. and Hawkes, A. G. (1995a). The principles of the stochastic interpretation of ion-channel mechanisms. In: *Single-Channel Recording* (B. Sakmann, and E. Neher, Eds.) pp. 397–482, Plenum Press, New York.

Colquhoun, D. and Hawkes, A. G. (1995b). A *Q*-matrix cookbook. How to write only one program to calculate the single-channel and macroscopic predictions for any kinetic mechanisms. In: *Single-Channel Recording* (B. Sakmann, and E. Neher, Eds.) pp. 589–633, Plenum Press, New York.

Colquhoun, D., Hawkes, A. G., and Srodzinski, K. (1996). Joint distributions of apparent open times and shut times of single ion channels and the maximum likelihood fitting of mechanisms. *Philos Trans R Soc Lond A* 354, 2555–2590.

Colquhoun, D. and Sakmann, B. (1981). Fluctuations in the microsecond time range of the current through single acetylcholine receptor ion channels. *Nature* 294, 464–466.

Colquhoun, D. and Sakmann, B. (1985). Fast events in single-channel currents activated by acetylcholine and its analogues at the frog muscle end-plate. *J Physiol* 369, 501–557.

Colquhoun, D. and Sigworth, F. J. (1995). Fitting and statistical analysis of single-channel records. In: *Single-Channel Recording* (B. Sakmann, and E. Neher, Eds.). Plenum Press, pp. 483–587, New York.

Dahan, M., Levi, S., Luccardini, C., Rostaing, P., Riveau, B., and Triller, A. (2003). Diffusion dynamics of glycine receptors revealed by single-quantum dot tracking. *Science* 302, 442–445.

del Castillo, J. and Katz, B. (1957). Interaction at end-plate receptors between different choline derivatives. *Proc R Soc Lond B Biol Sci* 146, 369–381.

Demuro, A. and Parker, I. (2003). Optical single-channel recording: imaging Ca^{2+} flux through individual N-type voltage-gated channels expressed in *Xenopus* oocytes. *Cell Calcium* 34, 499–509.

Demuro, A. and Parker, I. (2005). "Optical patch-clamping": single-channel recording by imaging Ca^{2+} flux through individual muscle acetylcholine receptor channels. *J Gen Physiol* 126, 179–192.

Gandhi, C. S. and Isacoff, E. Y. (2005). Shedding light on membrane proteins. *Trends Neurosci* 28, 472–479.

Ghosh, I. and Wirth, M. J. (2007). Parsing the motion of single molecules: a novel algorithm for deconvoluting the dynamics of individual receptors at the cell surface. *Sci STKE*, pe28.

Groc, L., Heine, M., Cousins, S. L., Stephenson, F. A., Lounis, B., Cognet, L., and Choquet, D. (2006). NMDA receptor surface mobility depends on NR2A-2B subunits. *Proc Natl Acad Sci U S A* 103, 18769–18774.

Grosman, C., Zhou, M., and Auerbach, A. (2000). Mapping the conformational wave of acetylcholine receptor channel gating. *Nature* 403, 773–776.

Ha, T. (2001). Single molecule fluorescence resonance energy transfer. *Methods* 25, 78–86.

Ha, T., Enderle, T., Ogletree, D. F., Chemla, D. S., Selvin, P. R., and Weiss, S. (1996). Probing the interaction between two single molecules: fluorescence resonance energy transfer between a single donor and a single acceptor. *Proc Natl Acad Sci USA* 93, 6264–6268.

Harms, G. S., Orr, G., Montal, M., Thrall, B. D., Colson, S. D., and Lu, H. P. (2003). Probing conformational changes of gramicidin ion channels by single molecule patch-clamp fluorescence microscopy. *Biophys J* 85, 1826–1838.

Hatton, C. J., Shelley, C., Brydson, M., Beeson, D., and Colquhoun, D. (2003). Properties of the human muscle nicotinic receptor, and of the slow-channel myasthenic syndrome mutant εL221F, inferred from maximum likelihood fits. *J Physiol* 547, 729–760.

Hawkes, A. G., Jalali, A., and Colquhoun, D. (1990). The distributions of the apparent open times and shut times in a single channel record when brief events can not be detected. *Philos Trans R Soc Lond A* 332, 511–538.

Hawkes, A. G., Jalali, A., and Colquhoun, D. (1992). Asymptotic distributions of apparent open times and shut times in a single channel record allowing for the omission of brief events. *Philos Trans R Soc Lond B Biol Sci* 337, 383–404.

Hladky, S. B. and Haydon, D. A. (1970). Discreteness of conductance change in bimolecular lipid membranes in the presence of certain antibiotics. *Nature* 225, 451–453.

Hladky, S. B. and Haydon, D. A. (1972). Ion transfer across lipid membranes in the presence of gramicidin A. I. Studies of the unit conductance channel. *Biochim Biophys Acta* 274, 294–312.

Jin, R., Banke, T. G., Mayer, M. L., Traynelis, S. F., and Gouaux, E. (2003). Structural basis for partial agonist action at ionotropic glutamate receptors. *Nat Neurosci* 6, 803–810.

Jones, M. V. and Westbrook, G. L. (1995). Desensitized states prolong $GABA_A$ channel responses to brief agonist pulses. *Neuron* 15, 181–191.

Klemic, K. G., Klemic, J. F., Reed, M. A., and Sigworth, F. J. (2002). Micromolded PDMS planar electrode allows patch clamp electrical recordings from cells. *Biosens Bioelectron* 17, 597–604.

Lape, R., Colquhoun, D., and Sivilotti, L. G. (2008). On the nature of partial agonism in the nicotinic receptor superfamily. *Nature* 454, 722–728.

Latorre, R., Ehrenstein, G., and Lecar, H. (1972). Ion transport through excitability-inducing material (EIM) channels in lipid bilayer membranes. *J Gen Physiol* 60, 72–85.

Mannuzzu, L. M., Moronne, M. M., and Isacoff, E. Y. (1996). Direct physical measure of conformational rearrangement underlying potassium channel gating. *Science* 271, 213–216.

McKinney, S. A., Joo, C., and Ha, T. (2006). Analysis of single molecule FRET trajectories using hidden Markov modeling. *Biophys J* 91, 1941–1951.

Meier, J., Vannier, C., Serge, A., Triller, A., and Choquet, D. (2001). Fast and reversible trapping of surface glycine receptors by gephyrin. *Nat Neurosci* 4, 253–260.

Moerner, W. E. (2007). Single molecule chemistry and biology special feature: new directions in single molecule imaging and analysis. *Proc Natl Acad Sci U S A* 104, 12596–12602.

Neher, E. and Sakmann, B. (1976). Single channel currents recorded from membrane of denervated frog muscle fibres. *Nature* 260, 799–802.

O'connell, K. M., Rolig, A. S., Whitesell, J. D., and Tamkun, M. M. (2006). Kv2.1 potassium channels are retained within dynamic cell surface microdomains that are defined by a perimeter fence. *J Neurosci* 26, 9609–9618.

Peran, M., Hicks, B. W., Peterson, N. L., Hooper, H., and Salas, R. (2001). Lateral mobility and anchoring of recombinant $GABA_A$ receptors depend on subunit composition. *Cell Motil Cytoskeleton* 50, 89–100.

Plested, A. J., Groot-Kormelink, P. J., Colquhoun, D., and Sivilotti, L. G. (2007). Single channel study of the spasmodic mutation α 1A52S in recombinant rat glycine receptors. *J Physiol* 581, 51–73.

Qin, F., Auerbach, A., and Sachs, F. (2000). Hidden Markov modeling for single channel kinetics with filtering and correlated noise. *Biophys J* 79, 1928–1944.

Rothberg, B. S. and Magleby, K. L. (1999). Gating kinetics of single large-conductance Ca^{2+}-activated K^+ channels in high Ca^{2+} suggest a two-tiered allosteric gating mechanism. *J Gen Physiol* 114, 93–124.

Salamone, F. N., Zhou, M., and Auerbach, A. (1999). A re-examination of adult mouse nicotinic acetylcholine receptor channel activation kinetics. *J Physiol* 516, 315–330.

Shuai, J. and Parker, I. (2005). Optical single-channel recording by imaging Ca^{2+} flux through individual ion channels: theoretical considerations and limits to resolution. *Cell Calcium* 37, 283–299.

Sonnleitner, A. and Isacoff, E. Y. Single ion channel imaging. Marriott, G. and Parker, I. [361B], 304–319. 2003. Academic Press. Methods in Enzymology. Abelson, J. N. and Simon, M. I.

Sonnleitner, A., Mannuzzu, L. M., Terakawa, S., and Isacoff, E. Y. (2002). Structural rearrangements in single ion channels detected optically in living cells. *Proc Natl Acad Sci U S A* 99, 12759–12764.

Stephenson, R. P. (1956). A modification of receptor theory. *Br J Pharmacol* 11, 379–393.

Tardin, C., Cognet, L., Bats, C., Lounis, B., and Choquet, D. (2003). Direct imaging of lateral movements of AMPA receptors inside synapses. *EMBO J* 22, 4656–4665.

Ulbrich, M. H. and Isacoff, E. Y. (2007). Subunit counting in membrane-bound proteins. *Nat Methods* 4, 319–321.

Venkataramanan, L., Kue, R., and Sigworth, F. J. (1998a). Identification of hidden Markov models for ion channel currents – part II: sate-dependent excess noise. *IEEE Trans Signal Process* 46, 1916–1929.

Venkataramanan, L., Kuc, R., and Sigworth, F. J. (2000). Identification of hidden Markov models for ion channel currents – part III: band-limited, sampled data. *IEEE Trans Signal Process* 48, 376–385.

Venkataramanan, L., Walsh, J. L., Kue, R., and Sigworth, F. J. (1998b). Identification of hidden Markov models for ion channel currents – part I: colored background noise. *IEEE Trans Signal Process* 46, 1901–1915.

Weiss, S. (1999). Fluorescence spectroscopy of single biomolecules. *Science* 283, 1676–1683.

Wyman, J. and Allen, D. W. (1951). The problem of the heme interactions in hemoglobin and the basis of the Bohr effect. *J Polym Sci* VII, 499–518.

Zheng, J. (2006). Patch fluorometry: shedding new light on ion channels. *Physiol (Bethesda)* 21, 6–12.

Zhou, M., Nakatani, E., Gronenberg, L. S., Tokimoto, T., Wirth, M. J., Hruby, V. J., Roberts, A., Lynch, R. M., and Ghosh, I. (2007). Peptide-labeled quantum dots for imaging GPCRs in whole cells and as single molecules. *Bioconjug Chem* 18, 323–332.

CHAPTER 9

Single Molecule Fluorescence in Membrane Biology

Lydia M. Harriss

Chemistry Research Laboratory, University of Oxford, 12 Mansfield Road, Oxford, OX1 3TA, UK

Mark I. Wallace

Chemistry Research Laboratory, University of Oxford, 12 Mansfield Road, Oxford, OX1 3TA, UK

Summary

Single molecule fluorescence (SMF) methods have made valuable contributions to the understanding of membrane biology. We explore some of the insights afforded by SMF into the mechanisms that govern membrane structure, cell signalling, ion channel gating, and vesicle transport. We also consider how recent advances in single molecule imaging technology could be applied to membranes and identify technological limitations currently inhibiting the field.

Key Words

single molecule fluorescence; membrane; localization; receptor; signalling; lipid raft; ion channel; vesicle

The Structure and Function of Cellular Membranes

All cells are enveloped by a lipid membrane that forms a vital barrier between intra- and extra-cellular environments. Similarly, membranes enclose organelles inside the cell, creating an assortment of distinct compartments. Such spatial arrangement is essential for maintaining proper cellular function; it helps to form an organized 3D system that provides protection and storage for cell components, conserves different environments, and influences the rates at which different reactions occur by controlling local concentrations and inhibiting the diffusion of reactants. In addition, lipid vesicles provide the means to effect the directed transport of molecules within the cell. Such transport mechanisms are crucial to a wide number of cellular processes, including endocytosis, exocytosis, and neurotransmission (Lin and Scheller, 2000).

Proteins are a primary membrane constituent, comprising approximately one-third of the dry membrane volume (Zimmerberg and Gawrisch, 2006). They are responsible for a variety of essential cellular processes that include signalling, transport, and maintaining cellular structure and organization. For example, carrier and channel proteins facilitate the transport of small, inorganic ions and water-soluble molecules through the membrane. Cell surface receptors are critical to the transmission of signals into the cell, becoming activated upon the binding of extracellular signalling molecules and initiating internal signalling cascades that modify cellular behaviour. Some membrane-bound proteins catalyse membrane-associated reactions, such as the F_1F_o-ATPase motor protein responsible for ATP synthesis. Other membrane proteins act as structural links connecting the cytoskeleton to the external cellular matrix or to adjacent cells (Alberts *et al.*, 2002). Given this role as intermediaries between the cell and its environment, it is not surprising that membrane proteins comprise around three-quarters of all current drug targets (Watts, 2005). It is obvious that if single molecule methods can offer an improved understanding of membrane protein behaviour, they will prove extremely valuable in developing new drug technologies.

Why Apply Single Molecule Fluorescence to Membranes?

Over the past 15 years, developments in lasers, optics, and detection technologies have made possible the observation of individual molecules. Some of the greatest advances facilitated by single molecule fluorescence (SMF) have been in the study of biological systems, as these techniques offer a way of correlating biological function with the movement, organization, and stoichiometry of individual molecular components. The general strengths of single molecule methods are discussed in more detail in the Introduction to this book. In short, the main advantage is the ability to avoid the ensemble averaging inherent in bulk measurements that might otherwise obscure heterogeneous behaviour. A second benefit is that, unlike bulk experiments, synchronization of the population is not needed in order to measure kinetics. Kinetic parameters can be recovered by simply observing the changes of a single molecule, even while the system as a whole is in dynamic equilibrium (Lu *et al.*, 1998).

Membranes and their constituents have several key characteristics that make them well suited to single molecule approaches (García-Sáez and Schwille, 2007). These characteristics include the following: (1) Their location at the cell surface makes them easily accessible for *in vivo* studies, while the availability of model cell membranes makes them an attractive prospect for *in vitro* study. (2) The confinement of molecules to 2D motion provides a geometry compatible with many single molecule experiments that rely on 2D imaging. This is made even more favorable by the slower timescale of membrane diffusion compared to 3D diffusion in solution. Slower timescales are better suited to the limits of current imaging technology, although significant progress in the speed of 3D imaging is being made (Holtzer *et al.*, 2007). (3) The presence of many receptor proteins at low concentrations makes them easily accessible to SMF in contrast to conventional techniques such as fluorescence recovery after photobleaching (FRAP) (Lippincott-Schwartz *et al.*, 2003), which requires higher fluorophore concentrations for sufficient photon statistics. (4) The behaviour of membranes and membrane proteins is governed by a complex range of intra- and inter-molecular interactions, which are amenable to analysis by SMF. (5) Membrane proteins have also been shown to exhibit intermediate states and subpopulations, while membrane processes (such as signalling) are known to be influenced by the local heterogeneity of the intracellular structure and environment (Ichinose *et al.*, 2004), both of which can be revealed using SMF.

Having hopefully convinced the reader of the value of SMF approaches to the study of membranes and membrane proteins, we now consider recent developments in dye and SMF microscopy technology and highlight some of the membrane processes that they have been used to investigate.

Fluorescent Labels and Artificial Cell Membranes

SMF imaging of membranes *in vivo* requires the specific attachment of fluorescent labels to proteins or lipids of interest within the membrane of live cells. The biggest challenge lies in limiting labelling to only the protein of interest. This can be achieved by genetically engineering cells to express fluorescent protein conjugated to the membrane protein, immunostaining with fluorescently labelled antibodies, or providing a means of specifically attaching chemical labels or nanoparticles. If the object is not to track a membrane protein directly, but rather to obtain a measure of membrane fluidity, then fluorescently labelled lipids can also be introduced directly into the cell membrane.

Fluorescent fusion proteins have been widely used to introduce *in vivo* labels. Green fluorescent protein (GFP) has been engineered to emit over a wide range of wavelengths, exhibit enhanced photostability, and have rapid maturation (Lippincott-Schwartz and Patterson, 2003; Zhang *et al.*, 2003). Other fluorescent proteins have also been developed to emit at longer wavelengths, including Venus (Nagai *et al.*, 2002), dsRED (Matz *et al.*, 1999), mCherry (Shaner *et al.*, 2004), monomeric red fluorescent proteins (Mrfp) (Campbell *et al.*, 2002), and mKate (Shcherbo *et al.*, 2007). Small molecule dyes can also be used *in vivo* (Marks and Nolan, 2006), although it is difficult to chemically label a specific protein in live cells. Recent methods developed to overcome this challenge include those that use histidine tags on proteins as ligands for fluorophore binding (Lata *et al.*, 2006) and others that exploit arsenic chemistry to recognize specific multi-cysteine motifs (Griffin *et al.*, 1998).

In contrast to labelling *in vivo*, *in vitro* studies generally use artificial bilayers assembled from synthesized or purified phospholipids. Fluorescently labelled, purified proteins and lipids can then be introduced into the bilayer to create a model cell membrane. It is possible to control the permeability, fluidity, curvature, and rigidity of a bilayer by varying the types and proportions of proteins and lipids used (Janmey and Kinnunen, 2006). *In vitro* studies also avoid problems with cellular auto-fluorescence, which can contribute to a high background signal. The result is a far simpler system than a live cell, which is particularly useful for dissecting the contribution that individual protein components make to a particular process. In addition, *in vitro* experiments enable otherwise inaccessible proteins (such as those on internal organelle membranes) to be investigated.

Lipid bilayers can be formed by a variety of methods, either substrate supported or unsupported. For example, they may be created by depositing vesicles onto a glass substrate (Keller *et al.*, 2000; Sackmann, 1996), painting lipid layers over apertures in hydrophobic films (Cheng *et al.*, 2001), moving a lipid monolayer formed at an air–water interface across an aperture (Montal and Mueller, 1972) or between an aqueous droplet and a hydrogel support immersed in a lipid–oil solution (Heron *et al.*, 2007).

Artificial lipid vesicles also make useful model systems. For example, small unilamellar vesicles (SUVs), approximately 20–100 nm in diameter (Lasic, 1988), can be used to form planar lipid bilayers by vesicle fusion or used directly as ultra-small biomimetic containers (Okumus *et al.*, 2004). Giant unilamellar vesicles (GUVs) are typically 5–100 μm in size and therefore more closely resemble cells and cell organelles. GUVs have been used to explore membrane structure, protein dynamics (Doeven *et al.*, 2005; Kahya *et al.*, 2004), and lipid–protein interactions (Sánchez and Gratton, 2005).

Single Molecule Fluorescence Techniques

Since their inception in the late 1980s (Moerner and Kador, 1989; Nguyen *et al.*, 1987; Peck *et al.*, 1989; Wilkerson *et al.*, 1993), SMF techniques have undergone rapid growth, with the development of numerous imaging methods capable of detecting an extensive range of parameters. SMF detection usually requires a microscope in either a wide-field or a scanning configuration (Michalet *et al.*, 2007; Moerner and Fromm, 2003; Nie and Zare, 1997). All of these optical geometries work on the principle of reducing the excitation volume and thereby increasing the ratio of SMF to background noise.

Fluorescent parameters studied at the single molecule level (Michalet *et al.*, 2003) include methods based on the analysis of fluorescence intensity, lifetime (Duncan, 2006), polarization (Forkey *et al.*, 2003; Harms *et al.*, 1999; Sase *et al.*, 1997; Schütz *et al.*, 1997a), spectral diffusion (Wazawa *et al.*, 2000), and Förster resonance energy transfer (FRET) (Clegg, 1995; Ha *et al.*, 1996; Schütz *et al.*, 1998). By analyzing fluctuations in the fluorescence of small numbers of molecules, fluorescence correlation spectroscopy (FCS) methods can extract thermodynamic and kinetic parameters relating to the behaviour of individual molecules (Bacia *et al.*, 2006; Haustein and Schwille, 2007; Medina and Schwille, 2002; Schwille *et al.*, 1999b). For example, FCS can observe the diffusion of fluorescently labelled lipids in GUVs and membranes (Schwille *et al.*, 1999b). It has also been employed *in vivo* to explore the involvement of the T-cell surface glycoprotein, CD8, in the interaction between peptide class I major histocompatibility complexes (MHCs) and T-cells (Gakamsky *et al.*, 2005), as well as the initiation of mast cell signalling (Larson *et al.*, 2005). Dual colour FCS has probed the endocytic pathway of bacterial cholera toxin internalization (Bacia *et al.*, 2002). It has also been possible to reduce the amount of damage sustained by samples during FCS by employing two-photon excitation (Schwille *et al.*, 1999a).

The majority of single molecule imaging techniques are wide-field and therefore unable to detect objects smaller than the diffraction limit (approximately 250 nm for visible light). However, this limit can be overcome by exploiting the distribution of fluorescence intensity from a single molecule. When imaged, a fluorophore behaves as a point source with an Airy disc point spread function. The center of mass of the function, and therefore the position of the molecule, can be obtained by performing a least-squares fit of an appropriate function (such as a Gaussian distribution) to the measured fluorescence intensity profile of the spot (Cheezum *et al.*, 2001; Thompson *et al.*, 2002). With a sufficient number of photons, these methods can provide a localization of 1–2 nm, allowing the measurement of distances on the scale of individual proteins.

Early applications of such techniques to membranes included analyzing the movement and clustering of cell surface low-density lipoprotein receptors (LDL-R) (Gross and Webb, 1986). This was later extended to the automated localization (with 30 nm accuracy) and tracking of individual LDL-R on the surface of human skin fibroblasts (Anderson *et al.*, 1992; Ghosh and Webb, 1994), the observation of single fluorescent lipids diffusing in artificial lipid bilayers (also with 30 nm accuracy, Schmidt *et al.*, 1996a), and the development of new algorithms for simpler single-particle localization (Thompson *et al.*, 2002).

More recently, a plethora of localization-based imaging techniques has been developed (Toprak and Selvin, 2007), key examples of which are summarized in Table 9.1. Fluorescence

Table 9.1: Recent single molecule fluorescence techniques

	Spatial Resolution	Timescale	Description	Strengths	Limitations
smFRET (single molecule Förster resonance energy transfer) (Greenleaf et al., 2007; Ha, 2001; Ha et al., 1996)	~1 nm	~1 ms	The separation of two fluorophores coupled via the Förster dipole–dipole interaction can be determined from the efficiency of energy transfer between them.	Can detect intramolecular motions.	Difficult to calibrate. Does not give absolute position.
FIONA (fluorescence imaging with one nanometer accuracy) (Nan et al., 2005; Toprak and Selvin, 2007; Yildiz et al., 2003)	<1.5 nm	~0.3 ms	Determining the centroid position of a fluorophore by fitting its fluorescence intensity distribution with a point-spread function.	Can use wide-field illumination to simultaneously observe many flurophores.	Cannot resolve two or more objects of the same emission spectrum within a diffraction limited spot.
SHREC (single molecule high-resolution co-localization) (Churchman et al., 2005)	<10 nm	~0.5 s per frame	Dual-color FIONA. Two fluorescent probes with different spectra are imaged separately and then localized and mapped onto the plane of the microscope.	Can measure intra- and inter-molecular distances through time.	Using fluorophores of two colors complicates labelling.
SHRImP (single molecule high-resolution imaging with photobleaching) (Gordon et al., 2004)	~5 nm	~0.5 s per frame	The distance between a pair of fluorophores is calculated by comparing images before and after one of them has photobleached, to obtain the intensity profile of each fluorophore. A point-spread function is then applied to find the centroid positions of both fluorophores.	Avoids difficulties of two-color labelling, e.g., finding an appropriate pair of fluorophores and suitable attachment method.	Cannot make distance measurements through time (photobleaching is irreversible). Pairs of flurophores must be well separated.

Technique	Resolution	Time	Description	Advantages	Disadvantages
NALMS (nanometer-localized multiple single molecule fluorescence microscopy) (Qu et al., 2004)	~8 nm	~1 s per frame	Localization of multiple fluorophores within a diffraction-limited spot (a multiple molecule version of SHRImP). The fluorescence intensity profile of each flurophore is obtained by comparing images taken as it photobleaches.	Allows multiple fluorophores to be localized in the same diffraction-limited region using a single fluorophore color.	Requires a static sample.
PALM (photoactivatable localization microscopy) (Betzig et al., 2006)	~2 nm	~1 min	A small number of flurophores within a much larger population are excited to fluoresce, imaged, and then photobleached. Repeating this process enables all molecules to be localized.	Can be applied to samples using high fluorophore densities.	Requires a static sample.
STORM (sub-diffraction-limit imaging by stochastic optical reconstruction microscopy) (Rust et al., 2006)	~8 nm	~mins	Small sub-populations of photo-switchable fluorophores are turned on and off using light of different colors, permitting the localization of single molecules. Repeated activation cycles produce a composite image of the entire sample.	Allows imaging of high fluorophore density samples. Fluorophores can be probed repeatedly, but not permanently deactivated.	Requires a static sample.
STED (stimulated emission depletion) (Westphal and Hell, 2005)	~16 nm	~10 mins	Reduces the excitation volume below that dictated by the diffraction limit, by co-aligning one beam of light capable of fluorophore excitation with another that induces de-excitation by stimulated emission.	Can be used in a scanning mode to image a sample of square microns in size, with resolutions approaching those of localization techniques.	High-intensity pulsed lasers can induce sample damage.

imaging with one nanometer accuracy (FIONA) has demonstrated spatial and temporal resolutions of 1.5 nm and 0.5 s *in vitro*, respectively (Ökten *et al.*, 2004; Yildiz *et al.*, 2003, 2004a, b), with a 400-fold improvement in temporal resolution, down to 0.3–1.1 ms, reported *in vivo* (Kural *et al.*, 2005; Nan *et al.*, 2005). Additional 3D information about fluorophore orientation has been obtained by combining FIONA with defocused orientation and position imaging (DOPI), although this has been at the expense of temporal resolution, requiring exposure times of 0.6–0.8 s (Toprak *et al.*, 2006). Single molecule high-resolution co-localization (SHREC) has demonstrated that two spectrally distinct fluorophores with separations greater than 10 nm can be resolved, providing a tool capable of resolving distances at which FRET begins to lose effectiveness (Churchman *et al.*, 2005). As far as we are aware to date, no explicit attempt has been made to exploit the nanometer resolutions of FIONA or SHREC on membranes, beyond a simple Gaussian fitting of diffusing fluorescent molecules; however, both techniques have the potential to image on timescales and distance scales relevant to membranes.

Other recent methods can localize many individual fluorophores in an image, each to within approximately 8 nm, with complete imaging requiring minutes for tens of fluorophores or up to hours for $\sim 10^5$ fluorophores (SHRImP (Balci *et al.*, 2005; Gordon *et al.*, 2004), NALMS (Qu *et al.*, 2004), PALM (Betzig *et al.*, 2006), and STORM (Rust *et al.*, 2006)). Consequently, imaging membrane events, which typically occur on the order of milliseconds, would require dramatic improvements to the speed of these approaches.

The Role of SMF in Studying Membranes and Membrane Proteins

Most SMF membrane experiments focus on how the structure of the membrane relates to the function of the proteins embedded within it (Figure 9.1). Diffusion studies, where the motion of individual lipids or proteins is tracked, have the power to reveal much about membrane organization, signalling processes, and vesicular transport.

Probing the Structure and Behaviour of Cell Membranes

Classical models of cellular membranes, such as the fluid mosaic model (Singer and Nicolson, 1972), described the membrane as a 2D solution of proteins diffusing in a viscous phospholipid bilayer solvent. More recently, an increasing body of evidence suggests the presence of a compartmentalized substructure within cellular membranes (Kusumi *et al.*, 2005). One of the most compelling observations is that lipid diffusion in

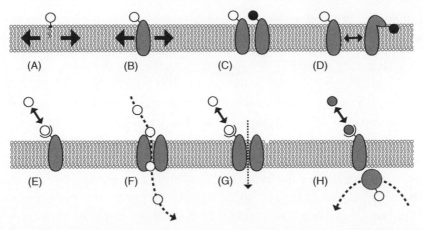

Figure 9.1: Probing membranes with single molecule fluorescence. (A) Lipids (shown in light grey) and (B) proteins (dark grey) can be labeled with fluorophores (white circles) and tracked as they diffuse within the membrane. (C) Protein–protein interactions can be probed by observing changes in the fluorescence intensity of pairs of fluorophores. FRET measurements are possible if labels of two colors are used. (D) Conformational changes of a protein can be detected by monitoring changes in fluorescence due to self-quenching. (E) The kinetics of fluorescently labelled ligands binding to membrane proteins can be measured. (F) The transport of small fluorescent molecules across the membrane can be monitored. (G) Observations of fluorescent ligands binding to ion channels could be coupled with electrophysiological measurements to better understand ion channel activation. (H) The recruitment of fluorescently labelled signalling proteins (large, dark grey circle) to the membrane after the activation of a membrane receptor by the binding of a ligand (small, dark grey circle) can be observed in live cells

artificial reconstituted membranes appears to be between 5 and 50 times faster than that in plasma membranes (Suzuki *et al.*, 2005), suggesting that constraints on diffusion exist in cells that are not present in simpler *in vitro* models.

The origins of this organization are not well understood, despite extensive experimental attempts to identify different categories of inhomogeneity. Some membrane domains have been attributed to lipid rafts (regions of membrane enriched in cholesterol and saturated acyl chain lipids that form an ordered liquid phase (Simons and Ikonen, 1997)), while others are thought to be due to interactions between the actin cytoskeleton and the cell membrane (Lommerse *et al.*, 2004b). SMF studies have made it possible to track individual lipids and proteins within the membrane, lending unique insight into membrane structure.

The earliest biologically related SMF experiments tracked the diffusion of individual, fluorescently labelled phospholipids within an artificial bilayer (Schmidt *et al.*, 1996a); for the first time, the trajectories of hundreds of lipids were visualized, from which the lateral diffusion constant of the bilayer was calculated. Unlike the ensemble-averaged measurements commonly obtained using FRAP, single molecule tracing does not mask different types of diffusion within the same measurement area. Such sensitivity to the inhomogeneity of lipid motion has been used to identify biphasic diffusion in fluid-supported bilayers (Schütz *et al.*, 1997b). Information about the rotational mobility of individual lipid molecules has been obtained by both single molecule anisotropy imaging (Harms *et al.*, 1999) and single molecule dichroism (Schütz *et al.*, 1997a).

More recently, the application of SMF techniques *in vivo* has facilitated some particularly interesting insights into the mobility and structure of membranes in live cells. Schütz *et al.* (2000a) introduced Cy5-labelled lipids to the plasma membrane of human coronary artery smooth muscle cells and used epifluorescence microscopy to track them. Membrane microdomains of 0.2–2 μm were observed, within which lipids were confined but still able to diffuse rapidly. Similar compartments of 0.03–0.23 μm were detected in a variety of other cell lines, using single-particle tracking (SPT) of labelled DOPE (1,2-Dioleoyl-s*n*-Glycero-3-phosphoethanolamine) lipids (Murase *et al.*, 2004). In this case, lipids were able to "hop" between domains with an average rate of between 1 and 17 ms.

Membrane–Cytoskeleton Interactions

SMF and SPT evidence also supports the idea that interactions between the lipid membrane and the underlying cytoskeleton create a barrier to membrane diffusion. For example, lipids labelled with 40 nm-diameter gold particles tracked in rat kidney fibroblasts and shown to be "corralled" within compartments (~230 nm) for many milliseconds before hopping to adjacent domains (Fujiwara *et al.*, 2002). Different levels of confinement were revealed, indicating that these "corrals" were located within larger 750 nm compartments (Figure 9.2). Cholesterol depletion had a minimal effect on diffusion, suggesting that lipid rafts were not responsible for obstructing lipid movement, whereas actin depolymerization increased the size of the observed compartments. Based

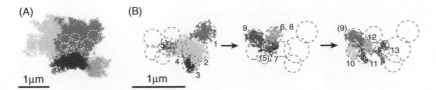

Figure 9.2: Normal rat kidney fibroblastic cell membrane is doubly compartmentalized with regard to translational diffusion of phospholipids. (A) A typical long-term trajectory of DOPE recorded with a 25 μs resolution for a period of 2 s (81 000 frames). Plausible compartments greater than 230 nm are shown in different shades of grey. Smaller 230 nm sub-compartments are marked by circles and shown enlarged (B). Over a period of 120 ms the lipid hops between compartments (numbered according to sequence of occupation). Note that compartments 5, 7, and 12 (also 6 and 8; 4 and 13) match with regard to their positions. Reproduced with permission from Fujiwara et al. (2002)

on these findings and the development of an earlier skeleton fence model (Kusumi and Sako, 1996), the anchored membrane protein picket model was devised. This postulated that transmembrane proteins anchored to the actin membrane skeleton behave as a row of pickets, temporarily fencing in lipids through steric hindrance and hydrodynamic effects (Figure 9.3). This interference of membrane proteins with lipid mobility also provided an explanation for how the cytoskeleton (which is only in contact with the cytoplasmic leaflet) can also induce confinement in the exoplasmic leaflet. Further support for this theory came from evidence showing that expression of ankyrin-G (another important constituent of the membrane skeleton) in polarized neurons correlated well with the suppression of lipid diffusion (Nakada *et al.*, 2003). Single molecule observations of the diffusion of μ-opioid receptors (a type of G-protein-coupled receptor) labelled with either GFP or gold particles have revealed that these proteins are also corralled within domains consistent with the anchored-protein picket model. It has been suggested that the bulk diffusion measured in cell membranes is slower than that in artificial membranes only because of compartmentalization causing a lower average rate, while the actual rate of lipid diffusion is the same (Suzuki *et al.*, 2005).

It is important to note that the involvement of the cytoskeleton in the restriction of protein and lipid diffusion is not universally accepted. Wieser *et al.* (2007) have reported that CD59, a glycosylphosphatidylinositol (GPI)-anchored protein, is not restricted by periodic cytoskeletal barriers while diffusing in T24 (ECV) cells.

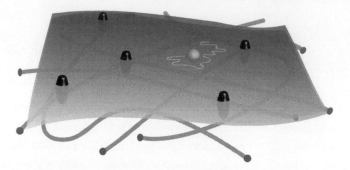

Figure 9.3: The anchored membrane protein picket model proposed by Fujiwara et al. (2002). The actin membrane skeleton (shown as intersecting tubes) is described as a fence and the transmembrane proteins (grey ellipsoids) that are anchored to it as rows of pickets, which temporarily confine diffusing biomolecules (white sphere) by hydrodynamic effects and steric hindrance

Lipid Rafts

Numerous investigations into the composition of lipid rafts and their influence on membrane dynamics have been conducted, but their effects remain unclear. This is in part due to the absence of a standard method for defining and quantifying lipid rafts, making it difficult to combine individual findings into an overall picture (Lommerse *et al.*, 2004b). Further difficulties arise from different studies producing seemingly contradictory results. For example, a study of the diffusion of two types of I-E^k MHC class II showed no evidence for the confinement of either native or GPI-anchored I-E^k, although domains larger than 0.3–4.0 μm^2 could not be ruled out (Vrljic *et al.*, 2002). This appears to contradict a separate study that used SPT to demonstrate the transient trapping of GPI-anchored proteins in cholesterol-enriched domains of ~ 0.03–0.07 μm^2 (Dietrich *et al.*, 2002). Reasons for this difference in behaviour are not yet understood, but may be due to variations between cell lines or experimental techniques. More recent work suggests that changes to cholesterol concentration affect the mobility of native I-E^k and GPI-linked proteins, which may be directly linked to changes in the size and stability of lipid rafts (Vrljic *et al.*, 2005).

Single molecule techniques have also produced evidence for other types of inhomogeneity. For example, by creating a fusion protein of the H-Ras membrane-targeting sequence and eYFP, Lommerse *et al.* (2004a) produced a fluorescent probe that, it was thought, would associate with lipid rafts in the cytoplasmic membrane. Around a third of the H-Ras membrane anchors expressed in both human embryonic kidney cells and mice fibroblast cells were found to be constrained to domains with an average size of 250 nm. Neither actin nor cholesterol extraction was found to significantly affect anchor mobility, suggesting that the H-Ras anchors were associating with a third type of lipid compartment, not due to lipid rafts or the membrane skeleton. In further investigations, Lommerse *et al.* (2006) fused eYFP to the membrane anchors of the GTPase K-Ras (with a supposed negligible affinity for lipid rafts) and the Src-kinase Lck (with an expected high affinity for lipid rafts). They tracked the diffusion of these proteins in the cytoplasmic leaflet of mice fibroblast cells and found that domains (~200 nm) confined a small proportion (~16–27%) of both types of anchor, while the majority of anchor proteins were able to diffuse freely. It was concluded that the 200 nm domains were unlikely to be lipid rafts, and that if rafts were present, they would have to be smaller than the resolution of the system (<137 nm) and have little influence on anchor protein mobility. A second study comparing the diffusion of wild-type H-Ras with constitutively inactive (S17N) and constitutively active (G12V) H-Ras mutants in mice fibroblast cells suggests that localization of H-Ras is dependent on its activation (Dadke and Chernoff, 2003). Major, fast-diffusing and smaller, slow-diffusing populations were identified for each of the three types of protein. For the inactive mutant and the GDP-bound wild type, both populations underwent free diffusion, while for the active mutant and the GTP loaded wild type (treated with insulin), the slow-diffusing fraction was confined to ~200 nm "corrals." It was suggested that these 200 nm domains were caused by the transient trapping of mobile 40 nm non-raft domains (as described by Rotblat *et al.*, 2004), perhaps due to interactions with the actin cytoskeleton, which has been shown to rearrange in the presence of insulin.

Further experimentation is obviously required to clarify seemingly contradictory evidence and to develop a thorough understanding of the mechanisms underpinning protein and lipid confinement. In particular, the conflicting nature of some of these findings may hint at a more interesting underlying mechanism not adequately described by existing models. An important first step would be the formulation of a generally accepted definition for lipid rafts, although this will undoubtedly be difficult to achieve if membrane confinement is a truly dynamic process. Greater spatial and temporal resolutions will probably be required to visualize protein and lipid mobility in enough detail to confirm or disprove the theories of lipid rafts and membrane–cytoskeleton interactions. Although FIONA (Kural *et al.*, 2005; Nan *et al.*, 2005) has not yet been applied to membranes, with reported resolutions of 1.5 nm and 0.3 ms, it has the potential to reveal this behaviour in far greater detail.

Understanding the Dynamics and Interactions of Membrane Proteins

SMF techniques are also providing insight into the stoichiometry, dynamics, and activation of protein complexes involved in a wide range of processes. Early proof-of-principle studies (Schmidt *et al.*, 1996b; Schütz *et al.*, 2000c) have heralded the exploration of many membrane-related processes on a level that allows us to see how individual molecules behave.

Signalling

Activation of Immunological Receptors

The MHC is present in the cellular membrane of antigen-presenting cells and is responsible for displaying antigens on the cell surface. When T-cells detect these antigens, they initiate a signalling cascade that results in the destruction of cells expressing foreign proteins (Alberts *et al.*, 2002). This system of antigen presentation and detection has proved a popular topic for single molecule experiments, due to its immunological significance and the relative ease with which antigens can be fluorescently labelled. We have already seen that Cy5-labelled peptides have been used to track the diffusion of GPI-linked and native MHC class II I-E^k protein complexes to investigate membrane confinement (Vrljic *et al.*, 2002, 2005). Anomalous diffusion of antibody-labelled MHC class I molecules has also been observed on cervical cancer cell membranes (Smith *et al.*, 1999). Several consequences of this diffusion were proposed, including a high local concentration of class I molecules, which might encourage T-cell binding, or the recruitment of class I molecules on antigen-presenting cells to the vicinity of a bound T-cell to enhance cell–cell interaction.

The events that follow antigen detection by T-cells have also been explored with SMF. When a T-cell receptor (TCR) present in the T-cell membrane detects an antigen, the cell becomes activated, and other TCRs, signalling molecules such as Lck, and adaptor proteins such as LAT are recruited to the region of antigen stimulation. Ike *et al.* (2003) have demonstrated that both Lck and the Lck anchor N10, localize in areas of induced TCR clustering in Jurkat T-cells. By analyzing trajectories of individual molecules, they discovered that average diffusion coefficients were lower for Lck and N10 in regions close to the TCR clusters, compared to molecules that were further away. Attempts to chemically disrupt membrane rafts and the actin membrane skeleton led to a significant increase in Lck diffusion throughout the cell. From this, it was proposed that TCR clustering causes a reorganization of actin filaments, increasing the density of the actin membrane skeleton and trapping Lck by a combination of the actin "fence" and the protein "pickets" bound to the actin. However, another study tracking the mobility of fluorescently labelled signalling proteins before and after Jurkat T-cell activation did not observe a significant change in the average diffusion coefficients of Lck and N10 molecules (Douglass and Vale, 2005). Only LAT exhibited a change in average diffusion rate. A post-activation clustering effect similar to that seen by Ike *et al.* (2003) was observed for Lck, LAT, and CD2 (a co-receptor in the membrane of antigen-presenting cells), but not for N10, suggesting that protein–protein interactions may be important for this rearrangement. Clusters remained undisrupted after cells were exposed to a cholesterol-depleting agent, and although actin was essential for their formation, once these domains had formed, they were unaffected by actin depolymerization. This led to the proposal that the microdomains formed by TCR activation were created by protein–protein interactions, not through lipid rafts or interactions between the actin cytoskeleton and the cell membrane.

A recently published study has visualized the motions of single-TCR complexes and the GPI proteins CD48 and CD59 in live T lymphocytes (Drbal *et al.*, 2007). Before initiation of TCR signalling, CD48 and CD59 were found to diffuse freely within the plasma membrane, while after T-cell activation (by initiating TCR clustering, using a coverslip coated with antibodies specific to the TCR subunit, $CD3_\varepsilon$), only CD48 became strongly immobilized and CD59 movement remained unaffected. It was concluded that this demonstrated the heterogeneous composition of lipid rafts, and that lipid raft components are not associated in resting T-cells until triggered by cell activation.

Cell Growth, Proliferation, and Differentiation

Epidermal growth factors (EGF) are important signalling molecules involved in communicating growth-related messages to the cell. EGF bind to epidermal growth factor receptors (EGFR) on the surface of the cell membrane and form signalling dimers (composed of two EGF and two EGFR molecules), which undergo auto-phosphorylation and initiate a signal transduction cascade (Schlessinger, 2002). This leads to changes inside the cell that promote proliferation, differentiation, or migration. Employing both TIRF and epifluorescence, Sako *et al.* (2000) have observed the binding of individual Cy3-labelled EGF molecules to EGFR on the surfaces of human carcinoma cells. When these experiments were combined with single molecule FRET measurements between Cy3-labelled EGF and Cy5-labelled EGF, they indicated that EGF–EGFR dimers form on the cell membrane, and that EGFR auto-phosphorylation occurs after dimerization.

Sako and co-workers also stimulated A431 human carcinoma cells with tetramethyl-rhodamine-conjugated mouse EGF (Rh-EGF) and then detected EGFR activation with fluorescently labelled antibody fragments directed against an activated form of EGFR. After stimulation, the number of phosphorylated EGFR was greater than the number of bound EGF molecules, indicating that some unliganded EGFR had also been activated, amplifying the EGF signal. Based on this, the following amplification mechanism was proposed: Two EGF bind to two EGFR to form a signalling homodimer, which after activation by auto-phosphorylation can dynamically cluster with unliganded, un-activated EGFR to form transient heterodimers that trigger the activation of the unliganded EGFR (Ichinose *et al.*, 2004). In this way, EGFR activation can propagate as more heterodimers form and more unliganded EGFR are activated. In accordance with this model, recent single-wavelength fluorescence cross-correlation spectroscopy (SW-FCCS) and bulk-fluorescence FRET experiments also indicated that EGFR and ErbB2 receptors (another type of receptor tyrosine kinase with a similar structure to EGFR (Schlessinger, 2002)) assemble into pre-formed dimeric structures before activation (Liu *et al.*, 2007).

Further single molecule observations of Rh-EGF have enabled the analysis of the kinetics of EGF binding and signalling dimer formation. Teramura *et al.* (2006) identified two rate constants for the association of EGF with the cell surface and the presence of a binding intermediate during EGFR dimerization. They proposed a model to explain how two possible mechanisms could promote the formation of signalling dimers at low concentrations of EGF. According to this model, a small fraction (1–2%) of EGFR formed "pre-dimers" (two EGFR molecules without bound EGF), to which EGF bound 100 times faster compared to monomeric EGFR, enhancing signalling dimer formation. Consistent with experimental evidence, the model also found that the association constant for the binding of the second EGF molecule to the EGFR dimer was 10 times greater than that for the first EGF molecule. This suggested that the first binding event induced a conformational change in the second binding site, forming an intermediate state with an increased association rate for the second EGF molecule, presenting a second method for accelerating signalling dimer formation.

The first event in the signal transduction sequence following the activation of EGFR has also been studied. The binding of growth factor receptor-binding protein 2 (Grb2) molecules to phosphorylated EGFR has been investigated by using EGF molecules immobilized on glass coverslips to activate EGFR. The bound EGF ensured that receptor stimulation was not interrupted by EGF diffusion or internalization, permitting extended observation of the association and dissociation of Cy3-labelled Grb2 with phosphorylated EGFR (Ichinose *et al.*, 2006).

GABA receptors (GABA$_A$R), instrumental in controlling growth cone steering and elongation, have been labelled with quantum dots and tracked across the membrane of nerve growth cones (Bouzigues and Dahan, 2007). A novel statistical method, the speed correlation index (SCI), was used to analyze the motion of individual GABA$_A$R. GABA$_A$R in the growth cone membrane were found to move with a conveyor-belt motion (displaying transient periods of directed movement), which was terminated upon microtubule depolymerization by nocodazole, and then reinstated after the cells were allowed to recover in fresh medium and microtubule polymerization restarted. It was suggested that the receptors may be carried along microtubules during polymerization, and that this coupling could even offer a means of feedback to regulate cytoskeleton rearrangement. It is possible that receptors that initiate microtubule growth by becoming activated are displaced as a result of the microtubule reorganization, halting the signalling of further growth.

SMF has also been used to investigate the role of histidine kinase PleC in the regulation of *Caulobacter* bacterial cell division. Images of eYFP-labelled PleC diffusing across *Caulobacter* cells were obtained under epifluorescence (Deich *et al.*, 2004). PleC was observed to localize at one of the cell poles, while other PleC molecules remained diffusing throughout the membrane.

Molecular Signalling Switches

Single molecule techniques have been used to visualize the activation of the important G protein Ras, which behaves as a switch in signalling pathways involved in cell migration, proliferation and death, and cytoskeleton reorganization and other processes. Ras is a proto-oncogene product, and its deregulation can coincide with the development of cancers and other diseases (Chien and Hoshijima, 2004; Hancock, 2003). Once activated, Ras transduces signals by binding effectors such as Raf1 kinase, which interact with other molecules, creating a signalling cascade. In an attempt to better understand this process, Hibino *et al.* (2003) used a combination of single molecule and ensemble techniques to observe the Ras–Raf1 interaction initiated by stimulation with EGF. HeLa cells were transfected with YFP-tagged H-Ras (an isoform of Ras) and GFP-tagged Raf1. After expression, these proteins were imaged while EGF was introduced to activate H-Ras. Bulk epifluorescence observations showed the recruitment of GFP-Raf1 from the cytoplasm to the plasma membrane after H-Ras activation. The binding and disassociation of individual GFP-Raf1 spots with the membrane were measured using TIRF and used to reconstruct dissociation curves from which binding periods could be calculated. Two binding states of Raf1 to the membrane were identified, which were suggested to be either due to multiple states of Ras activation, or Raf1 binding sites on proteins other than Ras.

Single molecule FRET has also been used to visualize Ras activation *in vivo*. A fluorescent GTP analogue (BodipyTR-GTP) acceptor was injected into human epidermoid mouth carcinoma cells expressing either H-Ras or K-Ras fused with a YFP donor (Murakoshi *et al.*, 2004). When stimulated with EGF, pre-bound GDP was released from Ras, permitting the binding of BodipyTR-GTP and inducing FRET. After activation, the diffusion of Ras over the bilayer appeared significantly reduced. Cholesterol depletion did not block Ras immobilization, suggesting that lipid rafts were unlikely to be the cause of impeded diffusion. From these findings, a model was proposed in which activated Ras specifically binds to activated Ras scaffolding proteins, initiating the formation of transient signalling complexes that include effector molecules (like Raf) and deactivating proteins. Large complexes would then bind to or become trapped in the actin membrane skeleton, forming membrane compartments.

Other Signalling Pathways

Several other signalling pathways have been probed with SMF methods, including chemotactic signalling in *Dictyostelium* cells (Ueda *et al.*, 2001) and the stoichiometry and turnover of MotB, a protein responsible for coupling ion flow to torque generation in the bacterial flagellar motor (Leake *et al.*, 2006).

Single glycine receptors in the neuronal membrane of live cells have been conjugated to quantum dots, enabling the characterization of distinct membrane diffusion regions (Dahan *et al.*, 2003). It has also been possible to observe the diffusion of AMPA glutamate receptors on both sides of the synapse in live rat hippocampal neurons, using fluorescently labelled antibodies. This has provided insight into the accumulation of these receptors at the synapses where they mediate rapid neurotransmission (Tardin *et al.*, 2003). Separately, TIRF has been used to visualize the binding and release of single pleckstrin homology domains fused with eGFP at the plasma membrane of mouse myoblast cells, increasing understanding of how these domains recruit proteins to the membrane (Mashanov *et al.*, 2004).

Membrane Transport

The mechanisms by which membrane proteins transport matter across both the outer and nuclear membranes regulate the cellular environment and consequently cell function. Ensemble techniques have been used extensively to investigate some of these proteins, while SMF has been applied only relatively recently.

The nuclear pore complex (NPC) is one such transporter system, which straddles the inner and outer nuclear membranes and mediates the movement of material between the nucleus and cytoplasm. The contribution made by single molecule studies to understanding the distinct transport modes supported by NPCs has been extensively reviewed elsewhere (Peters, 2007).

Ion Channels

Ion channels are transmembrane pores that comprise another significant group of membrane transport proteins by facilitating the movement of ions between the intra- and extra-cellular environments. Present in an enormous range of organisms, they help to regulate electrochemical gradients across the membrane, which are involved in many processes, from the transmission of nervous signals and regulation of the beating heart to the sensing of light and smell (Craven and Zagotta, 2006). Ion channel gating (opening and closing) can be triggered by changes in membrane voltage, ligand binding, or variations in membrane tension, depending on channel type. The application of single molecule techniques to ion channels is discussed in more detail in Chapter 8. Here we will focus on how SMF techniques in particular have added to our understanding of ion channel mobility and the conformational changes that occur during gating.

Several different ion channels have been fluorescently labelled and visualized *in vivo* on a single molecule level. $K_v1.3$ voltage gated potassium channels were found to be sparsely distributed throughout Jurkat cells when they were labelled by exposure to a fluorescently tagged hongotoxin and imaged in 3D (Schütz *et al.*, 2000b). When human cardiac L-type Ca^{2+} channels were labelled with eYFP and tracked individually in human embryonic kidney cells (HEK293), they were observed to aggregate, and measurements of their lateral and rotational movement across the bilayer became possible (Harms *et al.*, 2001). SMF has also been used to study the assembly of bacterial pore-forming toxins. Single-pair FRET observations have been made of the assembly of leukocidin fast fraction (LukF) and γ-hemolysin second component (HS) monomers into pore-forming oligomers (Nguyen *et al.*, 2003). Changes in the fluorescence intensity of an environment-sensitive fluorophore attached to LukF have also been monitored to demonstrate the subsequent insertion of these oligomers into erythrocyte membranes (Nguyen *et al.*, 2006).

Electrical recordings of ionic current flow through a single ion channel (also known as single-channel recording, SCR) were one of the earliest forms of single molecule measurement (Neher and Sakmann, 1976; Chapter 8). Over the last decade, efforts have been made to combine electrophysiological SCR with SMF imaging to better understand the mechanisms by which channel gating is controlled. Synchronous bulk-fluorescence and electrophysiological measurements of multiple olfactory-type CNGA2 channels in inside-out patches on the plasma membranes of *Xenopus laevis* oocytes have already been achieved (Biskup *et al.*, 2007), but work to recreate this on the single molecule level continues. Ide and Yanagida (1999) devised a TIRF system that demonstrated the potential for simultaneous measurement of channel position and electrical activity in artificial lipid bilayers. Initial studies showed Cy3-labelled alamethicin peptides assembling into conducting pores, which were later followed by observations of liposomes containing nystatins and ergosterols fusing with bilayers (Ide *et al.*, 2002). Single-pair FRET has also been employed by the Schütz group to show the heterodimerization of two gramicidin peptides (one labelled with a Cy3 donor and the other with a Cy5 acceptor) as they form a conducting channel in lipid bilayers. The sudden appearance of a FRET signal spot (due to two peptides moving sufficiently close for their fluorophores to undergo resonance energy transfer) was shown to coincide with an increase in electrical current, consistent with the formation of a gramicidin channel (Borisenko *et al.*, 2003). Some uncorrelated events were also noted, such as the presence of a FRET spot in the absence of an electrical channel measurement. One possible interpretation was that gramicidin channels may have nonconducting dimeric intermediates, illustrating how combined SMF and electrical measurements could potentially be used to probe the structural rearrangements experienced by ion channels.

Further evidence of the power of combining these techniques for probing channel conformation comes from the Lu group, who applied combined electrophysiology and single-pair FRET with fluorescence self-quenching to gramicidin ion channels (Harms *et al.*, 2003; Lu, 2005). They used a system similar to that of Borisenko *et al.* to measure the FRET efficiency distributions of gramicidin heterodimers that coincided with either "open" or "closed" electrical currents. Fluorescence intensity distributions corresponding to "open" and "closed" current levels were calculated from self-quenching measurements of gramicidin homodimers (the two TMR fluorophores on each ion channel experienced maximum self-quenching, and therefore minimum fluorescence, at their closest separation). Both FRET and self-quenching data indicated the presence of multiple channel conformations corresponding to the open and closed states. Some simultaneous measurements have also been made on single ion channels *in vivo*, including those to detect the voltage-driven structural rearrangements of fluorescently labelled Shaker K^+ channels (Sonnleitner *et al.*, 2002).

To date, few simultaneous SCR and SMF measurements have been published for any ion channel more complex than alamethicin or gramicidin. This may well be because unlike other channels with more complicated gating mechanisms, there is a clearly defined "activation" moment when two subunits in the top and bottom leaflets of the membrane align to form a pore. Similarly, FRET labelling of gramicidin monomers has proved a successful way of generating a clear energy transfer signal at the same "activation" instant, making it easier to distinguish between the fluorescence attributable to an open channel, compared to lone subunits and background.

An additional challenge to synchronizing SCR and SMF *in vitro* is confirming that only one channel is present in the bilayer, so that any electrical activity can be attributed to a single-channel system. It is often not possible to identify a single channel from SCR alone, as two channels opening at different times look much the same as one channel opening repeatedly. If the entire bilayer could be imaged at once, it would be possible to confirm the number of fluorescently labelled channels present, but this is not trivial. To improve experiments where only part of the bilayer is imaged, annexin A5 (which binds preferentially to negatively charged lipids) has been shown to reduce the lateral diffusion coefficients of ryanodine receptor type 2 ion channels by 200-fold (Ichikawa *et al.*, 2006). Annexin A5 could therefore be used to immobilize channel proteins within the microscope field of view to allow extended observation.

Vesicle Tracking, Docking, and Fusion

The varied nature of vesicle cargos makes them an important component of many biological processes. It is therefore vital to ensure that vesicles dock and fuse with the correct target membrane in a highly selective fashion, releasing their cargos at the intended destination. SNAREs (soluble *N*-ethylmaleimide-sensitive factor attachment protein receptors) provide one such mechanism and are particularly important in facilitating synaptic transmission (Wang and Tang, 2006). Each membrane-enclosed compartment within the cell has its own particular pair of SNAREs: a target t-SNARE (containing syntaxin and SNAP-25) in the membrane of the organelle and a complementary v-SNARE (containing synaptobrevin) located on the vesicle to be fused. Helical domains in both SNAREs wrap around each other to form a stable t-SNARE complex (SNAREpin) that forms a link between the two membranes. They are then thought to facilitate membrane fusion by drawing the membranes together, forming a new bilayer that finally ruptures. The exact mechanism by which this is achieved is not fully understood, but it is thought that energy released from the intertwining of the helices may contribute to overcoming the energy barrier to fusion.

In vivo, individual synaptic vesicles have been labelled with fluorescent lipid and tracked in the terminals of goldfish retinal bipolar neurons (Zenisek *et al.*, 2000). Preferred zones of exocytosis were identified and suggested to coincide with structures believed to capture and position vesicles against the plasma membrane. Similar tracking and fusion studies in the small terminals of rat hippocampal neurons showed that most fusion events involved only a partial loss of dye from synaptic vesicles, and that vesicles could fuse more than once when repeatedly stimulated by action potentials, consistent with the idea that vesicles are reused for neurotransmitter release (Aravanis *et al.*, 2003). Fusion events between viruses and target cells have also been visualized. Viruses containing avian sarcoma and leukosis virus envelope glycoprotein were labelled with a lipophilic dye to indicate lipid transfer between contacting membrane leaflets, and with palmitylated eYFP (which remains in the inner leaflet of the viral membrane until a fusion pore is formed between the viral inner leaflet and the cell plasma membrane) and gagYFP (which is incorporated into the virus nucleocapsid) to show content transfer after pore formation (Melikyan *et al.*, 2005). By imaging single virions, it was possible to follow the key virus fusion events of hemifusion, small pore formation, and pore enlargement. The influenza infection pathway has also been investigated by tracking single viruses from the periphery of CHO cells into the peri-nuclear region, where they finally fuse with the endosomal membrane (Lakadamyali *et al.*, 2003). Three distinct phases were identified within this movement: The first was slow, actin-dependent active transport in the cell periphery; the second was rapid, unidirectional movement likely to be driven by dynein along microtubules; and the third was an intermit-tent, bidirectional motion attributed to both dynein and kinesin motors.

In vitro, the formation of supported lipid bilayers has been investigated by imaging individual vesicles as they fuse and spread over a quartz support (Johnson *et al.*, 2002). Fluorescently labelled lipids, and a fluorescent dye encapsulated within the vesicles, made it possible to identify different fusion pathways. The fusion of individual vesicles, reconstituted with v-SNAREs, to planar lipid bilayers containing t-SNAREs has also been examined using TIRF (Fix *et al.*, 2004). Fluorescent lipid incorporated in the v-SNARE vesicles, excited only when within the range of the evanescent field, displayed a sudden increase in fluorescence when a vesicle approached and fused with the bilayer. As the lipid diffused out across the bilayer, the fluorescence intensity decreased. A similar method (Liu *et al.*, 2005) has been used to measure the rates of docking ($2.2 \times 10^7 \, \text{M}^{-1}\text{s}^{-1}$ at low SNARE densities and at $37\,^{\circ}\text{C}$) and fusion ($40 \pm 15 \, \text{s}^{-1}$) for individual v-SNARE-reconstituted vesicles and bilayers containing t-SNAREs. Single fusion events between vesicles reconstituted with yeast v-SNARE proteins and membrane-intercalating fluorescent donor dye, and vesicles reconstituted with t-SNARE proteins and a fluorescent acceptor have been observed using TIRF. The FRET value during a fusion event depended on the degree of lipid mixing at that moment, making it possible to detect fusion intermediates. This was a clear demonstration of how, as with single molecule studies, detecting individual fusion events can reveal more than an ensemble-averaged measurement.

To date, a limited number of SMF studies have been conducted on vesicle fusion. Those that have, highlight the potential for future studies to reveal the interactions and conformational changes of the proteins involved in far greater detail. FRET, lifetime, and anisotropy measurements have been made of individual syntaxin 1 proteins, which were double-labelled with acceptor and donor fluorophores (Margittai *et al.*, 2003). While freely diffusing in buffer, the protein was observed to switch between "open" and "closed" states, corresponding to SNARE conformations where the SNARE motif is either available or blocked and unable to form t-SNARE complexes. The SNARE mechanism has also been probed on a single molecule level in a lipid bilayer environment. Bowen *et al.* (2004) showed that docking and thermally induced fusion were possible in the absence of SNAP-25, by visualizing fusion between vesicles reconstituted with fluorescently labelled synaptobrevin and artificial bilayers reconstituted with syntaxin. Conformations of the SNARE complex after vesicle docking with the bilayer were observed by measuring the FRET between labelled syntaxin and synaptobrevin molecules. Experiments were conducted in the presence and absence of SNAP-25, and a combination of parallel and antiparallel configurations was identified for the complete three-protein complex. Intermediate FRET states were detected for syntaxin–synaptobrevin binary complexes, which may have been the result of partial folding in the absence of SNAP-25.

Conclusions

Several localization techniques have the potential to image membrane events, which typically occur on millisecond timescales, but have not (as far as we are aware) yet been used to do this. FIONA and SHREC are probably the two most applicable of the methods discussed here, as localizations with resolutions of up to 1.5 nm and 0.3–1.1 ms have already been reported for FIONA (Kural *et al.*, 2005; Nan *et al.*, 2005), while SHREC can localize fluorophores at separations of <10 nm (Churchman *et al.*, 2005). Techniques based on compiling a complete image of a sample by exciting only a subpopulation of fluorophores at one time (namely, SHRImP, NALMS, PALM, or STORM) would require significant improvements to imaging times, which are typically of the order of minutes. They could, however, be used to take snapshots of a membrane system from which information about the relative concentrations of two components could be obtained.

Although stimulated-emission-depletion fluorescence microscopy (STED) (Hell, 2003; Hell and Wichmann, 1994) and its generalization, reversible saturable optical (fluorescence) transitions (RESOLFT) (Schwentker *et al.*, 2007), are not currently capable of SMF, they may be developed to enable single molecule visualization in the future (Weiss,

2000). STED is capable of imaging with lateral resolutions of up to 16 nm (Westphal and Hell, 2005) and temporal resolutions of the order of tens of milliseconds. Although both temporal and spatial resolutions would need to be improved before STED could capture individual molecular membrane events in detail, it has already been shown to be compatible with live-cell imaging. Vacuolar membranes on live yeast cells (Klar *et al.*, 2000) and the distribution of F_1F_o–ATP synthase and Tom20 on the mitochondrial outer membrane of mammalian cells (Donnert *et al.*, 2007) have both been successfully imaged.

As with most biological applications of SMF, the primary challenges facing membrane studies are related to labelling (Foley and Burkart, 2007) and detection technology (Michalet *et al.*, 2007). Work is ongoing to engineer fluorescent labels, which will cause minimal interference with the systems being imaged and have improved photophysical properties. These properties include increased brightness for a larger signal-to-noise ratio, emission spectra shifted away from that of cellular auto-fluorescence, and reduced susceptibility to blinking, photobleaching, spectral jumps, dark and triplet states, and changes in quantum efficiency. In addition to the fluorescent dyes and proteins more commonly used, semiconductor quantum dots offer attractive advantages such as significantly increased brightness, greater resistance to photobleaching, and broad absorption and narrow emission spectra, which can be adjusted. Significant challenges to the use of quantum dots *in vivo* include their size (which is comparable to proteins) and surface properties (which render them sensitive to environmental changes, inclined to aggregate, and difficult to introduce into cells) (Jaiswal and Simon, 2004).

Given that the resolution achievable by localization techniques is ultimately limited only by the number of photons detected from a single fluorophore, it is highly likely that we will see improvements in the spatial and temporal resolutions attainable by existing and future SMF techniques. The applications of these methods to membranes will reveal membrane structure and function in far greater detail than currently possible. In this expanding field, it is likely that many of the most exciting and insightful findings are yet to come.

References

Alberts, B., Johnson, A., Lewis, J., Raff, M., Roberts, K., and Walter, P. (2002). *Molecular Biology of the Cell*. Taylor and Francis Group, New York.

Anderson, C. M., Georgiou, G. N., Morrison, I. E., Stevenson, G. V., and Cherry, R. J. (1992). Tracking of cell surface receptors by fluorescence digital imaging microscopy using a charge-coupled device camera. Low-density lipoprotein and influenza virus receptor mobility at 4 degrees C. *J Cell Sci* 101, 415–425.

Aravanis, A. M., Pyle, J. L., and Tsien, R. W. (2003). Single synaptic vesicles fusing transiently and successively without loss of identity. *Nature* 423, 643–647.

Bacia, K., Kim, S. A., and Schwille, P. (2006). Fluorescence cross-correlation spectroscopy in living cells. *Nat Methods* 3, 83–89.

Bacia, K., Majoul, I. V., and Schwille, P. (2002). Probing the endocytic pathway in live cells using dual-colour fluorescence cross-correlation analysis. *Biophys J* 83, 1184–1193.

Balci, H., Ha, T., Sweeney, H. L., and Selvin, P. R. (2005). Interhead distance measurements in myosin VI via SHRImP support a simplified hand-over-hand model. *Biophys J* 89, 413–417.

Betzig, E., Patterson, G. H., Sougrat, R., Lindwasser, O. W., Olenych, S., Bonifacino, J. S., Davidson, M. W., Lippincott-Schwartz, J., and Hess, H. F. (2006). Imaging intracellular fluorescent proteins at nanometer resolution. *Science* 313, 1642–1645.

Biskup, C., Kusch, J., Schulz, E., Nache, V., Schwede, F., Lehmann, F., Hagen, V., and Benndorf, K. (2007). Relating ligand binding to activation gating in CNGA2 channels. *Nature* 446, 440–443.

Borisenko, V., Lougheed, T., Hesse, J., Füreder-Kitzmüller, E., Fertig, N., Behrends, J. C., Woolley, G. A., and Schütz, G. J. (2003). Simultaneous optical and electrical recording of single gramicidin channels. *Biophys J* 84, 612–622.

Bouzigues, C. and Dahan, M. (2007). Transient directed motions of GABAA receptors in growth cones detected by a speed correlation index. *Biophys J* 92, 654–660.

Bowen, M. E., Weninger, K., Brunger, A. T., and Chu, S. (2004). Single molecule observation of liposome-bilayer fusion thermally induced by soluble N-ethyl maleimide sensitive-factor attachment protein receptors (SNAREs). *Biophys J* 87, 3569–3584.

Campbell, R. E., Tour, O., Palmer, A. E., Steinbach, P. A., Baird, G. S., Zacharias, D. A., and Tsien, R. Y. (2002). A monomeric red fluorescent protein. *Proc Natl Acad Sci U S A* 99, 7877–7882.

Cheezum, M. K., Walker, W. F., and Guilford, W. H. (2001). Quantitative comparison of algorithms for tracking single fluorescent particles. *Biophys J* 81, 2378–2388.

Cheng, Y. L., Bushby, R. J., Evans, S. D., Knowles, P. F., Miles, R. E., and Ogier, S. D. (2001). Single ion channel sensitivity in suspended bilayers on micromachined supports. *Langmuir* 17, 1240–1242.

Chien, K. R. and Hoshijima, M. (2004). Unravelling Ras signals in cardiovascular disease. *Nat Cell Biol* 6, 807–808.

Churchman, L. S., Ökten, Z., Rock, R. S., Dawson, J. F., and Spudich, J. A. (2005). Single molecule high-resolution colocalization of Cy3 and Cy5 attached to macromolecules measures intramolecular distances through time. *Proc Natl Acad Sci U S A* 102, 1419–1423.

Clegg, R. M. (1995). Fluorescence resonance energy transfer. *Curr Opin Biotechnol* 6, 103–110.

Craven, K. B. and Zagotta, W. N. (2006). CNG and HCN channels: two peas, one pod. *Annu Rev Physiol* 68, 375–401.

Dadke, S. and Chernoff, J. (2003). Protein-tyrosine phosphatase 1B mediates the effects of insulin on the actin cytoskeleton in immortalized fibroblasts. *J Biol Chem* 278, 40607–40611.

Dahan, M., Lévi, S., Luccardini, C., Rostaing, P., Riveau, B., and Triller, A. (2003). Diffusion dynamics of glycine receptors revealed by single-quantum dot tracking. *Science* 302, 442–445.

Deich, J., Judd, E. M., McAdams, H. H., and Moerner, W. E. (2004). Visualization of the movement of single histidine kinase molecules in live caulobacter cells. *Proc Natl Acad Sci U S A* 101, 15921–15926.

Dietrich, C., Yang, B., Fujiwara, T., Kusumi, A., and Jacobson, K. (2002). Relationship of lipid rafts to transient confinement zones detected by single particle tracking. *Biophys J* 82, 274–284.

Doeven, M. K., Folgering, J. H. A., Krasnikov, V., Geertsma, E. R., van den Bogaart, G., and Poolman, B. (2005). Distribution, lateral mobility and function of membrane proteins incorporated into giant unilamellar vesicles. *Biophys J* 88, 1134–1142.

Donnert, G., Keller, J., Wurm, C. A., Rizzoli, S. O., Westphal, V., Schönle, A., Jahn, R., Jakobs, S., Eggeling, C., and Hell, S. W. (2007). Two-colour far-field fluorescence nanoscopy. *Biophys J* 92, L67–L69.

Douglass, A. D. and Vale, R. D. (2005). Single molecule microscopy reveals plasma membrane microdomains created by protein-protein networks that exclude or trap signalling molecules in T cells. *Cell* 121, 937–950.

Drbal, K., Moertelmaier, M., Holzhauser, C., Muhammad, A., Fuertbauer, E., Howorka, S., Hinterberger, M., Stockinger, H., and Schütz, G. J. (2007). Single molecule microscopy reveals heterogeneous dynamics of lipid raft components upon TCR engagement. *Int Immunol* 19, 675–684.

Duncan, R. R. (2006). Fluorescence lifetime imaging microscopy (FLIM) to quantify protein-protein interactions inside cells. *Biochem Soc Trans* 34, 679–682.

Fix, M., Melia, T. J., Jaiswal, J. K., Rappoport, J. Z., You, D., Söllner, T. H., Rothman, J. E., and Simon, S. M. (2004). Imaging single membrane fusion events mediated by SNARE proteins. *Proc Natl Acad Sci U S A* 101, 7311–7316.

Foley, T. L. and Burkart, M. D. (2007). Site-specific protein modification: advances and applications. *Curr Opin Chem Biol* 11, 12–19.

Forkey, J. N., Quinlan, M. E., Shaw, M. A., Corrie, J. E. T., and Goldman, Y. E. (2003). Three-dimensional structural dynamics of myosin V by single molecule fluorescence polarization. *Nature* 422, 399–404.

Fujiwara, T., Ritchie, K., Murakoshi, H., Jacobson, K., and Kusumi, A. (2002). Phospholipids undergo hop diffusion in compartmentalized cell membrane. *J Cell Biol* 157, 1071–1082.

Gakamsky, D. M., Luescher, I. F., Pramanik, A., Kopito, R. B., Lemonnier, F., Vogel, H., Rigler, R., and Pecht, I. (2005). CD8 kinetically promotes ligand binding to the T cell antigen receptor. *Biophys J* 89, 2121–2133.

García-Sáez, A. J. and Schwille, P. (2007). Single molecule techniques for the study of membrane proteins. *Appl Microbiol Biotechnol* 76 (2), 257–266.

Ghosh, R. N. and Webb, W. W. (1994). Automated detection and tracking of individual and clustered cell surface low density lipoprotein receptor molecules. *Biophys J* 66, 1301–1318.

Gordon, M. P., Ha, T., and Selvin, P. R. (2004). Single molecule high-resolution imaging with photobleaching. *Proc Natl Acad Sci U S A* 101, 6462–6465.

Greenleaf, W. J., Woodside, M. T., and Block, S. M. (2007). High-resolution, single molecule measurements of biomolecular motion. *Annu Rev Biophys Biomol Struct* 36, 171–190.

Griffin, B. A., Adams, S. R., and Tsien, R. Y. (1998). Specific covalent labelling of recombinant protein molecules inside live cells. *Science* 281, 269–272.

Gross, D. and Webb, W. W. (1986). Molecular counting of low-density-lipoprotein particles as individuals and small clusters on cell surfaces. *Biophys J* 49, 901–911.

Ha, T. (2001). Single molecule fluorescence resonance energy transfer. *Methods* 25, 78–86.

Ha, T., Enderle, T., Ogletree, D. F., Chemla, D. S., Selvin, P. R., and Weiss, S. (1996). Probing the interaction between two single molecules: fluorescence resonance energy transfer between a single donor and a single acceptor. *Proc Natl Acad Sci U S A* 93, 6264–6268.

Hancock, J. F. (2003). Ras proteins: different signals from different locations. *Nat Rev Mol Cell Biol* 4, 373–385.

Harms, G. S., Cognet, L., Lommerse, P. H. M., Blab, G. A., Kahr, H., Gamsjäger, R., Spaink, H. P., Soldatov, N. M., Romanin, C., and Schmidt, T. (2001). Single molecule imaging of L-type Ca^{2+} channels in live cells. *Biophys J* 81, 2639–2646.

Harms, G. S., Orr, G., Montal, M., Thrall, B. D., Colson, S. D., and Lu, H. P. (2003). Probing conformational changes of gramicidin ion channels by single molecule patch-clamp fluorescence microscopy. *Biophys J* 85, 1826–1838.

Harms, G. S., Sonnleitner, M., Schütz, G. J., Gruber, H. J., and Schmidt, T. (1999). Single molecule anisotropy imaging. *Biophys J* 77, 2864–2870.

Haustein, E. and Schwille, P. (2007). Fluorescence correlation spectroscopy: novel variations of an established technique. *Annu Rev Biophys Biomol Struct* 36, 151–169.

Hell, S. W. (2003). Toward fluorescence nanoscopy. *Nat Biotechnol* 21, 1347–1355.

Hell, S. W. and Wichmann, J. (1994). Breaking the diffraction resolution limit by stimulated-emission: stimulated-emission-depletion fluorescence microscopy. *Opt Lett* 19, 780–782.

Heron, A. J., Thompson, J. R., Mason, A. E., and Wallace, M. I. (2007). Direct detection of membrane channels from gels using water-in-oil droplet bilayers. *J Am Chem Soc* 129, 16042–16047.

Hibino, K., Watanabe, T. M., Kozuka, J., Iwane, A. H., Okada, T., Kataoka, T., Yanagida, T., and Sako, Y. (2003). Single- and multiple-molecule dynamics of the signalling from H-Ras to cRaf-1 visualized on the plasma membrane of living cells. *Chemphyschem* 4, 748–753.

Holtzer, L., Meckel, T., and Schmidt, T. (2007). Nanometric three-dimensional tracking of individual quantum dots in cells. *Appl Phys Lett* 90, 053902.

Ichikawa, T., Aoki, T., Takeuchi, Y., Yanagida, T., and Ide, T. (2006). Immobilizing single lipid and channel molecules in artificial lipid bilayers with annexin A5. *Langmuir* 22, 6302–6307.

Ichinose, J., Morimatsu, M., Yanagida, T., and Sako, Y. (2006). Covalent immobilization of epidermal growth factor molecules for single molecule imaging analysis of intracellular signalling. *Biomaterials* 27, 3343–3350.

Ichinose, J., Murata, M., Yanagida, T., and Sako, Y. (2004). EGF signalling amplification induced by dynamic clustering of EGFR. *Biochem Biophys Res Commun* 324, 1143–1149.

Ide, T., Takeuchi, Y., and Yanagida, T. (2002). Development of an experimental apparatus for simultaneous observation of optical and electrical signals from single ion channels. *Single Mol* 3, 33–42.

Ide, T. and Yanagida, T. (1999). An artificial lipid bilayer formed on an agarose-coated glass for simultaneous electrical and optical measurement of single ion channels. *Biochem Biophys Res Commun* 265, 595–599.

Ike, H., Kosugi, A., Kato, A., Iino, R., Hirano, H., Fujiwara, T., Ritchie, K., and Kusumi, A. (2003). Mechanism of Lck recruitment to the T-cell receptor cluster as studied by single molecule-fluorescence video imaging. *Chemphyschem* 4, 620–626.

Jaiswal, J. K. and Simon, S. M. (2004). Potentials and pitfalls of fluorescent quantum dots for biological imaging. *Trends Cell Biol* 14, 497–504.

Janmey, P. A. and Kinnunen, P. K. J. (2006). Biophysical properties of lipids and dynamic membranes. *Trends Cell Biol* 16, 538–546.

Johnson, J. M., Ha, T., Chu, S., and Boxer, S. G. (2002). Early steps of supported bilayer formation probed by single vesicle fluorescence assays. *Biophys J* 83, 3371–3379.

Kahya, N., Scherfeld, D., Bacia, K., and Schwille, P. (2004). Lipid domain formation and dynamics in giant unilamellar vesicles explored by fluorescence correlation spectroscopy. *J Struct Biol* 147, 77–89.

Keller, C. A., Glasmästar, K., Zhdanov, V. P., and Kasemo, B. (2000). Formation of supported membranes from vesicles. *Phys Rev Lett* 84, 5443–5446.

Klar, T. A., Jakobs, S., Dyba, M., Egner, A., and Hell, S. W. (2000). Fluorescence microscopy with diffraction resolution barrier broken by stimulated emission. *Proc Natl Acad Sci U S A* 97, 8206–8210.

Kural, C., Kim, H., Syed, S., Goshima, G., Gelfand, V. I., and Selvin, P. R. (2005). Kinesin and dynein move a peroxisome *in vivo*: a tug-of-war or coordinated movement? *Science* 308, 1469–1472.

Kusumi, A., Nakada, C., Ritchie, K., Murase, K., Suzuki, K., Murakoshi, H., Kasai, R., Kondo, J., and Fujiwara, T. (2005). Paradigm shift of the plasma membrane concept from the two-dimensional continuum fluid to the partitioned fluid: high speed single molecule tracking of membrane molecules. *Annu Rev Biophys Biomol Struct* 34, 351–378.

Kusumi, A. and Sako, Y. (1996). Cell surface organization by the membrane skeleton. *Curr Opin Cell Biol* 8, 566–574.

Lakadamyali, M., Rust, M. J., Babcock, H. P., and Zhuang, X. (2003). Visualizing infection of individual influenza viruses. *Proc Natl Acad Sci U S A* 100, 9280–9285.

Larson, D. R., Gosse, J. A., Holowka, D. A., Baird, B. A., and Webb, W. W. (2005). Temporally resolved interactions between antigen-stimulated IgE receptors and Lyn kinase on living cells. *J Cell Biol* 171, 527–536.

Lasic, D. D. (1988). The mechanism of vesicle formation. *Biochem J* 256, 1–11.

Lata, S., Gavutis, M., Tampe, R., and Piehler, J. (2006). Specific and stable fluorescence labelling of histidine-tagged proteins for dissecting multi-protein complex formation. *J Am Chem Soc* 128, 2365–2372.

Leake, M. C., Chandler, J. H., Wadhams, G. H., Bai, F., Berry, R. M., and Armitage, J. P. (2006). Stoichiometry and turnover in single, functioning membrane protein complexes. *Nature* 443, 355–358.

Lin, R. C. and Scheller, R. H. (2000). Mechanisms of synaptic vesicle exocytosis. *Annu Rev Cell Dev Biol* 16, 19–49.

Lippincott-Schwartz, J., Altan-Bonnet, N., and Patterson, G. H. (2003). Review: Photobleaching and photoactivation: following protein dynamics in living cells. *Nat Cell Biol* 5, S7–S14.

Lippincott-Schwartz, J. and Patterson, G. H. (2003). Development and use of fluorescent protein markers in living cells. *Science* 300, 87–91.

Liu, P., Sudhaharan, T., Koh, R. M. L., Hwang, L. C., Ahmed, S., Maruyama, I. N., and Wohland, T. (2007). Investigation of the dimerization of proteins from the epidermal growth factor receptor family by single wavelength fluorescence cross-correlation spectroscopy. *Biophys J* 93, 684–698.

Liu, T., Tucker, W. C., Bhalla, A., Chapman, E. R., and Weisshaar, J. C. (2005). SNARE-driven, 25-millisecond vesicle fusion *in vitro*. *Biophys J* 89, 2458–2472.

Lommerse, P. H. M., Blab, G. A., Cognet, L., Harms, G. S., Snaar-Jagalska, B. E., Spaink, H. P., and Schmidt, T. (2004a). Single molecule imaging of the H-Ras membrane-anchor reveals domains in the cytoplasmic leaflet of the cell membrane. *Biophys J* 86, 609–616.

Lommerse, P. H. M., Spaink, H. P., and Schmidt, T. (2004b). *In vivo* plasma membrane organization: results of biophysical approaches. *Biochim Biophys Acta Biomembr* 1664, 119–131.

Lommerse, P. H. M., Vastenhoud, K., Pirinen, N. J., Magee, A. I., Spaink, H. P., and Schmidt, T. (2006). Single molecule diffusion reveals similar mobility for the Lck, H-Ras, and K-Ras membrane anchors. *Biophys J* 91, 1090–1097.

Lu, H. P. (2005). Probing single molecule protein conformational dynamics. *Acc Chem Res* 38, 557–565.

Lu, H. P., Xun, L. Y., and Xie, X. S. (1998). Single molecule enzymatic dynamics. *Science* 282, 1877–1882.

Margittai, M., Widengren, J., Schweinberger, E., Schroder, G. F., Felekyan, S., Haustein, E., Konig, M., Fasshauer, D., Grubmuller, H., Jahn, R., and Seidel, C. A. M. (2003). Single molecule fluorescence resonance energy transfer reveals a dynamic equilibrium between closed and open conformations of syntaxin 1. *Proc Natl Acad Sci U S A* 100, 15516–15521.

Marks, K. M. and Nolan, G. P. (2006). Chemical labelling strategies for cell biology. *Nat Methods* 3, 591–596.

Mashanov, G. I., Tacon, D., Peckham, M., and Molloy, J. F. (2004). The spatial and temporal dynamics of pleckstrin homology domain binding at the plasma membrane measured by imaging single molecules in live mouse myoblasts. *J Biol Chem* 279, 15274–15280.

Matz, M. V., Fradkov, A. F., Labas, Y. A., Savitsky, A. P., Zaraisky, A. G., Markelov, M. L., and Lukyanov, S. A. (1999). Fluorescent proteins from nonbioluminescent *Anthozoa* species. *Nat Biotechnol* 17, 969–973.

Medina, M.Á. and Schwille, P. (2002). Fluorescence correlation spectroscopy for the detection and study of single molecules in biology. *Bioessays* 24, 758–764.

Melikyan, G. B., Barnard, R. J. O., Abrahamyan, L. G., Mothes, W., and Young, J. A. T. (2005). Imaging individual retroviral fusion events: from hemifusion to pore formation and growth. *Proc Natl Acad Sci U S A* 102, 8728–8733.

Michalet, X., Kapanidis, A. N., Laurence, T., Pinaud, F., Doose, S., Pflughoefft, M., and Weiss, S. (2003). The power and prospects of fluorescence microscopies and spectroscopies. *Annu Rev Biophys Biomol Struct* 32, 161–182.

Michalet, X., Siegmund, O. H. W., Vallerga, J. V., Jelinsky, P., Millaud, J. E., and Weiss, S. (2007). Detectors for single molecule fluorescence imaging and spectroscopy. *J Mod Opt* 54, 239–281.

Moerner, W. E. and Fromm, D. P. (2003). Methods of single molecule fluorescence spectroscopy and microscopy. *Rev Sci Instrum* 74, 3597–3619.

Moerner, W. E. and Kador, L. (1989). Optical-detection and spectroscopy of single molecules in a solid. *Phys Rev Lett* 62, 2535–2538.

Montal, M. and Mueller, P. (1972). Formation of bimolecular membranes from lipid monolayers and a study of their electrical properties. *Proc Natl Acad Sci U S A* 69, 3561–3566.

Murakoshi, H., Iino, R., Kobayashi, T., Fujiwara, T., Ohshima, C., Yoshimura, A., and Kusumi, A. (2004). Single molecule imaging analysis of Ras activation in living cells. *Proc Natl Acad Sci U S A* 101, 7317–7322.

Murase, K., Fujiwara, T., Umemura, Y., Suzuki, K., Iino, R., Yamashita, H., Saito, M., Murakoshi, H., Ritchie, K., and Kusumi, A. (2004). Ultrafine membrane compartments for molecular diffusion as revealed by single molecule techniques. *Biophys J* 86, 4075–4093.

Nagai, T., Ibata, K., Park, E. S., Kubota, M., Mikoshiba, K., and Miyawaki, A. (2002). A variant of yellow fluorescent protein with fast and efficient maturation for cell-biological applications. *Nat Biotechnol* 20, 87–90.

Nakada, C., Ritchie, K., Oba, Y., Nakamura, M., Hotta, Y., Iino, R., Kasai, R. S., Yamaguchi, K., Fujiwara, T., and Kusumi, A. (2003). Accumulation of anchored proteins forms membrane diffusion barriers during neuronal polarization. *Nat Cell Biol* 5, 626–632.

Nan, X. L., Sims, P. A., Chen, P., and Xie, X. S. (2005). Observation of individual microtubule motor steps in living cells with endocytosed quantum dots. *J Phys Chem B* 109, 24220–24224.

Neher, E. and Sakmann, B. (1976). Single-channel currents recorded from membrane of denervated frog muscle fibres. *Nature* 260, 799–802.

Nguyen, A. H., Nguyen, V. T., Kamio, Y., and Higuchi, H. (2006). Single molecule visualization of environment-sensitive fluorophores inserted into cell membranes by staphylococcal γ-hemolysin. *Biochemistry* 45, 2570–2576.

Nguyen, D. C., Keller, R. A., Jett, J. H., and Martin, J. C. (1987). Detection of single molecules of phycoerythrin in hydrodynamically focused flows by laser-induced fluorescence. *Anal Chem* 59, 2158–2161.

Nguyen, V. T., Kamio, Y., and Higuchi, H. (2003). Single molecule imaging of cooperative assembly of gamma-hemolysin on erythrocyte membranes. *EMBO J* 22, 4968–4979.

Nie, S. M. and Zare, R. N. (1997). Optical detection of single molecules. *Annu Rev Biophys Biomol Struct* 26, 567–596.

Ökten, Z., Churchman, L. S., Rock, R. S., and Spudich, J. A. (2004). Myosin VI walks hand-over-hand along actin. *Nat Struct Mol Biol* 11, 884–887.

Okumus, B., Wilson, T. J., Lilley, D. M. J., and Ha, T. (2004). Vesicle encapsulation studies reveal that single molecule ribozyme heterogeneities are intrinsic. *Biophys J* 87, 2798–2806.

Peck, K., Stryer, L., Glazer, A. N., and Mathies, R. A. (1989). Single molecule fluorescence detection-auto-correlation criterion and experimental realization with phycoerythrin. *Proc Natl Acad Sci U S A* 86, 4087–4091.

Peters, R. (2007). Single molecule fluorescence analysis of cellular nanomachinery components. *Annu Rev Biophys Biomol Struct* 36, 371–394.

Qu, X., Wu, D., Mets, L., and Scherer, N. F. (2004). Nanometer-localized multiple single molecule fluorescence microscopy. *Proc Natl Acad Sci U S A* 101, 11298–11303.

Rotblat, B., Prior, I. A., Muncke, C., Parton, R. G., Kloog, Y., Henis, Y. I., and Hancock, J. F. (2004). Three separable domains regulate GTP-dependent association of H-Ras with the plasma membrane. *Mol Cell Biol* 24, 6799–6810.

Rust, M. J., Bates, M., and Zhuang, X. W. (2006). Sub-diffraction-limit imaging by stochastic optical reconstruction microscopy (STORM). *Nat Methods* 3, 793–795.

Sackmann, E. (1996). Supported membranes: scientific and practical applications. *Science* 271, 43–48.

Sako, Y., Minoghchi, S., and Yanagida, T. (2000). Single molecule imaging of EGFR signalling on the surface of living cells. *Nat Cell Biol* 2, 168–172.

Sánchez, S. A. and Gratton, E. (2005). Lipid-protein interactions revealed by two-photon microscopy and fluorescence correlation spectroscopy. *Acc Chem Res* 38, 469–477.

Sase, I., Miyata, H., Ishiwata, S.i., and Kinosita Jr., K. (1997). Axial rotation of sliding actin filaments revealed by single-fluorophore imaging. *Proc Natl Acad Sci U S A* 94, 5646–5650.

Schlessinger, J. (2002). Ligand-induced, receptor-mediated dimerization and activation of EGF receptor. *Cell* 110, 669–672.

Schmidt, T., Schütz, G. J., Baumgartner, W., Gruber, H. J., and Schindler, H. (1996a). Imaging of single molecule diffusion. *Proc Natl Acad Sci U S A* 93, 2926–2929.

Schmidt, T., Schütz, G. J., Gruber, H. J., and Schindler, H. (1996b). Local stoichiometries determined by counting individual molecules. *Anal Chem* 68, 4397–4401.

Schütz, G. J., Kada, G., Pastushenko, V. P., and Schindler, H. (2000a). Properties of lipid microdomains in a muscle cell membrane visualized by single molecule microscopy. *EMBO J* 19, 892–901.

Schütz, G. J., Pastushenko, V. P., Gruber, H. J., Knaus, H. G., Pragl, B., and Schindler, H. (2000b). 3D imaging of individual ion channels in live cells at 40 nm resolution. *Single Mol* 1, 25–31.

Schütz, G. J., Schindler, H., and Schmidt, T. (1997a). Imaging single molecule dichroism. *Opt Lett* 22, 651–653.

Schütz, G. J., Schindler, H., and Schmidt, T. (1997b). Single molecule microscopy on model membranes reveals anomalous diffusion. *Biophys J* 73, 1073–1080.

Schütz, G. J., Sonnleitner, M., Hinterdorfer, P., and Schindler, H. (2000c). Single molecule microscopy of biomembranes (review). *Mol Membr Biol* 17, 17–29.

Schütz, G. J., Trabesinger, W., and Schmidt, T. (1998). Direct observation of ligand colocalization on individual receptor molecules. *Biophys J* 74, 2223–2226.

Schwentker, M. A., Bock, H., Hofmann, M., Jakobs, S., Bewersdorf, J., Eggeling, C., and Hell, S. W. (2007). Wide-field subdiffraction RESOLFT microscopy using fluorescent protein photoswitching. *Microsc Res Tech* 70, 269–280.

Schwille, P., Haupts, U., Maiti, S., and Webb, W. W. (1999a). Molecular dynamics in living cells observed by fluorescence correlation spectroscopy with one- and two-photon excitation. *Biophys J* 77, 2251–2265.

Schwille, P., Korlach, J., and Webb, W. W. (1999b). Fluorescence correlation spectroscopy with single molecule sensitivity on cell and model membranes. *Cytometry* 36, 176–182.

Shaner, N. C., Campbell, R. E., Steinbach, P. A., Giepmans, B. N. G., Palmer, A. E., and Tsien, R. Y. (2004). Improved monomeric red, orange and yellow fluorescent proteins derived from *Discosoma* sp. red fluorescent protein. *Nat Biotechnol* 22, 1567–1572.

Shcherbo, D., Merzlyak, E. M., Chepurnykh, T. V., Fradkov, A. F., Ermakova, G. V., Solovieva, E. A., Lukyanov, K. A., Bogdanova, E. A., Zaraisky, A. G., Lukyanov, S., and Chudakov, D. M. (2007). Bright far-red fluorescent protein for whole-body imaging. *Nat Methods* 4, 741–746.

Simons, K. and Ikonen, E. (1997). Functional rafts in cell membranes. *Nature* 387, 569–572.

Singer, S. J. and Nicolson, G. L. (1972). The fluid mosaic model of the structure of cell membranes. *Science* 175, 720–731.

Smith, P. R., Morrison, I. E., Wilson, K. M., Fernández, N., and Cherry, R. J. (1999). Anomalous diffusion of major histocompatibility complex class I molecules on HeLa cells determined by single particle tracking. *Biophys J* 76, 3331–3344.

Sonnleitner, A., Mannuzzu, L. M., Terakawa, S., and Isacoff, E. Y. (2002). Structural rearrangements in single ion channels detected optically in living cells. *Proc Natl Acad Sci U S A* 99, 12759–12764.

Suzuki, K., Ritchie, K., Kajikawa, E., Fujiwara, T., and Kusumi, A. (2005). Rapid hop diffusion of a G-protein-coupled receptor in the plasma membrane as revealed by single molecule techniques. *Biophys J* 88, 3659–3680.

Tardin, C., Cognet, L., Bats, C., Lounis, B., and Choquet, D. (2003). Direct imaging of lateral movements of AMPA receptors inside synapses. *EMBO J* 22, 4656–4665.

Teramura, Y., Ichinose, J., Takagi, H., Nishida, K., Yanagida, T., and Sako, Y. (2006). Single molecule analysis of epidermal growth factor binding on the surface of living cells. *EMBO J* 25, 4215–4222.

Thompson, R. E., Larson, D. R., and Webb, W. W. (2002). Precise nanometer localization analysis for individual fluorescent probes. *Biophys J* 82, 2775–2783.

Toprak, E., Enderlein, J., Syed, S., McKinney, S. A., Petschek, R. G., Ha, T., Goldman, Y. E., and Selvin, P. R. (2006). Defocused orientation and position imaging (DOPI) of myosin V. *Proc Natl Acad Sci U S A* 103, 6495–6499.

Toprak, E. and Selvin, P. R. (2007). New fluorescent tools for watching nanometer-scale conformational changes of single molecules. *Annu Rev Biophys Biomol Struct* 36, 349–369.

Ueda, M., Sako, Y., Tanaka, T., Devreotes, P., and Yanagida, T. (2001). Single molecule analysis of chemotactic signalling in *Dictyostelium* cells. *Science* 294, 864–867.

Vrljic, M., Nishimura, S. Y., Brasselet, S., Moerner, W. E., and McConnell, H. M. (2002). Translational diffusion of individual class II MHC membrane proteins in cells. *Biophys J* 83, 2681–2692.

Vrljic, M., Nishimura, S. Y., Moerner, W. E., and McConnell, H. M. (2005). Cholesterol depletion suppresses the translational diffusion of class II major histocompatibility complex proteins in the plasma membrane. *Biophys J* 88, 334–347.

Wang, Y. and Tang, B. L. (2006). SNAREs in neurons-beyond synaptic vesicle exocytosis. *Mol Membr Biol* 23, 377–384. Review

Watts, A. (2005). Solid-state NMR in drug design and discovery for membrane-embedded targets. *Nat Rev Drug Discov* 4, 555–568.

Wazawa, T., Ishii, Y., Funatsu, T., and Yanagida, T. (2000). Spectral fluctuation of a single fluorophore conjugated to a protein molecule. *Biophys J* 78, 1561–1569.

Weiss, S. (2000). Shattering the diffraction limit of light: a revolution in fluorescence microscopy? *Proc Natl Acad Sci U S A* 97, 8747–8749.

Westphal, V. and Hell, S. W. (2005). Nanoscale resolution in the focal plane of an optical microscope. *Phys Rev Lett* 94, 143903.

Wieser, S., Moertelmaier, M., Fuertbauer, E., Stockinger, H., and Schütz, G. J. (2007). (Un)confined diffusion of CD59 in the plasma membrane determined by high-resolution single molecule microscopy. *Biophys J* 92, 3719–3728.

Wilkerson, C. W., Goodwin, P. M., Ambrose, W. P., Martin, J. C., and Keller, R. A. (1993). Detection and lifetime measurement of single molecules in flowing sample streams by laser-induced fluorescence. *Appl Phys Lett* 62, 2030–2032.

Yildiz, A., Forkey, J. N., McKinney, S. A., Ha, T., Goldman, Y. E., and Selvin, P. R. (2003). Myosin V walks hand-over-hand: single fluorophore imaging with 1.5-nm localization. *Science* 300, 2061–2065.

Yildiz, A., Park, H., Safer, D., Yang, Z., Chen, L.-Q., Selvin, P. R., and Sweeney, H. L. (2004a). Myosin VI steps via a hand-over-hand mechanism with its lever arm undergoing fluctuations when attached to actin. *J Biol Chem* 279, 37223–37226.

Yildiz, A., Tomishige, M., Vale, R. D., and Selvin, P. R. (2004b). Kinesin walks hand-over-hand. *Science* 303, 676–678.

Zenisek, D., Steyer, J. A., and Almers, W. (2000). Transport, capture and exocytosis of single synaptic vesicles at active zones. *Nature* 406, 849–854.

Zhang, J., Campbell, R. E., Ting, A. Y., and Tsien, R. Y. (2003). Creating new fluorescent probes for cell biology. *Nat Rev Mol Cell Biol* 4, 80.

Zimmerberg, J. and Gawrisch, K. (2006). The physical chemistry of biological membranes. *Nat Chem Biol* 2, 564–567.

CHAPTER 10

Single Molecule Microarray Analysis

Jan Hesse

Center for Biomedical Nanotechnology, Upper Austrian Research GmbH, Scharitzerstr. 6-8, A-4020 Linz, Austria

Thomas Haselgrübler

Biophysics Institute, Johannes Kepler University Linz, Altenbergerstr. 69, A-4040 Linz, Austria

Christian Wechselberger

Center for Biomedical Nanotechnology, Upper Austrian Research GmbH, Scharitzerstr. 6-8, A-4020 Linz, Austria

Gerhard J. Schütz

Biophysics Institute, Johannes Kepler University Linz, Altenbergerstr. 69, A-4040 Linz, Austria

Summary

Microarrays have become the paramount tool for comparative analysis of biological samples. Alterations in the RNA expression profiles allow for rapid identification of pathway up- or downregulation, thereby enabling a new discovery-driven approach for biomedical research. Yet, current technology demands large amounts of sample material, which hampers the specific analysis of complex and heterogeneous biological tissue. In this chapter, we explore the limits of sensitivity reachable by microarray technology. In particular, applications based on the detection of single nucleic acid molecules on a microarray format are discussed, which offer additional access to new types of information not obtainable from ensemble-based studies.

Key Words

single molecule microscopy; spectroscopy; fluorescence; genomics; DNA; RNA expression profiling; DNA sequencing

Introduction

The success story of DNA microarrays is closely linked to the recognition of the diagnostic quality of gene and protein expression profiles. In recent years, microarrays have become a standard tool not only for expression analysis in molecular biology and genomic research, but also for pharmacogenomics, and infectious and genetic disease and cancer diagnostics (Heller, 2002). Analysis of a whole genome in a single experiment has enabled large-scale surveys of gene expression patterns, with the outcome of, for example, functional gene dependencies (Brown and Botstein, 1999), the identification of novel microRNAs (Bentwich et al., 2005), precise tumor classification (Beer et al., 2002; Nielsen et al., 2002; van't Veer et al., 2002), or drug target validation (Marton et al., 1998). Recently, a note of caution was sounded concerning the reliability of microarray data, in particular upon a publication from Margaret Cam, who compared gene expression levels on three different platforms and found hardly any overlap (Tan et al., 2003). A rigorous test by the MicroArray Quality Control consortium, however, reported a high level of intra- and interplatform concordance (Shi et al., 2006), further confirming previous interpretations.

There is an increasing demand for understanding gene regulation in more and more refined samples. Many tissue preparations were found to be heterogeneous in composition, yielding emerging interest in the specific microarray analysis of cellular subpopulations. A certain cell type can be isolated routinely from heterogeneous fluid or tissue samples using fluorescence-activated cell sorting (FACS) or laser capture microdissection (LCM). Purification and biochemical processing can be performed with minimum loss of material via microfluidic devices (Hong et al., 2004; Paegel et al., 2003). Still, analysis of such minute samples remains a challenging task. Current protocols include an RNA/DNA amplification step. PCR-based approaches, however, often lead to complications due to cross-contamination of sample material, and distorted statistical outputs (Mutter and Boynton, 1995). Linear amplification is less prone to distortions of the relative abundance of mRNA species in the original sample; however, the method is time-consuming and requires sophisticated protocols that are difficult to establish (Baugh et al., 2001; Hu et al., 2002; Kacharmina et al., 1999; Li et al., 2005; Mahadevappa and Warrington, 1999; Nygaard et al., 2005; Taylor et al., 2004; Van Gelder et al., 1990; Wang et al., 2000).

The rising demand for minute sample analysis matches recent developments in spectroscopy, which provide numerous tools for analyzing even single molecules. Pioneering work for the detection and characterization of single fluorescent molecules was performed in the early 1990s by groups studying the spectroscopic behavior of single chromophores embedded in condensed matter at low temperature (Moerner and Kador, 1989; Moerner and Orrit, 1999). Milestones on the way to applicability in bioscience include single molecule detection at room temperature (Shera et al., 1990), in buffer (Funatsu et al., 1995; Sase et al., 1995; Schmidt et al., 1996a), and in living cells (Sako et al., 2000; Schütz et al., 2000a,b).

In this chapter, we will highlight the progress toward single molecule detection on microarray platforms. Our group has recently reported the development of an mRNA expression profiling system operating at single molecule sensitivity (Hesse et al., 2004, 2006), demonstrating the applicability of single molecule approaches for large-scale screening problems. After discussing technical considerations for highly sensitive microarray analysis, we will go beyond merely quantitative approaches and raise additional topics that may bring a qualitative improvement of single molecule readout over classical "ensemble"-based methodologies.

Speed and Sensitivity: Mutually Exclusive Demands?

Two strategies have been introduced for approaching parallel screening at single-molecule sensitivity: first, "fluorescent molecules in solution" can be detected by a stationary excitation and detection system. Most prominently, confocal devices have been applied to reduce the detection volume down to a few femtoliters (Nie et al., 1994). Using avalanche photodiodes (APDs) as detectors allows fluorescence brightness, spectral properties, and spatial mobility to be measured. By combining multiple detectors, additional information about spectral properties can be accessed. Such devices have been extensively used to measure binding kinetics (Bismuto et al., 2001), conformational fluctuations (Wennmalm et al., 1997), spectroscopic transitions (Eggeling et al., 2001; Haupts et al., 1998), and molecular mobility (Korlach et al., 1999) using fluorescence correlation spectroscopy (FCS) (for a review see Krichevski and Bonnet, 2002). The information can be assessed within a few seconds, rendering FCS an applicable tool for miniaturized high-throughput screening (Auer, 2001). Also, cameras have been used to image fluorescent molecules in solution, as they passed through the defocused excitation beam. The greatly increased observation area, compared to a point detector as used for FCS, markedly increases detection speed, which has been utilized, for example, for DNA fragment sizing (Van Orden et al., 2000). Constant flow of the solution and hydrodynamic focusing ensure that most molecules will pass through the detection volume, making this method applicable for detection of trace amounts of molecules.

The second strategy, which aims at the detection of "surface-immobilized molecules," is particularly relevant to microarray analysis. Historically, the first studies used confocal microscopy to efficiently discriminate single molecule signals from background. The method is based on scanning a diffraction-limited laser focus over the sample; the fluorescence emitted from the sample is imaged via a pinhole onto a point detector (Minsky, 1988; Pawley, 1995). Using APDs as detectors allows additional information about the excited state lifetime to be obtained. Combination of several APDs placed after a stack of dichroic and polarization beam splitters was used to maximize the information content acquired from a single molecule (Eggeling et al., 2001). However, due to the serial data acquisition, the inspection of large sample areas remains a time-consuming task.

In a different approach, several groups utilized intensified or back-illuminated CCD cameras for imaging single molecules on surfaces (Funatsu et al., 1995; Sase et al., 1995; Schmidt et al., 1996a). In this case, sample areas of several square micrometers are illuminated with a defocused laser spot (Köhler illumination) and imaged directly. Background signal can be efficiently suppressed by exciting the sample via an evanescent wave, by a technique known as total internal reflection fluorescence microscopy (TIRFM) (Axelrod et al., 1984). The principal advantage of using pixel arrays (like CCDs) as detectors is parallel signal acquisition. Still, the pixel array covers only a small part of the sample area, making additional scanning inevitable. Motorized scanning stages equipped with automated positioning systems enable repeatability of ~100 nm; however, inertia of the moving parts requires time-consuming feedback loops for precise stops.

We have recently developed a single molecule microarray or "biochip" reader, which makes use of an alternative readout scheme: it allows the reduction of overhead times due to stage positioning and signal integration by synchronizing continuous stage motion and camera readout (Hesse et al., 2004). When operating the camera in time-delayed-integration (TDI) mode, charges are integrated on the CCD chip during the line-shifts that accompany the readout process (Figure 10.1A). Stationary point emitters on the sample would therefore be imaged as lines. When the sample is moved synchronously with the charges on the CCD chip, point emitters are imaged as diffraction-limited spots (Figure 10.1B). With this system, single Cy5 molecules immobilized on a glass surface can be easily detected with a signal-to-background-noise ratio of ~37. No eccentricity of the signals was observed, indicating sufficient synchronization of stage scanning and camera readout (Hesse et al., 2004).

Single molecule detection demands precise control of the focal plane. Fluctuations of the order of ~100 nm would yield a significant peak broadening, impeding single molecule identification. We therefore implemented a focus-hold system, which keeps the sample

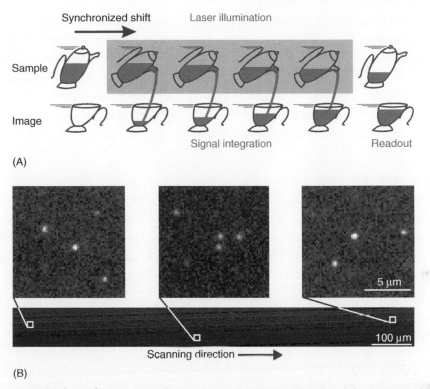

Figure 10.1: Single molecule detection on large areas. (A) Sketch of the TDI-mode data acquisition. In this mode, the parallel register shift of the camera is synchronized with the object motion so that charge packets on the CCD pixels (cups) always correspond to the same sample region (pots) as they move across the parallel register. Charge accumulates and signal strength increases as the pixels approach the readout register. (B) Image of a supported lipid bilayer of dipalmitoylphosphatidylcholine (DPPC) containing trace amounts of fluorescent lipid, Cy5-dipalmitoylphosphoethanolamine (molar ratio Cy5-DPPE/DPPC = 10^{-8}). The stripe shows a 1 mm × 0.1 mm region of the bilayer; arbitrarily chosen details demonstrate the homogeneity of the bilayer over the full scanning region. Individual fluorescent peaks can be easily discriminated over the background and represent individual Cy5-DPPE molecules. Reprinted with permission from Hesse et al. (2004). Copyright (2004) by ACS

in the focal plane of the objective during the scanning process. It makes use of the back-reflection of laser light from the biochip surface (Hellen and Axelrod, 1990). Slight changes in the distance to the focal plane can be detected on a two-segment photodiode; the differential signal is used to control a z-shift of the objective in a feedback loop.

In the direction of scanning, the intensity profile of the excitation beam need not be constant. In practice, it is convenient to use a Gaussian profile with a width corresponding to the field of view. This ensures maximum excitation of dye molecules within the region of observation, and minimum photobleaching of dyes in regions still to be scanned. Perpendicular to the scanning direction, a homogenous illumination profile is preferable, which can be achieved by Köhler illumination; a field stop introduced for restricting the excitation to the field of view imaged on the chip prevents photobleaching of adjacent areas during a scan.

The device operates at a much higher data acquisition rate than alternative imaging devices (Hesse et al., 2002). Scanning an area of $1\,cm^2$ with single molecule sensitivity and 320 nm pixel resolution would take ~13 days with confocal scanning microscopy; even an imaging approach based on recording the whole area sequentially would still take ~10 h. Using the TDI approach, the overall scanning time can be reduced to just 10 min. This rapid scanning procedure enabled ultrasensitive and high-resolution readout of microarrays on a reasonable timescale.

Microarrays: Attempting a Definition

A biochip is generally defined as a carrier for biological samples. The most common chips are based on glass slides for the detection of fluorescently labeled material ranging from tissues or cells to biomolecules. A microarray can be defined as a biochip surface containing miniaturized spots that react specifically with different components of the sample. Ideally, each microarray spot is *pure* and of *known* composition. DNA microarrays are currently fabricated via in situ synthesis or via printing, yielding spots with a size of 10–150 μm; current technology enables the analysis of tens of thousands of different reactants on a single such "biochip" (Dufva, 2005). If not further specified, we will assume the usage of such classical microarrays throughout the manuscript.

From the beginning, attempts were made to reduce the size of microarray spots in order to maximize the information content of the biochip. Using the tip of an atomic force microscope as a nanoscopic pen, dip-pen nanolithography has been introduced for patterning DNA on surfaces at feature sizes less than 100 nm (Demers et al., 2002). While such small features contain only a few hundred DNA molecules and are therefore not applicable to conventional microarray analysis at high dynamic range (Dufva, 2005), they bring new levels of miniaturization into reach, which in combination with advances in lab-on-a-chip technologies and microfluidics (Sims and Allbritton, 2007) may well be useful for analysis of minute amounts of sample material.

A regular array of spots, each containing just a single molecule, can be regarded as the ultimate limit of a microarray. Recently, a method for producing high-density single DNA molecule surface patterns has been reported (Schwartz and Quake, 2007). Such spots are intrinsically *pure*, yet generally *unknown* in composition. How to make use of such "microarrays" for DNA sequencing will be discussed below.

Microarray Surfaces: There is Plenty to Groom at the Bottom

One critical component of microarray technology is the surface chemistry of the substrate. Glass slides are extensively used as support material due to their favorable optical characteristics. Glass surfaces may be chemically modified either by adsorption of positively charged compounds on the negatively charged substrate or by covalent attachment exploiting the selective reactivity of glass with compounds known as silane coupling reagents (Figure 10.2).

The most common surfaces used for microarray analysis bear chemically reactive amine, epoxy, or aldehyde groups. Amine groups are formed either by adsorption of poly-L-lysine (Kusnezow et al., 2003) or by reaction with amine-containing silane derivatives such as 3-aminopropyltriethoxysilane (APTES) (Howarter and Youngblood, 2006; Kim et al., 2007a; Kusnezow et al., 2003; Mahajan et al., 2006) or *N*-[3-(trimethoxysilyl)-propyl] ethylenediamine (Groll et al., 2004). Epoxy surfaces are prepared by reaction with 3-glycidoxypropyl-trimethoxysilane (GPS) (Kim et al., 2007a; Kusnezow et al., 2003; Piehler et al., 2000). The epoxy groups may be converted into aldehyde groups by a two-step procedure: (i) hydrolysis of the epoxy group under mild acidic conditions and (ii) oxidation of the formed vicinal diol using sodium periodate (Schlapak et al., 2006).

DNA Microarrays

Binding of DNA to amine surfaces occurs via electrostatic interaction. After spotting, unreacted amino groups on the substrate are commonly blocked using succinic anhydride/*N*-methylimidazol in *N*-methylpyrrolidone/borate buffer. When a nonaqueous solvent such as 1,2-dichloroethane is used instead, the background is markedly diminished (Diehl et al., 2001).

For attachment of DNA to an epoxy or aldehyde surface, an amine group has to be synthetically added to the 5′-terminus of the DNA of interest. Amino-terminated DNA is covalently coupled to epoxy and aldehyde surfaces by formation of secondary amines and imines (Schiff bases), respectively. The imine bond is prone to hydrolysis (i.e., reverse reaction) in the subsequent hybridization step. In order to prevent hydrolysis,

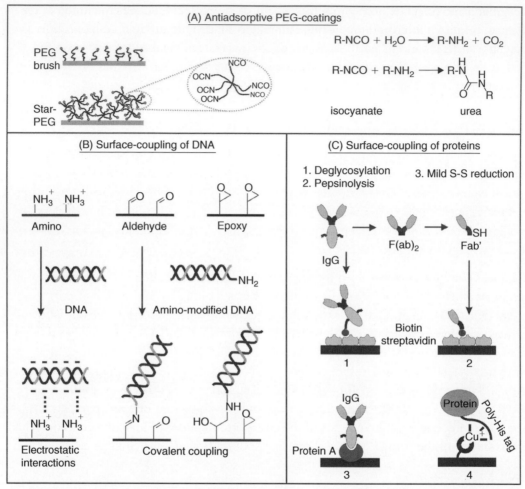

Figure 10.2: (A) Antiadsorptive coatings for biochips using variants of PEG. The PEG brush causes the formation of partially structured water and has a high exclusion volume; both effects minimize direct interactions between the biochip surface and biomolecules in solution. Star-PEG consists of 6 PEG arms bearing terminal isocyanate groups. The hexavalent molecules can cross-link via urea groups, thereby forming a 3D network on the substrate surface that has superior antiadsorptive properties. (B) Ways of binding DNA to chemically modified surfaces. DNA binds to positively charged amine surfaces via electrostatic interaction with the negatively charged DNA backbone. For more flexible attachment, an amine group is synthetically added to the 5'-terminus of the DNA. Upon contact with aldehyde- or epoxy-modified surfaces, the amine group is bound via the formation of secondary amines and imines. This bond can be stabilized against hydrolysis by chemical

the spotted surface is treated with sodium borohydride, which reduces the imine into a stable secondary amine and unreacted, still reactive aldehyde groups into nonreactive alcohol groups (blocking step). Sodium dodecyl sulfate (SDS) is commonly added to hybridization reactions to decrease background (Schena, 2003).

We have used aldehyde-functionalized glass slides for DNA microarray analysis at the single molecule level (Hesse et al., 2006). Sodium cyanoborohydride ($NaBH_3CN$) and 4-aminobutanoic acid were selected for imine reduction and blocking of unreacted aldehyde groups. Low nonspecific signals of approximately 0.001 molecules/μm^2 were measured for target oligonucleotides at concentrations below 1 pM. For concentrations above 1 pM, the nonspecific binding of oligonucleotides resulted in a nonspecific signal that increased linearly with the target oligonucleotide concentration, approximately four orders of magnitude below the specific signal.

Glass surfaces grafted with poly(ethylene glycol) (PEG), widely used to prevent nonspecific adsorption of proteins (see below), may also be used for DNA microarrays. Both linear and branched polymers have been investigated (Ameringer et al., 2005; Schlapak et al., 2006). PEG helps to resist the nonspecific adsorption of DNA, thus lowering the background signal, and functions as a spacer to avoid steric hindrance. A dense PEG layer also masks inhomogeneities of the underlying glass surface (Schlapak et al., 2006).

High-density linear structures, so-called PEG "brushes," were prepared by grafting of PEG-diamine onto glass slides bearing aldehyde or epoxy groups (Piehler et al., 2000; Schlapak et al., 2006). A thiol-modified DNA was linked to the maleimide-derivatized PEG layer. Fluorescence imaging showed that for target DNA concentrations in excess of 1 nM, the PEG layer improved hybridization up to fourfold and reduced nonspecific adsorption of DNA by a factor of up to 13 (Schlapak et al., 2006).

Protein Microarrays

Protein microarrays are the continuation of the DNA array approach (Piehler, 2005). However, owing to fundamental biochemical differences between DNAs and proteins, the chemical aspects of DNA microarrays cannot be simply applied to protein microarrays, which require a more sophisticated surface and immobilization chemistry (Kusnezow

reduction using sodium borohydride. (C) Oriented coupling of antibodies via (1) their two carbohydrate chains, (2) their disulfide bond in the hinge region, and (3) their Fc region. (4) Recombinant proteins are immobilized on chelated metal surfaces via their poly-His tag

and Hoheisel, 2003). DNAs are relatively simple polyanions and can be chemically modified and immobilized on solid surfaces based on electrostatic interaction or covalent coupling (see the previous section). Proteins are chemically and structurally much more complex and heterogeneous. The activity and function is critically dependent on their delicate three-dimensional structure. Proteins tend to adsorb nonspecifically to most surfaces, which often results in the disruption of their structure. This strong tendency can be attributed to hydrophobic, ionic, and hydrogen bonding interactions with the solid surface. Keeping proteins immobilized in their native conformation, with their 3D structures intact and their active domains accessible, is one of the challenging aspects of protein microarray technology.

Immobilization

Ideally, the linking chemistry should meet the following criteria: (i) the protein molecule of interest is immobilized irreversibly in a straightforward fashion; (ii) the immobilization does not disrupt the 3D structure of the protein; (iii) the coupling chemistry allows control of protein orientation; and (iv) the immobilization chemistry is highly specific, such that prepurification of protein samples is not required (Guo, 2006).

Proteins may be linked to the surface with random or controlled orientation. A wide variation in orientation gives rise to a distribution of binding affinities and kinetics (Guo, 2006). Covalent coupling between the amine groups of the protein molecules and aldehyde, epoxy, or *N*-hydroxysuccinimide (NHS)-ester groups on the substrate leads to random orientation. So does the coupling via biotin–streptavidin interaction. Biotin-tagged proteins are conjugated to the biotinylated surface via the intermediate streptavidin. The biotin–streptavidin interaction is the strongest among all noncovalent interactions and shows strong affinities and high specificity. Selective coupling of proteins with controlled orientation is achieved by using the recombinant polyhistidine (His) tag. A chelating group such as nitrilotriacetic acid (NTA) attached to the substrate strongly binds Ni^{2+} or Cu^{2+} ions via three sites of coordination, leaving three vacant sites for coordination by the imidazole groups of polyhistidine. The His tag may be attached to either the C- or N-terminus and generally does not interfere with the function of the protein (Cha et al., 2004).

For antibody microarrays, there are three additional approaches for chemical attachment with controlled coupling (Guo, 2006; Kusnezow and Hoheisel, 2003; Seong and Choi, 2003). Orientation may be achieved via selective coupling (i) to the Fc portion of the antibody, (ii) to the carbohydrate side chain in the Fc region, or (iii) the disulfide bond in the hinge region (see Figure 10.2).

Nonspecific Adsorption

In order to prevent nonspecific adsorption, high background signals, and loss of protein activity, protein-resistant coatings have been developed. The most inert and efficient polymer against nonspecific protein adsorption is PEG (Guo, 2006). The PEG backbone is extensively hydrogen-bonded to water molecules, resulting in the formation of partially structured water extending into the aqueous phase. Adsorption of a protein molecule requires the disruption of this structured water layer. In addition, PEG has a high exclusion volume due to high conformational entropy. Protein adsorption leads to the compression of the PEG layer toward the solid surface and thus is entropically unfavorable (Guo, 2006).

Glass substrates with a high-density PEG brush are commercially available. Recently, a dense network of isocyanate-terminated star-shaped poly(ethylene glycol) (star-PEG), cross-linked at their chain ends via urea groups, was shown to be extremely resistant to nonspecific adsorption of proteins. Its advantageous properties for biofunctional and biomimetic surfaces have been demonstrated by single molecule fluorescence microscopy (Groll et al., 2004). The protein RNase H labeled with a fluorescence resonance energy transfer (FRET) pair was coupled to the star-PEG surface via the biotin–streptavidin interaction. FRET analysis showed completely reversible denaturation/renaturation behavior. The comparison of star-PEG surfaces with linear PEG surfaces (PEG brushes) and bovine serum albumin (BSA)-coated surfaces showed the superior properties of star-PEG. The latter showed negligible nonspecific adsorption (0.01 molecules/μm^2). Significantly higher levels were observed for linear PEG-grafted surfaces (0.1 molecules/μm^2) and for BSA-coated surfaces (10 molecules/μm^2). The star-PEG, however, is not yet commercially available.

Star-PEG surface coatings have also been used to study specific cell adhesion (Groll et al., 2005b), the patterning of proteins (Groll et al., 2005c), the nanostructured ordering on substrates (Groll et al., 2005a), and protein dynamics under denaturing conditions (Kuzmenkina et al., 2005). To the authors' knowledge, there is not yet a publication describing protein microarray analysis at the single molecule level.

Applications: Learning from Singles

In the following section, we will highlight successful applications of single molecule detection on biochip surfaces, and promising routes to pursue.

mRNA Expression Profiling

In a typical microarray experiment for RNA expression profiling, sample RNA is reverse transcribed to cDNA; during the transcription, fluorescently labeled bases are incorporated into the cDNA. The sample expression profile is inferred from the fluorescence signals obtained on the individual spots upon hybridization. Commercial microarray platforms are optimized for rapid and cheap analysis, with the consequence that they operate at moderate detection sensitivity. In an attempt to explore the potential of ultrasensitive microarray analysis (Hesse et al., 2006), we developed a biochip readout device operating at single molecule sensitivity (as discussed in previous section). Microarray surfaces were optimized for low autofluorescence background and high antiadsorptivity so that the weak signals of single dye molecules could be discriminated with high reliability from nonspecific background. By counting individual fluorescent cDNA molecules specifically hybridized to their complementary sequence on the chip surface, mRNA expression profiles of an equivalent of only 10^4 cells could be determined, which is two orders of magnitude less material than typically required when using conventional amplification-free methods.

A single molecule approach for microarray analysis differs fundamentally from classical ensemble-based methods, as not the spot brightness but the actual number of bound molecules is determined as the readout signal. Besides the apparent advantage of increased sensitivity, the decoupling of brightness information from the number of bound cDNA molecules allows for precise determination of the expression level, independent of variations in the molecular brightness. Intrinsically, the approach is only suitable for low surface densities at which individual molecules can be resolved as isolated signals; when working at surface densities <1 molecule per $5\,\mu m^2$, the probability of overlapping signals is kept below 10%. Some spots on the microarray may well contain a higher amount of hybridized cDNA; in such cases, classical ensemble analysis is employed.

We have validated the sensitivity and dynamic range of the system by hybridizing a 60-mer fluorescent sense-oligonucleotide to its complementary antisense and noncomplementary nonsense oligonucleotide on the microarray (Hesse et al., 2006) (Figure 10.3). All experiments were performed in a hybridization volume of $50\,\mu l$. At high sample concentrations $\gg 1\,pM$, individual spots show a homogenous signal, reminiscent of images well known from conventional ensemble assays. Around $1\,pM$, however, individual peaks emerge in the high-resolution images, which become more and more dispersed when reducing the concentration further; these peaks can be attributed to individual oligonucleotide molecules specifically hybridized to their complementary sequence on the chip surface. As expected, the single molecule brightness is not affected by changes in the sample concentration.

The detection limit of the readout device per se is given by a single dye molecule bound to the chip surface. In practice, the stochastic nature of the binding process and background

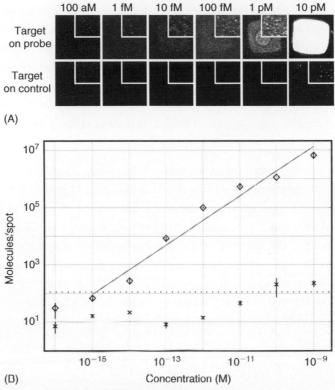

(A)

(B)

Figure 10.3: Oligonucleotide detection at the single molecule level. (A) The first row shows fluorescence images of individual microarray spots specifically hybridized with different concentrations of a 60-mer fluorescent oligonucleotide. The insets show spot details at 3.5-fold higher magnification. At concentrations below ~1 pM, individual fluorescent peaks can be resolved, corresponding to single oligonucleotide molecules hybridized to their complementary sequence on the microarray. At concentrations higher than 1 pM, overlapping single molecule signals generate homogeneous fluorescence. In the second row, the corresponding images of noncomplementary control spots are shown. (B) Number of fluorescent oligonucleotides detected per spot versus concentration in the hybridization solution ($\diamond$ specific signal, $\times$ nonspecific signal measured on noncomplementary spots). The detected nucleotide density increases linearly with the applied oligonucleotide concentration over 6 orders of magnitude. Assuming a signal-to-background-noise ratio of 10 as detection limit (dashed line), a concentration of ~1.3 fM is required for unambiguous analysis, corresponding to ~39 000 molecules in the sample volume of 50 μl. A dynamic range of 4.7 orders of magnitude can be specified (Hesse et al., 2006). Reprinted with permission from Genome Res. (Hesse et al., 2006). Copyright (2006) by Cold Spring Harbor Laboratory Press

signals due to surface impurities define the limit of detection. By optimizing the surface chemistry, we could reduce the amount of background signals to ~13 peaks within a circular spot area of 100 μm diameter. Taking into account Poissonian-binding statistics, reliable microarray analysis with a signal-to-noise ratio of 10 therefore demands an average number of 112 molecules hybridized to a spot. Using this definition, the limit of detection for hybridization of 60-mer oligonucleotides was reached at a concentration of 1.3 fM. Considering the assay volume of 50 μl, this limit corresponds to a total number of ~39 000 oligonucleotide molecules within the sample.

In an mRNA expression profiling experiment, the relative abundances in the sample may vary over orders of magnitude, yielding massive differences in the number of hybridized molecules per spot. In the same experiment, we can expect spots containing just a few well-separated cDNA molecules and other spots with highly overlapping single molecule signals – equivalent to the conventional ensemble approach. In order to enable both single molecule and ensemble analysis, we set up the reader with a 16-bit camera system, which allows for imaging at extremely high dynamic range (Hesse et al., 2004; Sonnleitner et al., 2005). As described, nonspecific background and binding statistics demand ~112 molecules per spot as the lower signal limit for reliable analysis. The upper limit (n_{max}) is given by:

$$n_{max} = 2^{16} \frac{A_{spot}}{r_{SB} \, N \, A_{pixel}} \tag{10.1}$$

where r_{SB} is the signal-to-background-noise ratio, N the background noise, and A_{spot} and A_{pixel} the spot area and pixel area, respectively. Assuming $A_{spot} = 10^4 \, μm^2$, $A_{pixel} = 0.04 \, μm^2$, $r_{SB} = 37$, and $N = 5$, we estimate $n_{max} \approx 5 \times 10^7$ molecules per spot, which is equivalent to a dynamic range close to 6 orders of magnitude. It should be noted that the operational dynamic range on the microarray platform is, in general, lower due to nonspecific binding of fluorescent oligonucleotides to the target control spots. At a concentration of 67 pM, the nonspecific signal equals the specific signal recorded at the detection limit of 1.3 fM, yielding an operational dynamic range of 4.7 orders of magnitude (Hesse et al., 2006).

For comprehensive mRNA expression profiling, a large number of genes have to be analyzed; in cancer research, for example, specialized subsets of genes contain a few thousand different capture probes. To address this need for larger arrays, we analyzed samples containing ~10 000 spots on a biochip surface of ~2.2 cm × 1.4 cm. A complex mixture of labeled cDNA was produced via reverse transcription of 5 μg total RNA extracted from HaCaT cells; 4% of that cDNA (corresponding to 200 ng total RNA) were used for expression profiling without further amplification (Figure 10.4). The low nonspecific background of $\sim$1 cDNA molecule per 200 μm^2, corresponding to $\sim$40 molecules per spot area, allowed reliable quantification of expressed genes down to a detection limit of 170 molecules per spot. The two-color image was recorded within 6 h at a pixel size of 200 nm.

The chosen ultrasensitive and high-resolution imaging approach not only allows for microarray analysis of low amounts of sample material at high dynamic range, but further provides additional advantages over standard measurements. First, spot morphology can be determined with high precision. This allows, in particular, the identification of bad spots containing clumps, scratches, or dye separation, which would severely impair standard

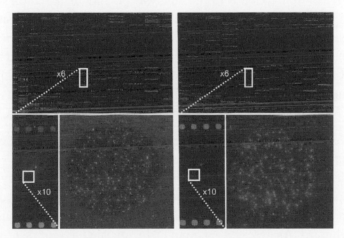

Figure 10.4: Readout of large biochips with single molecule sensitivity. Alexa555- (left column) and Alexa647-conjugated (right column) cDNA was produced by reverse transcription from 5 μg total RNA from two different HaCaT cell populations. Competitive hybridization was performed on a microarray containing 10 430 spots using 4% of the produced cDNA (corresponding to 200 ng total RNA) for each sample/color. The $\sim$2.2 cm $\times$ 1.4 cm microarray was read out in two colors within $\sim$6 h at a pixel size of 200 nm $\times$ 200 nm. The images show the feasibility of high-resolution imaging over several square centimeters: no defocusing occurs in the images; single molecules can be detected in both color channels; and no shift between the two images is observed. The low nonspecific background of $\sim$1 cDNA molecule per 200 μm^2, corresponding to $\sim$40 molecules per spot area, allows reliable quantification of expressed genes down to the detection limit of 170 molecules per spot

microarray analysis (Brown et al., 2001). Second, the single molecule brightness measured for different genes was found to vary significantly (Hesse et al., 2006, 2007). Differences in the degree of labeling, as well as in the local environment of the fluorophores, may contribute to this effect. In particular, it was shown that neighboring nucleobases can lead to sequence-specific quenching of fluorescent dyes (Seidel et al., 1996). This effect may well distort the apparent expression profiles obtained in ensemble measurements, even in ratiometric two-color recordings; on the quantification by single molecule counting, it has no effect.

DNA Methylation Analysis

Mechanisms for the regulation of gene expression include epigenetic modifications such as cytosine methylation (Ferguson-Smith and Surani, 2001; Issa, 2003; Lee, 2003; Robertson, 2005). Multiple methods for the analysis of CpG methylation status have been developed including chromatographic separation, use of methylation-sensitive restriction enzymes, and bisulfite-driven conversion of nonmethylated cytosine to uracil (Ushijima, 2005). Despite the widespread application of the latter, there are still limitations concerning DNA degradation during sodium bisulfite treatment, depurination under the required acidic and thermal conditions, as well as inconsistent conversion of cytosine to uracil in a given DNA sample (Derks et al., 2004; Shiraishi and Hayatsu, 2004).

In the approach reported in Proll et al. (2006), we quantified the methylation of genomic DNA hybridized to an oligonucleotide microarray. 5′-Methyl-cytosine (5′mC) was specifically labeled using monoclonal anti-5′mC antibody and a secondary Cy3-conjugated antibody. For target oligonucleotides, a detection limit of 600 aM (1.5 fg DNA per hybridization volume of 3 μl), corresponding to the DNA content of nuclei from 540 cells, was achieved. Analyses of genomic DNA revealed a clear discrimination between methylated and unmethylated promoter DNA for all selected genes.

DNA Fragment Sizing

As discussed earlier, single molecule approaches make expression profiling independent of brightness variations of the individual targets, rendering results less prone to distortions. Moreover, information on the single molecule brightness can be utilized for further characterization of the investigated sample. In general, a single dye molecule emits a well-defined number of photons during a given illumination time, which is a characteristic parameter of the molecule. The fluorescence brightness obtained from a single biomolecule can thus be used as a measure of the number of bound fluorophores (Schmidt et al., 1996b).

DNA can be homogenously stained using intercalating dyes, yielding a linear dependence of the brightness with the fragment size (Filippova et al., 2003; Laib et al., 2003). Several groups utilized this linearity for determination of the DNA fragment sizes present in the sample (Filippova et al., 2003; Foquet et al., 2002; Goodwin et al., 1993; Laib et al., 2003; Meng et al., 1995; Schlapak et al., 2007). Rapid data analysis, in combination with microfluidic devices, enabled the construction of a single molecule DNA sorting device (Chou et al., 1999).

DNA fragment sizing was developed as the ultrasensitive counterpart to classical restriction analysis, a standard method for DNA typing by determining restriction fragment length polymorphisms. It has also been used to quantify ionizing radiation-induced double-strand breaks (Filippova et al., 2003). A recent application indicates nicely the diagnostic relevance of this approach. By random incorporation of fluorescent nucleotides in a primer-extension reaction, the number of CAG repeats present in the original single-stranded DNA sample could be determined; CAG repeats encode polyglutamine tracts, which are characteristic signatures of at least nine inherited neurodegenerative diseases such as Huntington's disease (Li and Li, 2004). By single molecule brightness analysis, pathologically high (>40) and normal low repeat numbers in Huntingtin-derived sequences could be discriminated.

DNA (Combing and) Mapping

Detailed sequence information directly on individual high-molecular-weight DNA molecules can be obtained by physical mapping approaches using the "molecular combing" technique. For this purpose, individual DNA molecules are covalently bound to a solid substrate and elongated by a receding meniscus of solute (Kim et al., 2007b; Lebofsky and Bensimon, 2003). This procedure allows stretching and aligning individual strands of DNA molecules on planar surfaces at adjustable densities. Numerous reports have been published in the past describing the optimization of this technology and characterizing factors that influence, for example, the immobilization capabilities of DNA strands (depending on surface chemistry and solvent properties) and the extent of the resultant DNA stretching (Bensimon et al., 1995; Herrick and Bensimon, 1999; Yokota et al., 1997).

Once individual DNA strands have been immobilized, several techniques are available for high-resolution mapping based on fluorescence microscopy readout. By performing in situ hybridization, it is possible to directly visualize the precise position of the genomic probes on the combed DNA molecules. Exemplary in this respect are reports describing the analysis of combed yeast and human genomic DNA (Michalet et al., 1997), characterizing the human apolipoprotein(a) gene by direct visualization of the kringle-IV repeats (Erdel et al., 1999), quantification of fluorescent probe length on combed genomic DNA to detect subtle gains and losses of genomic DNA as small as ~50 kb in size (Herrick et al., 2000), and the high-resolution mapping of human and mouse chemokine clusters (Erdel et al., 2001). A different approach for DNA mapping to localize multiple sequence motifs on combed DNA used the incorporation of fluorescent dye molecules by combining nicking endonuclease and DNA polymerase enzymes. With this approach, the authors generated a sequence motif map of λ-phage DNA, a strain of human adenovirus and several strains of human rhinoviruses (Xiao et al., 2007).

Further development of this technique combined fluorescent in situ hybridization with antibody-based detection of newly synthesized DNA stretches from cells in S-phase. Incorporation of the nucleotide analogue bromodeoxyuridine (BrdU) was used to characterize the molecular regulation of activation or silencing of DNA origins of replication (ORI). This allowed for the first quantitative analysis of the distribution of ORIs in the repeated ribosomal DNA array of *Saccharomyces cerevisiae* (Pasero et al., 2002). Further characterization of the spatial and temporal regulation of ORI "firing" along eukaryotic chromosomes allowed for measurements with an impressive ~1 kb resolution (Patel et al., 2006). By establishing a "genomic Morse code" by use of fluorescence hybridization probes and combining this with the technique of pulse-labeling with BrdU analogues, ORI characterization has recently been achieved on a 1.5 Mb region of human chromosome 14q11.2 and allowed the correlation of initiation events with specific sequence tracts (Lebofsky et al., 2006).

Single Molecule Sequencing

Since the early 1990s, it has been noted that single molecule detection might be useful to obtain sequence information from individual DNA molecules. It was the call for strategies to sequence the whole genome at a price below $1000, which, however, boosted novel creative sequencing approaches (Service, 2006).

In one approach (Braslavsky et al., 2003), single-stranded target molecules are bound to a universal primer that is immobilized on a solid substrate with an intermolecular distance larger than the optical diffraction limit. The chip is then incubated with a buffer containing one type of labeled nucleotides and DNA polymerase. The reaction conditions are chosen such that the second strand synthesis can be stopped after 1–2 incorporation reactions, followed by fluorescence readout and silencing of the reporter dye. Repeating this scheme for all four bases yields digital information (yes/no) for each site (molecule) and each nucleotide. If each type of nucleotide carries a chromatically different fluorophore, in the reaction step all four bases can be present. Repeating this scheme several times yields sequence information for each molecule bound to the substrate.

In a similar approach, Rigler proposes to immobilize exonuclease molecules on the surface and observe the cleavage of individual bases from bound DNA or RNA molecules (Rigler, 2005). If each nucleotide is labeled, the cleavage results in a change in fluorescence upon progression of the cleavage process.

Another sequencing approach utilizes the perfect match hybridization of short oligonucleotides to obtain sequence information from the target molecules (Jacobson, 2005). A library of oligonucleotides of M residues (M-mers) is sequentially hybridized to the surface-immobilized targets and the reaction conditions are chosen such that only perfect match sequences can form a stable bond. After each hybridization, the fluorescence signals are detected, the target–oligo complexes are melted, and the oligonucleotides are washed away. Depending on the length of the M-mer, the procedure is repeated M^4 times to cover the whole library. Although each oligonucleotide provides information only on a short part of the target sequence, the alignment of all nucleotides that could bind to the target yields the complete sequence of the target.

Conclusions

The technological advances in microarray analysis have supported a paradigm shift in the scientific approach in the life sciences, away from hypothesis-driven toward discovery-driven research. At the moment, the new approach is restricted to the most accessible readout parameter, the gene expression profile. However, more and more researchers are attempting to use microarrays to perform more complex experiments, for example, by growing and analyzing cells in an array format (Ziauddin and Sabatini, 2001). In this chapter, we discussed why the opposite strategy – a reduction in the complexity by analyzing single biomolecules in a spot – can also be envisioned to have a major impact on future bioscience.

Acknowledgments

This work was supported by the GEN-AU project of the Austrian Federal Ministry for Science and Research, the Austrian Science Fund (FWF; L422-N-20 and Y250-B10), and the state of Upper Austria.

References

Ameringer, T., Hinz, M., Mourran, C., Seliger, H., Groll, J., and Moeller, M. (2005). Ultrathin functional star PEG coatings for DNA microarrays. *Biomacromolecules* 6, 1819–1823.

Auer, M. (2001). HTS: understanding the physiology of life. *Drug Discov Today* 6, 935–936.

Axelrod, D., Burghardt, T. P., and Thompson, N. L. (1984). Total internal reflection fluorescence. *Annu Rev Biophys Bioeng* 13, 247–268.

Baugh, L. R., Hill, A. A., Brown, E. L., and Hunter, C. P. (2001). Quantitative analysis of mRNA amplification by in vitro transcription. *Nucleic Acids Res* 29, E29.

Beer, D. G., Kardia, S. L., Huang, C. C., Giordano, T. J., Levin, A. M., Misek, D. E., Lin, L., Chen, G., Gharib, T. G., Thomas, D. G., Lizyness, M. L., Kuick, R., Hayasaka, S., Taylor, J. M., Iannettoni, M. D., Orringer, M. B., and Hanash, S. (2002). Gene-expression profiles predict survival of patients with lung adenocarcinoma. *Nat Med* 8, 816–824.

Bensimon, D., Simon, A. J., Croquette, V. V., and Bensimon, A. (1995). Stretching DNA with a receding meniscus: experiments and models. *Phys Rev Lett* 74, 4754–4757.

Bentwich, I., Avniel, A., Karov, Y., Aharonov, R., Gilad, S., Barad, O., Barzilai, A., Einat, P., Einav, U., Meiri, E., Sharon, E., Spector, Y., and Bentwich, Z. (2005). Identification of hundreds of conserved and nonconserved human microRNAs. *Nat Genet* 37, 766–770.

Bismuto, E., Gratton, E., and Lamb, D. C. (2001). Dynamics of ANS binding to tuna apomyoglobin measured with fluorescence correlation spectroscopy. *Biophys J* 81, 3510–3521.

Braslavsky, I., Hebert, B., Kartalov, E., and Quake, S. R. (2003). Sequence information can be obtained from single DNA molecules. *Proc Natl Acad Sci U S A* 100, 3960–3964.

Brown, C. S., Goodwin, P. C., and Sorger, P. K. (2001). Image metrics in the statistical analysis of DNA microarray data. *Proc Natl Acad Sci U S A* 98, 8944–8949.

Brown, P. O. and Botstein, D. (1999). Exploring the new world of the genome with DNA microarrays. *Nat Genet* 21, 33–37.

Cha, T., Guo, A., Jun, Y., Pei, D., and Zhu, X. Y. (2004). Immobilization of oriented protein molecules on poly(ethylene glycol)-coated Si(111). *Proteomics* 4, 1965–1976.

Chou, H. P., Spence, C., Scherer, A., and Quake, S. (1999). A microfabricated device for sizing and sorting DNA molecules. *Proc Natl Acad Sci U S A* 96, 11–13.

Demers, L. M., Ginger, D. S., Park, S. J., Li, Z., Chung, S. W., and Mirkin, C. A. (2002). Direct patterning of modified oligonucleotides on metals and insulators by dip-pen nanolithography. *Science* 296, 1836–1838.

Derks, S., Lentjes, M. H., Hellebrekers, D. M., de Bruine, A. P., Herman, J. G., and van Engeland, M. (2004). Methylation-specific PCR unraveled. *Cell Oncol* 26, 291–299.

Diehl, F., Grahlmann, S., Beier, M., and Hoheisel, J. D. (2001). Manufacturing DNA microarrays of high spot homogeneity and reduced background signal. *Nucleic Acids Res* 29, E38.

Dufva, M. (2005). Fabrication of high quality microarrays. *Biomol Eng* 22, 173–184.

Eggeling, C., Berger, S., Brand, L., Fries, J. R., Schaffer, J., Volkmer, A., and Seidel, C. A. (2001). Data registration and selective single molecule analysis using multi-parameter fluorescence detection. *J Biotechnol* 86, 163–180.

Erdel, M., Hubalek, M., Lingenhel, A., Kofler, K., Duba, H. C., and Utermann, G. (1999). Counting the repetitive kringle-IV repeats in the gene encoding human apolipoprotein(a) by fibre-FISH. *Nat Genet* 21, 357–358.

Erdel, M., Theurl, M., Meyer, M., Duba, H. C., Utermann, G., and Werner-Felmayer, G. (2001). High-resolution mapping of the human 4q21 and the mouse 5E3 SCYB chemokine cluster by fiber-fluorescence in situ hybridization. *Immunogenetics* 53, 611–615.

Ferguson-Smith, A. C. and Surani, M. A. (2001). Imprinting and the epigenetic asymmetry between parental genomes. *Science* 293, 1086–1089.

Filippova, E. M., Monteleone, D. C., Trunk, J. G., Sutherland, B. M., Quake, S. R., and Sutherland, J. C. (2003). Quantifying double-strand breaks and clustered damages in DNA by single molecule laser fluorescence sizing. *Biophys J* 84, 1281–1290.

Foquet, M., Korlach, J., Zipfel, W., Webb, W. W., and Craighead, H. G. (2002). DNA fragment sizing by single molecule detection in submicrometer-sized closed fluidic channels. *Anal Chem* 74, 1415–1422.

Funatsu, T., Harada, Y., Tokunaga, M., Saito, K., and Yanagida, T. (1995). Imaging of single fluorescent molecules and individual ATP turnovers by single myosin molecules in aqueous solution. *Nature* 374, 555–559.

Goodwin, P. M., Johnson, M. E., Martin, J. C., Ambrose, W. P., Marrone, B. L., Jett, J. H., and Keller, R. A. (1993). Rapid sizing of individual fluorescently stained DNA fragments by flow cytometry. *Nucleic Acids Res* 21, 803–806.

Groll, J., Albrecht, K., Gasteier, P., Riethmueller, S., Ziener, U., and Moeller, M. (2005a). Nanostructured ordering of fluorescent markers and single proteins on substrates. *Chembiochem* 6, 1782–1787.

Groll, J., Amirgoulova, E. V., Ameringer, T., Heyes, C. D., Rocker, C., Nienhaus, G. U., and Moller, M. (2004). Biofunctionalized, ultrathin coatings of cross-linked star-shaped poly(ethylene oxide) allow reversible folding of immobilized proteins. *J Am Chem Soc* 126, 4234–4239.

Groll, J., Fiedler, J., Engelhard, E., Ameringer, T., Tugulu, S., Klok, H. A., Brenner, R. E., and Moeller, M. (2005b). A novel star PEG-derived surface coating for specific cell adhesion. *J Biomed Mater Res A* 74, 607–617.

Groll, J., Haubensak, W., Ameringer, T., and Moeller, M. (2005c). Ultrathin coatings from isocyanate terminated star PEG prepolymers: patterning of proteins on the layers. *Langmuir* 21, 3076–3083.

Guo, A. and Zhu, X.-Y. (2007). The critical role of surface chemistry in protein microarrays. In: *Functional Protein Microarrays: Pathways to Discovery* (P. Predki, Ed.), pp. 53–71. CRC Press, Boca Raton.

Haupts, U., Maiti, S., Schwille, P., and Webb, W. W. (1998). Dynamics of fluorescence fluctuations in green fluorescent protein observed by fluorescence correlation spectroscopy. *Proc Natl Acad Sci U S A* 95, 13573–13578.

Hellen, E. H. and Axelrod, D. (1990). An automatic focus/hold system for optical microscopes. *Rev Sci Instrum* 61, 3722–3725.

Heller, M. J. (2002). DNA microarray technology: devices, systems, and applications. *Annu Rev Biomed Eng* 4, 129–153.

Herrick, J. and Bensimon, A. (1999). Imaging of single DNA molecule: applications to high-resolution genomic studies. *Chromosome Res* 7, 409–423.

Herrick, J., Michalet, X., Conti, C., Schurra, C., and Bensimon, A. (2000). Quantifying single gene copy number by measuring fluorescent probe lengths on combed genomic DNA. *Proc Natl Acad Sci U S A* 97, 222–227.

Hesse, J., Jacak, J., Kasper, M., Regl, G., Eichberger, T., Winklmayr, M., Aberger, F., Sonnleitner, M., Schlapak, R., Howorka, S., Muresan, L., Frischauf, A. M., and Schutz, G. J. (2006). RNA expression profiling at the single molecule level. *Genome Res* 16, 1041–1045.

Hesse, J., Jacak, J., Regl, G., Eichberger, T., Aberger, F., Schlapak, R., Howorka, S., Muresan, L., Frischauf, A., and Schütz, G. J. (2007). Single molecule fluorescence microscopy for ultra-sensitive RNA expression profiling. In: *Proc SPIE* (E. Jorg, K. G. Zygmunt, Eds.), Vol. 6444, pp. 64440F.

Hesse, J., Sonnleitner, M., Sonnleitner, A., Freudenthaler, G., Jacak, J., Hoglinger, O., Schindler, H., and Schutz, G. J. (2004). Single molecule reader for high-throughput bioanalysis. *Anal Chem* 76, 5960–5964.

Hesse, J., Wechselberger, C., Sonnleitner, M., Schindler, H., and Schütz, G. J. (2002). Single molecule reader for proteomics and genomics. *J Chromatogr B Analyt Technol Biomed Life Sci* 782, 127–135.

Hong, J. W., Studer, V., Hang, G., Anderson, W. F., and Quake, S. R. (2004). A nanoliter-scale nucleic acid processor with parallel architecture. *Nat Biotechnol* 22, 435–439.

Howarter, J. A. and Youngblood, J. P. (2006). Optimization of silica silanization by 3-aminopropyltriethoxysilane. *Langmuir* 22, 11142–11147.

Hu, L., Wang, J., Baggerly, K., Wang, H., Fuller, G. N., Hamilton, S. R., Coombes, K. R., and Zhang, W. (2002). Obtaining reliable information from minute amounts of RNA using cDNA microarrays. *BMC Genomics* 3, 16.

Issa, J. P. (2003). Age-related epigenetic changes and the immune system. *Clin Immunol* 109, 103–108.

Jacobson, J. M. (2005). Nucleotide sequencing via repetitive single molecule hybridization. US Patent US20050153324.

Kacharmina, J. E., Crino, P. B., and Eberwine, J. (1999). Preparation of cDNA from single cells and subcellular regions. *Methods Enzymol* 303, 3–18.

Kim, D., Lee, H. G., Jung, H., and Kang, S. H. (2007a). Single-protein molecular interactions on polymer-modified glass substrates for nanoarray chip application using dual-color TIRFM. *Bull Korean Chem Soc* 28 (5), 783–790.

Kim, J. H., Shi, W. X., and Larson, R. G. (2007b). Methods of stretching DNA molecules using flow fields. *Langmuir* 23, 755–764.

Korlach, J., Schwille, P., Webb, W. W., and Feigenson, G. W. (1999). Characterization of lipid bilayer phases by confocal microscopy and fluorescence correlation spectroscopy. *Proc Natl Acad Sci U S A* 96, 8461–8466.

Krichevski, O. and Bonnet, G. (2002). Fluorescence correlation spectroscopy: the technique and its applications. *Rep Prog Phys* 65, 251–297.

Kusnezow, W. and Hoheisel, J. D. (2003). Solid supports for microarray immunoassays. *J Mol Recognit* 16, 165–176.

Kusnezow, W., Jacob, A., Walijew, A., Diehl, F., and Hoheisel, J. D. (2003). Antibody microarrays: an evaluation of production parameters. *Proteomics* 3, 254–264.

Kuzmenkina, E. V., Heyes, C. D., and Nienhaus, G. U. (2005). Single molecule Forster resonance energy transfer study of protein dynamics under denaturing conditions. *Proc Natl Acad Sci U S A* 102, 15471–15476.

Laib, S., Rankl, M., Ruckstuhl, T., and Seeger, S. (2003). Sizing of single fluorescently stained DNA fragments by scanning microscopy. *Nucleic Acids Res* 31, e138.

Lebofsky, R. and Bensimon, A. (2003). Single DNA molecule analysis: applications of molecular combing. *Brief Funct Genomic Proteomic* 1, 385–396.

Lebofsky, R., Heilig, R., Sonnleitner, M., Weissenbach, J., and Bensimon, A. (2006). DNA replication origin interference increases the spacing between initiation events in human cells. *Mol Biol Cell* 17, 5337–5345.

Lee, J. T. (2003). Molecular links between X-inactivation and autosomal imprinting: X-inactivation as a driving force for the evolution of imprinting?. *Curr Biol* 13, R242–R254.

Li, L., Roden, J., Shapiro, B. E., Wold, B. J., Bhatia, S., Forman, S. J., and Bhatia, R. (2005). Reproducibility, fidelity, and discriminant validity of mRNA amplification for microarray analysis from primary hematopoietic cells. *J Mol Diagn* 7, 48–56.

Li, S. H. and Li, X. J. (2004). Huntingtin-protein interactions and the pathogenesis of Huntington's disease. *Trends Genet* 20, 146–154.

Mahadevappa, M. and Warrington, J. A. (1999). A high-density probe array sample preparation method using 10- to 100-fold fewer cells. *Nat Biotechnol* 17, 1134–1136.

Mahajan, S., Kumar, P., and Gupta, K. C. (2006). Oligonucleotide microarrays: immobilization of phosphorylated oligonucleotides on epoxylated surface. *Bioconjug Chem* 17, 1184–1189.

Marton, M. J., DeRisi, J. L., Bennett, H. A., Iyer, V. R., Meyer, M. R., Roberts, C. J., Stoughton, R., Burchard, J., Slade, D., Dai, H., Bassett Jr., D. E., Hartwell, L. H., Brown, P. O., and Friend, S. H. (1998). Drug target validation and identification of secondary drug target effects using DNA microarrays. *Nat Med* 4, 1293–1301.

Meng, X., Benson, K., Chada, K., Huff, E. J., and Schwartz, D. C. (1995). Optical mapping of lambda bacteriophage clones using restriction endonucleases. *Nat Genet* 9, 432–438.

Michalet, X., Ekong, R., Fougerousse, F., Rousseaux, S., Schurra, C., Hornigold, N., van Slegtenhorst, M., Wolfe, J., Povey, S., Beckmann, J. S., and Bensimon, A. (1997). Dynamic molecular combing: stretching the whole human genome for high-resolution studies. *Science* 277, 1518–1523.

Minsky, M. (1988). Memoir on inventing the confocal scanning microscope. *Scanning* 10, 128–138.

Moerner, W. E. and Kador, L. (1989). Optical detection and spectroscopy of single molecules in a solid. *Phys Rev Lett* 62, 2535–2538.

Moerner, W. E. and Orrit, M. (1999). Illuminating single molecules in condensed matter. *Science* 283, 1670–1676.

Mutter, G. L. and Boynton, K. A. (1995). PCR bias in amplification of androgen receptor alleles, a trinucleotide repeat marker used in clonality studies. *Nucleic Acids Res* 23, 1411–1418.

Nie, S., Chiu, D. T., and Zare, R. N. (1994). Probing individual molecules with confocal fluorescence microscopy. *Science* 266, 1018–1021.

Nielsen, T. O., West, R. B., Linn, S. C., Alter, O., Knowling, M. A., O'Connell, J. X., Zhu, S., Fero, M., Sherlock, G., Pollack, J. R., Brown, P. O., Botstein, D., and van de Rijn, M. (2002). Molecular characterisation of soft tissue tumours: a gene expression study. *Lancet* 359, 1301–1307.

Nygaard, V., Holden, M., Loland, A., Langaas, M., Myklebost, O., and Hovig, E. (2005). Limitations of mRNA amplification from small-size cell samples. *BMC Genomics* 6, 147.

Paegel, B. M., Blazej, R. G., and Mathies, R. A. (2003). Microfluidic devices for DNA sequencing: sample preparation and electrophoretic analysis. *Curr Opin Biotechnol* 14, 42–50.

Pasero, P., Bensimon, A., and Schwob, E. (2002). Single molecule analysis reveals clustering and epigenetic regulation of replication origins at the yeast rDNA locus. *Genes Dev* 16, 2479–2484.

Patel, P. K., Arcangioli, B., Baker, S. P., Bensimon, A., and Rhind, N. (2006). DNA replication origins fire stochastically in fission yeast. *Mol Biol Cell* 17, 308–316.

Pawley, J. B. (1995). Handbook of Biological Confocal Microscopy. Plenum, New York.

Piehler, J. (2005). New methodologies for measuring protein interactions in vivo and in vitro. *Curr Opin Struct Biol* 15, 4–14.

Piehler, J., Brecht, A., Valiokas, R., Liedberg, B., and Gauglitz, G. (2000). A high-density poly(ethylene glycol) polymer brush for immobilization on glass-type surfaces. *Biosens Bioelectron* 15, 473–481.

Proll, J., Fodermayr, M., Wechselberger, C., Pammer, P., Sonnleitner, M., Zach, O., and Lutz, D. (2006). Ultra-sensitive immunodetection of 5′methyl cytosine for DNA methylation analysis on oligonucleotide microarrays. *DNA Res* 13, 37–42.

Rigler, R. (2005). Parallel high throughput single molecule sequencing process. European Patent EP2005008511.

Robertson, K. D. (2005). DNA methylation and human disease. *Nat Rev Genet* 6, 597–610.

Sako, Y., Minoghchi, S., and Yanagida, T. (2000). Single molecule imaging of EGFR signalling on the surface of living cells. *Nat Cell Biol* 2, 168–172.

Sase, I., Miyata, H., Corrie, J. E., Craik, J. S., and Kinosita Jr., K. (1995). Real time imaging of single fluorophores on moving actin with an epifluorescence microscope. *Biophys J* 69, 323–328.

Schena, M. (2003). *Microarray Analysis*. John Wiley and Sons, Hobeken.

Schlapak, R., Kinns, H., Wechselberger, C., Hesse, J., and Howorka, S. (2007). Sizing trinucleotide repeat sequences by single molecule analysis of fluorescence brightness. *Chemphyschem* 8, 1618–1621.

Schlapak, R., Pammer, P., Armitage, D., Zhu, R., Hinterdorfer, P., Vaupel, M., Fruhwirth, T., and Howorka, S. (2006). Glass surfaces grafted with high-density poly(ethylene glycol) as substrates for DNA oligonucleotide microarrays. *Langmuir* 22, 277–285.

Schmidt, T., Schutz, G. J., Baumgartner, W., Gruber, H. J., and Schindler, H. (1996a). Imaging of single molecule diffusion. *Proc Natl Acad Sci U S A* 93, 2926–2929.

Schmidt, T., Schütz, G. J., Gruber, H. J., and Schindler, H. (1996b). Local stoichiometries determined by counting individual molecules. *Anal Chem* 68, 4397–4401.

Schütz, G. J., Kada, G., Pastushenko, V. P., and Schindler, H. (2000a). Properties of lipid microdomains in a muscle cell membrane visualized by single molecule microscopy. *EMBO J* 19, 892–901.

Schütz, G. J., Pastushenko, V. P., Gruber, H. J., Knaus, H.-G., Pragl, B., and Schindler, H. (2000b). 3D imaging of individual ion channels in live cells at 40nm resolution. *Single Mol* 1, 25–31.

Schwartz, J. J. and Quake, S. R. (2007). High density single molecule surface patterning with colloidal epitaxy. *Appl Phys Lett* 91, 083902.

Seidel, C. A. M., Schulz, A., and Sauer, M. H. M. (1996). Nucleobase-specific quenching of fluorescent dyes. 1. Nucleobase one-electron redox potentials and their correlation with static and dynamic quenching efficiencies. *J Phys Chem* 100, 5541–5553.

Seong, S. Y. and Choi, C. Y. (2003). Current status of protein chip development in terms of fabrication and application. *Proteomics* 3, 2176–2189.

Service, R. F. (2006). Gene sequencing: the race for the $1000 genome. *Science* 311, 1544–1546.

Shera, E. B., Seitzinger, N. K., Davis, L. M., Keller, R. A., and Soper, S. A. (1990). Detection of single flourescent molecules. *Chem Phys Lett* 174, 553–557.

Shi, L., Reid, L. H., Jones, W. D., Shippy, R., Warrington, J. A., Baker, S. C., Collins, P. J., de Longueville, F., Kawasaki, E. S., Lee, K. Y., Luo, Y., Sun, Y. A., Willey, J. C., Setterquist, R. A., Fischer, G. M., Tong, W., Dragan, Y. P., Dix, D. J., Frueh, F. W., Goodsaid, F. M., Herman, D., Jensen, R. V., Johnson, C. D., Lobenhofer, E. K., Puri, R. K., Schrf, U., Thierry-Mieg, J., Wang, C., Wilson, M., Wolber, P. K., Zhang, L., Amur, S., Bao, W., Barbacioru, C. C., Lucas, A. B., Bertholet, V., Boysen, C., Bromley, B., Brown, D., Brunner, A., Canales, R., Cao, X. M., Cebula, T. A., Chen, J. J., Cheng, J., Chu, T. M., Chudin, E., Corson, J., Corton, J. C., Croner, L. J., Davies, C., Davison, T. S., Delenstarr, G., Deng, X., Dorris, D., Eklund, A. C., Fan, X. H., Fang, H., Fulmer-Smentek, S., Fuscoe, J. C., Gallagher, K., Ge, W., Guo, L., Guo, X., Hager, J., Haje, P. K., Han, J., Han, T., Harbottle, H. C., Harris, S. C., Hatchwell, E., Hauser, C. A., Hester, S., Hong, H., Hurban, P., Jackson, S. A., Ji, H., Knight, C. R., Kuo, W. P., LeClerc, J. E., Levy, S., Li, Q. Z., Liu, C., Liu, Y., Lombardi, M. J., Ma, Y., Magnuson, S. R., Maqsodi, B., McDaniel, T., Mei, N., Myklebost, O., Ning, B., Novoradovskaya, N., Orr, M. S., Osborn, T. W., Papallo, A., Patterson, T. A., Perkins, R. G., Peters, E. H., Peterson, R. et al. (2006). The MicroArray Quality Control (MAQC) project shows inter- and intraplatform reproducibility of gene expression measurements. *Nat Biotechnol* 24, 1151–1161.

Shiraishi, M. and Hayatsu, H. (2004). High-speed conversion of cytosine to uracil in bisulfite genomic sequencing analysis of DNA methylation. *DNA Res* 11, 409–415.

Sims, C. E. and Allbritton, N. L. (2007). Analysis of single mammalian cells on-chip. *Lab Chip* 7, 423–440.

Sonnleitner, M., Freudenthaler, G., Hesse, J., and Schütz, G. J. (2005). High-throughput scanning with single molecule sensitivity. In: *Proc. SPIE* (D. V. Nicolau, J. Enderlein, R. C. Leif, D. L. Farkas, and R. Raghavachari, Eds.), Vol. 5699, pp. 202–210.

Tan, P. K., Downey, T. J., Spitznagel Jr., E. L., Xu, P., Fu, D., Dimitrov, D. S., Lempicki, R. A., Raaka, B. M., and Cam, M. C. (2003). Evaluation of gene expression measurements from commercial microarray platforms. *Nucleic Acids Res* 31, 5676–5684.

Taylor, B. T., Nambiar, P. R., Raja, R., Cheung, E., Rosenberg, D. W., and Anderegg, B. (2004). Microgenomics: identification of new expression profiles via small and single-cell sample analyses. *Cytometry Part A* 59A, 254–261.

Ushijima, T. (2005). Detection and interpretation of altered methylation patterns in cancer cells. *Nat Rev Cancer* 5, 223–231.

Van Gelder, R. N., von Zastrow, M. E., Yool, A., Dement, W. C., Barchas, J. D., and Eberwine, J. H. (1990). Amplified RNA synthesized from limited quantities of heterogeneous cDNA. *Proc Natl Acad Sci U S A* 87, 1663–1667.

Van Orden, A., Keller, R. A., and Ambrose, W. P. (2000). High-throughput flow cytometric DNA fragment sizing. *Anal Chem* 72, 37–41.

van't Veer, L. J., Dai, H., van de Vijver, M. J., He, Y. D., Hart, A. A., Mao, M., Peterse, H. L., van der Kooy, K., Marton, M. J., Witteveen, A. T., Schreiber, G. J., Kerkhoven, R. M., Roberts, C., Linsley, P. S., Bernards, R., and Friend, S. H. (2002). Gene expression profiling predicts clinical outcome of breast cancer. *Nature* 415, 530–536.

Wang, E., Miller, L. D., Ohnmacht, G. A., Liu, E. T., and Marincola, F. M. (2000). High-fidelity mRNA amplification for gene profiling. *Nat Biotechnol* 18, 457–459.

Wennmalm, S., Edman, L., and Rigler, R. (1997). Conformational fluctuations in single DNA molecules. *Proc Natl Acad Sci U S A* 94, 10641–10646.

Xiao, M., Phong, A., Ha, C., Chan, T. F., Cai, D., Leung, L., Wan, E., Kistler, A. L., DeRisi, J. L., Selvin, P. R., and Kwok, P. Y. (2007). Rapid DNA mapping by fluorescent single molecule detection. *Nucleic Acids Res* 35, e16.

Yokota, H., Johnson, F., Lu, H., Robinson, R. M., Belu, A. M., Garrison, M. D., Ratner, B. D., Trask, B. J., and Miller, D. L. (1997). A new method for straightening DNA molecules for optical restriction mapping. *Nucleic Acids Res* 25, 1064–1070.

Ziauddin, J. and Sabatini, D. M. (2001). Microarrays of cells expressing defined cDNAs. *Nature* 411, 107–110.

Abbreviations

The following abbreviations may be encountered in the text.

Abbreviation	Term
ADP	Adenosine diphosphate
AFM	Atomic force microscopy
AOD	Acousto-optic deflector
APD	Avalanche photodiode
ATP	Adenosine triphosphate
bp	Base pairs
BSA	Bovine serum albumin
CCD	Charge-coupled device
CGS	Centimeter-gram-second (a metric system which predated the SI)
Cy3, Cy5	Fluorescent cyanine dyes
DOPE	Dioleoyl phosphatidylethanolamine
DOPI	Defocused orientation and position imaging
DPPC	Dipalmitoylphosphatidylcholine
DPPE	Dipalmitoylphosphoethanolamine
FACS	Fluorescence-activated cell sorting
FCS	Fluorescence correlation spectroscopy
FIONA	Fluorescence imaging with one nanometer accuracy
FRAP	Fluorescence recovery after photobleaching
FRET	Förster/fluorescence resonance energy transfer
GFP	Green fluorescent protein
GPI	Glycosylphosphatidylinositol
GTP	Guanosine triphosphate
GUV	Giant unilamellar vesicles

(Continued)

Abbreviation	Term
kb	Kilobase pairs
LCM	Laser capture microdissection
MHC	Major histocompatibility complex
NALMS	Nanometer-localized multiple single molecule fluorescence microscopy
Nd:YAG	Neodymium:yttrium aluminum garnet
PALM	Photoactivated localization microscopy
PCR	Polymerase chain reaction
PEG	Polyethylene glycol
PSF	Point spread function
SD	Shine–Dalgarno (sequence)
SHREC	Single molecule high-resolution co-localization
SHRImP	Single molecule high-resolution imaging with photobleaching
SI	Système International d'Unités (International System of Units)
SMF	Single molecule fluorescence
SPT	Single-particle tracking
STORM	Stochastic optical reconstruction microscopy
SUV	Small unilamellar vesicles
TDI	Time-delayed integration
TIRF(M)	Total internal reflection fluorescence (microscopy)
TIR-FRET	Total internal reflection – Förster resonance energy transfer
TMR	Tetramethylrhodamine
WLC	Worm-like chain
(e)YFP	(Enhanced) Yellow fluorescent protein
λ	Bacteriophage lambda

The Amino Acids

These are the codes that are commonly used to represent the 20 amino acids that are normally found in proteins.

Single-Letter Code	Three-Letter Code	Name
A	Ala	Alanine
C	Cys	Cysteine
D	Asp	Aspartic acid
E	Glu	Glutamate
F	Phe	Phenylalanine
G	Gly	Glycine
H	His	Histidine
I	Ile	Isoleucine
K	Lys	Lysine
L	Leu	Leucine
M	Met	Methionine
N	Asn	Asparagine
P	Pro	Proline
Q	Gln	Glutamine
R	Arg	Arginine
S	Ser	Serine
T	Thr	Threonine
V	Val	Valine
W	Trp	Tryptophan
Y	Tyr	Tyrosine
Refer to IUPAC (1984).		

The Bases

These are the abbreviations for the most common bases found in DNA and RNA molecules.

Abbreviation	Base
A	Adenine
T	Threonine
C	Cytosine
G	Guanine
U	Uracil

SI Units

SI Prefixes

Power of 10	Power of 10^3	Factor	Name	Symbol	Adoption
24	8	10^{24}	yotta	y	§
21	7	10^{21}	zetta	z	§
18	6	10^{18}	exa	e	‡
15	5	10^{15}	peta	p	‡
12	4	10^{12}	tera	t	*
9	3	10^9	giga	g	*
6	2	10^6	mega	m	*
3	1	10^3	kilo	k	*
2		10^2	hecto	h	*
1		10^1	deka (deca)	da	*
0	0	10^0	–		
−1		10^{-1}	deci	d	*
−2		10^{-2}	centi	c	*
−3	−1	10^{-3}	milli	m	*
−6	−2	10^{-6}	micro	μ	*
−9	−3	10^{-9}	nano	n	*
−12	−4	10^{-12}	pico	p	*
−15	−5	10^{-15}	femto	f	†
−18	−6	10^{-18}	atto	a	†
−21	−7	10^{-21}	zepto	z	§
−24	−8	10^{-24}	yocto	y	§

The prefixes recognized for use with the SI system. The number of prefixes has been added to at intervals over the years, so that some of them (particularly Y, Z, y, and z) may be unfamiliar. The symbols in the rightmost column indicate the date of official adoption into the SI: *adopted by 11th CGPM (1960, 1961) (although many were in wide use before this, see Boer, 1966); † adopted by the 12th CGPM (1964); ‡ adopted by the 15th CGPM (1975) (Terrien, 1975); § adopted by the 19th CGPM (1991) (Quinn, 1992).

SI Base Units

There are seven base units in the SI (BIPM, 2006)

Unit	Symbol	Definition
meter (metre)	m	The meter is the length of the path traveled by light in vacuum during a time interval of 1/299 792 458 of a second.
kilogram	kg	The kilogram is the unit of mass; it is equal to the mass of the international prototype of the kilogram.
second	s	The second is the duration of 9 192 631 770 periods of the radiation corresponding to the transition between the two hyperfine levels of the ground state of the cesium 133 atom.
ampere	A	The ampere is that constant current which, if maintained in two straight parallel conductors of infinite length, of negligible circular cross-section, and placed 1 m apart in vacuum, would produce between these conductors a force equal to 2×10^{-7} newton per meter of length.
kelvin	K	The Kelvin, unit of thermodynamic temperature, is the fraction 1/273.16 of the thermodynamic temperature of the triple point of water.
mole	mol	1. The mole is the amount of substance of a system that contains as many elementary entities as there are atoms in 0.012 kilogram of carbon 12.[1] 2. When the mole is used, the elementary entities must be specified and may be atoms, molecules, ions, electrons, other particles, or specified groups of such particles.
candela	cd	The candela is the luminous intensity, in a given direction, of a source that emits monochromatic radiation of frequency 540×10^{12} hertz and that has a radiant intensity in that direction of 1/683 watt per steradian.

[1] This number is the Avogadro constant, N_A, which has the value $6.022 141 79(30) \times 10^{23}\,\mathrm{mol}^{-1}$ (see http://www.physics.nist.gov/cuu/Constants/index.html).

SI Derived Units

There are many SI units that are derived from the above units but have special names for convenience. A few relevant units, many of which are used in the text, are presented in the following table (BIPM, 2006).

Derived Quantity	Name	Symbol	Expressed in terms of Other SI Units	Expressed in terms of SI Base Units
Plane angle	radian	rad	1	m/m
Solid angle	steradian	sr	1	m^2/m^2
Frequency	hertz	Hz		s^{-1}
Force	newton	N		$m\,kg\,s^{-2}$
Energy, work, amount of heat	joule	J	$N\,m$	$m^2\,kg\,s^{-2}$
Power, radiant flux	watt	W	J/s	$m^2\,kg\,s^{-3}$
Electric potential difference, electromotive force	volt	V	W/A	$m^2\,kg\,s^{-3}\,A^{-1}$
Electric resistance	ohm	Ω	V/A	$m^2\,kg\,s^{-3}\,A^{-2}$
Electric conductance	siemens	S	A/V	$m^{-2}\,kg^{-1}\,s^3\,A^2$
Celsius temperature	degree Celsius	°C		K
Catalytic activity	katal	kat		$s^{-1}\,mol$

Non-SI Units Popular in Biology and Chemistry

A number of non-SI units are so widely used in biological and chemical sciences that it would simply have confused the reader to avoid their use in this book. In most cases, these units are related to the SI in a simple way. These units (other than the svedberg and angström) are often used in combination with the SI prefixes, for example, millimolar, kilodalton, kilobase pair. Note that the svedberg, confusingly, has the same symbol as the siemens, the SI unit of conductance.

Unit Name	Symbol	Description	Relation to SI
Svedberg	S	Used to describe the sedimentation coefficient of particles in ultracentrifugation	$1\,S = 10^{-13}\,s$
Ångström	Å	Unit of length popular among crystallographers; of the order of chemical bond lengths	$1\,Å = 10^{-10}\,m$; $1\,nm = 10\,Å$
Molarity	M	Unit of molar concentration, more widely adopted than the SI equivalent, $mol\,m^{-3}$	$1\,M = 1\,mol\,dm^{-3}$ or $1000\,mol\,m^{-3}$
Dalton	Da	Non-SI unit of molecular weight, also known as the unified atomic mass unit, 1/12 the mass of an atom of carbon-12	$1\,Da = 1/N_A\,g$ $\approx 1.66 \times 10^{-27}\,kg$ (BIPM)
Base pair	bp	Convenience unit for describing lengths and distances along DNA molecules. For single-stranded molecules, the term nucleotide (nt) is used instead	In B-form DNA, $1\,bp \approx 0.34\,nm$. The average mass of a base pair of DNA is $\approx 660\,Da$
"Unit"	U	Widespread unit for the measurement of enzyme activity. The katal is preferred	$1\,U = 1\,\mu mol\,min^{-1}$ $\approx 16.67\,nkat$ (Dybkaer, 2000)

References

BIPM (2006). *The International System of Units (SI)*. Bureau International des Poids et Mesures, Paris. http://www.bipm.org/en/si/.

de Boer, J. (1966). Short history of the prefixes. *Metrologia*, 2(4), 165–166.

CGPM (1961). Comptes Rendus de la 11e CGPM (1960).

CGPM (1965). Comptes Rendus de la 12e CGPM (1964), p. 94.

CGPM (1976). Comptes Rendus de la 15e CGPM (1975), p. 106.

IUPAC (1984). Nomenclature and symbolism for amino acids and peptides. *Pure Appl Chem* 56, 595–624.

CGPM (1992). Comptes Rendus de la 19e CGPM (1991), p. 185.

Dybkaer, R. (2000). The special name "katal" for the SI derived unit, mole per second, when expressing catalytic activity. *Metrologia* 37, 671.

Quinn, T. J. (1992). News from the BIPM. *Metrologia*, 29(1), 1–7.

Terrien, J. (1975). News from the Bureau International des Poids et Mesures. *Metrologia*, 11(4), 179–183.

Index

Printed and bound by CPI Group (UK) Ltd, Croydon, CR0 4YY

21/10/2024

01777130-0003